Lecture Notes in Physics

Springer-Verlag Berlin Heidelberg GmbH

The Editorial Policy for Proceedings

The series Lecture Notes in Physics reports new developments in physical research and teaching – quickly, informally, and at a high level. The proceedings to be considered for publication in this series should be limited to only a few areas of research, and these should be closely related to each other. The contributions should be of a high standard and should avoid lengthy redraftings of papers already published or about to be published elsewhere. As a whole, the proceedings should aim for a balanced presentation of the theme of the conference including a description of the techniques used and enough motivation for a broad readership. It should not be assumed that the published proceedings must reflect the conference in its entirety. (A listing or abstracts of papers presented at the meeting but not included in the proceedings could be added as an appendix.)

When applying for publication in the series Lecture Notes in Physics the volume's editor(s) should submit sufficient material to enable the series editors and their referees to make a fairly accurate evaluation (e.g. a complete list of speakers and titles of papers to be presented and abstracts). If, based on this information, the proceedings are (tentatively) accepted, the volume's editor(s), whose name(s) will appear on the title pages, should select the papers suitable for publication and have them refereed (as for a journal) when appropriate. As a rule discussions will not be accepted. The series editors and Springer-Verlag will normally not interfere with the detailed editing except in fairly obvious cases or on technical matters.

Final acceptance is expressed by the series editor in charge, in consultation with Springer-Verlag only after receiving the complete manuscript. It might help to send a copy of the authors' manuscripts in advance to the editor in charge to discuss possible revisions with him. As a general rule, the series editor will confirm his tentative acceptance if the final manuscript corresponds to the original concept discussed, if the quality of the contribution meets the requirements of the series, and if the final size of the manuscript does not greatly exceed the number of pages originally agreed upon. The manuscript should be forwarded to Springer-Verlag shortly after the meeting. In cases of extreme delay (more than six months after the conference) the series editors will check once more the timeliness of the papers. Therefore, the volume's editor(s) should establish strict deadlines, or collect the articles during the conference and have them revised on the spot. If a delay is unavoidable, one should encourage the authors to update their contributions if appropriate. The editors of proceedings are strongly advised to inform contributors about these points at an early stage.

The final manuscript should contain a table of contents and an informative introduction accessible also to readers not particularly familiar with the topic of the conference. The contributions should be in English. The volume's editor(s) should check the contributions for the correct use of language. At Springer-Verlag only the prefaces will be checked by a copy-editor for language and style. Grave linguistic or technical shortcomings may lead to the rejection of contributions by the series editors. A conference report should not exceed a total of 500 pages. Keeping the size within this bound should be achieved by a stricter selection of articles and not by imposing an upper limit to the length of the individual papers. Editors receive jointly 30 complimentary copies of their book. They are entitled to purchase further copies of their book at a reduced rate. As a rule no reprints of individual contributions can be supplied. No royalty is paid on Lecture Notes in Physics volumes. Commitment to publish is made by letter of interest rather than by signing a formal contract. Springer-Verlag secures the copyright for each volume.

The Production Process

The books are hardbound, and the publisher will select quality paper appropriate to the needs of the author(s). Publication time is about ten weeks. More than twenty years of experience guarantee authors the best possible service. To reach the goal of rapid publication at a low price the technique of photographic reproduction from a camera-ready manuscript was chosen. This process shifts the main responsibility for the technical quality considerably from the publisher to the authors. We therefore urge all authors and editors of proceedings to observe very carefully the essentials for the preparation of camera-ready manuscripts, which we will supply on request. This applies especially to the quality of figures and halftones submitted for publication. In addition, it might be useful to look at some of the volumes already published. As a special service, we offer free of charge LATEX and TEX macro packages to format the text according to Springer-Verlag's quality requirements. We strongly recommend that you make use of this offer, since the result will be a book of considerably improved technical quality. To avoid mistakes and time-consuming correspondence during the production period the conference editors should request special instructions from the publisher well before the beginning of the conference. Manuscripts not meeting the technical standard of the series will have to be returned for improvement.

For further information please contact Springer-Verlag, Physics Editorial Department II, Tiergartenstrasse 17, D-69121 Heidelberg, Germany

Aron M. Bernstein   Barry R. Holstein (Eds.)

Chiral Dynamics: Theory and Experiment

Proceedings of the Workshop
Held at MIT, Cambridge, MA, USA,
25-29 July 1994

 Springer

Editors

Aron M. Bernstein
Barry R. Holstein
Department of Physics
Massachusetts Institute of Technology
Cambridge, MA 02139, USA

Library of Congress Cataloging-in-Publication Data

Chiral dynamics : theory and experiment : proceedings of the workshop
  held at MIT, Cambridge, MA, USA, 25-29 July 1994 / Aron M.
Bernstein, Barry R. Holstein, eds.
       p.   cm. -- (Lecture notes in physics ; 452)
   Includes bibliographical references (p.).
   ISBN 978-3-662-13994-3   ISBN 978-3-540-49227-6 (eBook)
   DOI 10.1007/978-3-540-49227-6
   1. Particles (Nuclear physics)--Chirality--Congresses. 2. Quantum
chromodynamics--Congresses. 3. Nuclear interactions--Congresses.
I. Bernstein, Aron M., 1931-   . II. Holstein, Barry R., 1943-   .
III. Workshop on Chiral Dynamics: Theory and Experiment (1994 : MIT)
IV. Series.
QC793.3.C54C48   1995
539.7'25--dc20   95-17871
                                                                   CIP

ISBN 978-3-662-13994-3

© Springer-Verlag Berlin Heidelberg 1995
Originally published by Springer-Verlag Berlin Heidelberg New York in 1995
Softcover reprint of the hardcover 1st edition 1995

Typesetting: Camera-ready by the editors
SPIN: 10501383   55/3142-543210 - Printed on acid-free paper

Introduction

This volume presents the proceedings of the Workshop on Chiral Dynamics: Theory and Experiment, which was held at MIT, July 25-29, 1994. The main purpose of the workshop was to bring together the active participants in the field, both theorists and experimentalists, to assess where we are and to explore the most fruitful future directions. It was extremely gratifying that after deciding to organize this workshop we received an enthusiastic response from the community. This was reflected by a 100% acceptance of our initial invitations to speak.

The main attraction of chiral dynamics is that it represents a serious effort to provide a rigorous and model-independent methodology by which to make QCD predictions at the confinement scale. This is one of the forefront arenas of the standard model and as such it is essential to confront these reliable theoretical predictions with correspondingly precise experiments. The unique feature of this workshop was the equal mixture of experiment and theory. In order to explore the optimum future direction of this field we organized working groups on many of the important areas. We believe that the discussions which took place at the workshop have inspired several new initiatives in the field and have also led to new collaborations.

The function of the working groups was to focus on the predictions and measurements of physical processes, and the phenomenology that is often required to bridge the gap between what is predicted and what is measured. For example, there exist solid chiral predictions for low energy $\pi - \pi$ scattering but many relevant experiments involve the $\pi N \rightarrow \pi\pi N$ process. The phenomenological models used to extract the $\pi - \pi$ scattering lengths are far less rigorous than are the chiral predictions. Improvements in this situation are urgently needed in order to make progress. This is just one of the issues addressed.

The workshop schedule had morning and evening talks, with the afternoons left open for the working groups and for informal discussions. The πN and photopion working groups held a joint session. The last afternoon of the workshop was devoted to the reports of the working groups and to the summary talks.

This volume is organized to reflect the range of topics that were discussed. The reports of the working groups are included in the relevant topic areas. It is difficult in a book to convey the spirit of the workshop, which was one of optimism and enthusiasm. The discussions were extremely lively and lengthy, but, of course, the workshop reports are limited so that much of this excitement remains only in the minds of the participants. Nevertheless we believe that this volume can provide the reader with a useful overview of the field ca. 1994.

MIT, July 24-29, 1994

A.M. Bernstein
B.R. Holstein

Acknowledgements

The success of the workshop was made possible by the hard work and enthusiasm of all of the participants. We particularly want to thank the speakers and the working group coordinators and participants. We would also like to thank Jean Flanagan, Joanne Gregory, and the staff of the Laboratory for Nuclear Science of MIT, for their skillful and unflagging logistic support. We would like to thank CEBAF and the National Science Foundation for supporting the workshop and for providing support for young scientists to attend. The workshop would not have been possible except through the financial support of MIT; in particular we gratefully acknowledge support from the Department of Physics, the Dean of Science, the Center for Theoretical Physics, the Medium Energy Physics Group, The Bates Linear Accelerator Center, and the Laboratory for Nuclear Science.

Contents

Part I: Chiral Dynamics and QCD

Strong Interactions at Low Energies

*Steven Weinberg**

Theory Group, Department of Physics,
University of Texas
Austin, TX, 78712

Effective field theories are playing an increasing role in the study of a wide variety of physical phenomena, from W and Z interactions to superconductivity. Regarding the subject of this talk, we have known for years that the low energy strong interactions of nucleons and pions are well described in the tree approximation[2] by an effective field theory, with Lagrangian[1]

$$\mathcal{L}_{\text{eff}} = -\frac{\partial_\mu \vec{\pi} \cdot \partial^\mu \vec{\pi}}{2(1 + \vec{\pi}^2/F_\pi^2)^2} - \frac{m_\pi^2 \, \vec{\pi}^2}{2(1 + \vec{\pi}^2/F_\pi^2)^2}$$
$$+\bar{N} \left[i\partial_0 - \frac{2\vec{t} \cdot (\vec{\pi} \times \partial_0 \vec{\pi})}{F_\pi^2 (1 + \vec{\pi}^2/F_\pi^2)^2} - m_N - \frac{2g_A \vec{t} \cdot (\sigma \cdot \nabla)\vec{\pi}}{F_\pi(1 + \vec{\pi}^2/F_\pi^2)} \right] N$$
$$-\frac{1}{2}C_S(\bar{N}N)^2 - \frac{1}{2}C_T(\bar{N}\sigma N)^2 \tag{1}$$

where $g_A = 1.25$ and $F_\pi = 190$ MeV, and C_S and C_T are constants whose values can be fit to the two nucleon-nucleon scattering lengths. (Spatial vectors are boldface; arrows denote isovectors.) In this talk I will describe some current research that takes us beyond the leading terms provided by Eq. (1), in three different directions.

1 Isospin Breaking Corrections

Using Eq. (1) in the tree approximation gives just the first term in an expansion in powers of q, the typical value of the pion and nucleon three-momenta and pion mass. Higher terms in the expansion are generated[2] by including more derivatives in $\mathcal{L}_{\text{eff}}$, each of which contributes a factor of order q, or more nucleon fields, each of which contributes a factor of order $q^{1/2}$, or more factors of u and d quark masses, each of which contributes a factor of order q^2 (because $m_\pi^2 \propto m_u + m_d$), or loops in the Feynman graphs, each of which contributes a factor q^2. These corrections have been explored in great detail (including also strange

---

* Research supported in part by the Robert A. Welch Foundation and NSF Grant PHY 9009850. E-mail address: weinberg@physics.utexas.edu.

[2] In dealing with nuclear forces, the tree approximation must be applied to the nucleon-nucleon potential, rather than to the scattering amplitude.

particles), especially by Gasser and Leutwyler[3] for pions and single nucleons, and more recently by Bernard, Kaiser, and Meissner[4] for pion photoproduction and by Ordoñez and van Kolck[5] for multinucleon problems. Here I want to concentrate on the quark mass corrections, which produce violations of isospin conservation. This is of renewed interest now, because as I learned from Aron Bernstein there are plans to measure the π^0-nucleon scattering length, which is sensitive to isospin violating terms in the effective Lagrangian.

The quark mass terms in quantum chromodynamics may be put in the form

$$\mathcal{L}_{\text{mass}} = -(m_u + m_d)V_4 - (m_u - m_d)A_3 \tag{2}$$

where

$$V_4 = \frac{1}{2}(\bar{u}u + \bar{d}d) \qquad A_3 = \frac{1}{2}(\bar{u}u - \bar{d}d) . \tag{3}$$

The operators V_4 and A_3 are spatial scalars, and components of independent chiral four-vectors A_α and V_α. We must add terms to the effective Lagrangian with these transformation properties. From just the pion field alone (with no derivatives) we can construct no term A_3 and just one term V_4, the term in (1) proportional to m_π^2. From pion fields and a nucleon bilinear (but no derivatives) we can construct only one term of each type

$$V_4 \propto \left(\frac{1 - \vec{\pi}^2/F_\pi^2}{1 + \vec{\pi}^2/F_\pi^2}\right) \bar{N}N$$

$$A_3 \propto \bar{N}t_3N - \frac{2}{F_\pi^2}\left(\frac{\pi_3}{1 + \vec{\pi}^2/F_\pi^2}\right) \bar{N}\vec{t}\cdot\vec{\pi}N$$

where N is the nucleon doublet. Therefore in the effective Lagrangian we must include a term

$$\delta\mathcal{L}_{eff} = -\frac{A}{2}\left(\frac{1 - \vec{\pi}^2/F_\pi^2}{1 + \vec{\pi}^2/F_\pi^2}\right) \bar{N}N$$

$$-B\left[\bar{N}t_3N - \frac{2}{F_\pi^2}\left(\frac{\pi_3}{1 + \vec{\pi}^2/F_\pi^2}\right) \bar{N}\vec{t}\cdot\vec{\pi}N\right] \tag{4}$$

where A and B are constants proportional to the coefficients in (2):

$$A \propto m_u + m_d \qquad B \propto m_u - m_d . \tag{5}$$

The pion-nucleon terms in (1) have single derivatives, which contribute factors of order q, while the quark masses in (4) contribute factors of order q^2, so the effects of (4) are leading corrections, suppressed by just *one* factor of q.

The terms in (4) make a contribution to the scattering length for the pion-nucleon scattering process $\pi_a + N \rightarrow \pi_b + N$ (written as a matrix in the isospin space of the nucleon):

$$\delta a_{ba} = \frac{1}{4\pi[1 + m_\pi/m_N]} \times \left[\frac{2A}{F_\pi^2}\delta_{ab} + \frac{2B}{F_\pi^2}(t_a\delta_{3b} + t_b\delta_{3a})\right] . \tag{6}$$

The A term is the notorious σ-term. The B term was also described years ago[6]. What (I think) is new here is the full isospin-breaking term[7] in the effective action (4), which allows us easily to calculate the effect of isospin violations in other processes, such as $\pi + N \to \pi + \pi + N$. One immediate consequence of (4) is that isospin violation never appears in any process that does not involve at least one neutral pion.

Inspection of (4) shows that the constants A and B are related to the shifts δm_n and δm_p in the nucleon masses due to the quark masses:

$$A = \delta m_p + \delta m_n = (m_u + m_d) < p|(\bar{u}u + \bar{d}d|p >$$
$$B = \delta m_p - \delta m_n = (m_u - m_d) < p|(\bar{u}u - \bar{d}d|p > . \tag{7}$$

Unfortunately the nucleon expectation value of $\bar{u}u + \bar{d}d$ is not related in any simple way to observable quantities, so it is not possible to calculate A without dynamical assumptions. On the other hand, B is given by $SU(3)$ symmetry as:

$$B \simeq \left(\frac{m_u - m_d}{m_s}\right)(m_\Xi - m_\Sigma) \approx -2.5 \text{ MeV} . \tag{8}$$

This satisfies an important consistency condition. The full proton-neutron mass difference is the sum of B and an electromagnetic term, which is almost certainly positive, so we must have $B < m_p - m_n = -1.3$ MeV, and we do. It will be very interesting to see if experiments on low energy π^0-nucleon interactions confirm these predictions.

2 General Effective Lagrangians

The structure of the effective Lagrangian (1) is dictated by its invariance under $SU(2) \times SU(2)$ spontaneously broken to $SU(2)$, which induces on the pion field the non- linear symmetry transformation[8]:

$$\delta\vec{\pi} = \vec{\epsilon}(1 - \vec{\pi}^2) + 2\vec{\pi}(\epsilon \cdot \vec{\pi}) . \tag{9}$$

This was generalized by Callan, Coleman, Wess, and Zumino[9] to any group G broken to any subgroup $H \subset G$, in which case an element $g \in G$ induces on the general Goldstone boson fields π^a the transformation $\pi \to \pi'$, defined by

$$gU(\pi) = U(\pi')h(\pi, g) , \tag{10}$$

where $h \in H$ and $U(\pi)$ is a representative of the coset space G/H, parameterized by the Goldstone boson fields. Now, we know how to construct G-invariant Lagrangian densities our of covariant derivatives of π^a, but this is not the most general possibility. We also can have a Lagrangian density that under G transformations changes by a derivative

$$\mathcal{L} \to \mathcal{L} + \partial_\mu \mathcal{F}^\mu$$

so that the action is still invariant. Wess and Zumino[10] pointed out that the ABJ anomaly from fermion loops yields such a term in the effective Lagrangian for the case of $SU(3) \times SU(3)$ spontaneously broken to $SU(3)$:

$$\mathcal{L}_{\text{WZ}} = \frac{1}{6\pi^2} \epsilon^{\mu\nu\rho\sigma} \operatorname{Tr} \{ \Pi \, \partial_\mu \Pi \, \partial_\nu \Pi \, \partial_\rho \Pi \, \partial_\sigma \Pi \} + O(\Pi^6) \tag{11}$$

where

$$\Pi \equiv \frac{1}{2} \lambda_a \pi^a ,$$

λ_a are the Gell-Mann matrices (with $\operatorname{Tr} \lambda_a^2 = 2$), and the coset representatives $U(\pi)$ are chosen as

$$U(\pi) = e^{i\Pi} . \tag{12}$$

Witten[11] then showed that although the correction $\int d^4x \mathcal{L}_{\text{WZ}}$ to the action is not the integral of a G-invariant Lagrangian density over spacetime, it *is* the integral of a G-invariant Lagrangian density $\mathcal{L}_{\text{WZW}}$ over a five-dimensional ball that has four-dimensional spacetime (Euclideanized and compactified to a four-sphere) as its boundary. This raises the question whether there are any other terms in the effective Lagrangian density, not necessarily related to ABJ anomalies, that although not invariant under G nevertheless yield G-invariant contributions to the action. Have we been missing something?

This question has now been answered by Eric D'Hoker and myself[12], with help at the start from Eddie Farhi. Our analysis is in four steps:

(a) As in ref. [11], we first compactify spacetime to a sphere S_4 by assuming that all fields approach definite limits for $x^\mu \to \infty$. If the homotopy group $\pi_4(G/H)$ is trivial (as is the case for $SU(N) \otimes SU(N)$ spontaneously broken to $SU(N)$), or if $U(\pi(x))$ belongs to the trivial element of $\pi_4(G/H)$, then we may introduce a smooth function $\tilde{\pi}^a(x, t_1)$, such that

$$\tilde{\pi}^a(x, 1) = \pi^a(x) \qquad \tilde{\pi}^a(x, 0) = 0.$$

In this way spacetime is extended to a five-ball B_5 with boundary S_4 and coordinates x^μ and t_1. The action may then be written in the five-dimensional form

$$S[\pi] = \int_{B_5} d^4x \, dt_1 \, \mathcal{L}_1$$

where

$$\mathcal{L}_1 \equiv \frac{\delta I[\tilde{\pi}]}{\delta \tilde{\pi}^a(x, t_1)} \frac{\partial \tilde{\pi}^a(x, t_1)}{\partial t_1} .$$

(b) It is straightforward to show that if the action I is invariant under G, then the density $\mathcal{L}_1$ is also invariant under G. Thus any G-invariant term in the action can be written in the Witten form, as a five-dimensonal integral of an invariant density.

(c) From the definition of $\mathcal{L}_1$ in terms of $\delta I/\delta\pi$, we learn not only that it is G-invariant, but also that it satisfies an integrability condition, which implies that

$\mathcal{L}_1$ is a component of a G-invariant closed five-form Ω_5. That is, in the language of differential forms:

$$I = \int_{B_5} \Omega_5 \qquad d\Omega_5 = 0 . \tag{13}$$

Now, if Ω_5 is exact, then the four-form F_4 satisfying $\Omega_5 = dF_4$ can be chosen to be G-invariant, in which case I is the four-dimensional integral $\int_{S_4} F_4$ over S_4 of a G-invariant Lagrangian density. Hence the allowed terms in the four- dimensional Lagrangian density that are *not* G- invariant are in one-to-one correspondence with closed five-forms, modulo exact five-forms. These are the generators of the fifth de Rham cohomology $H^5(G/H; \mathbf{R})$ of the space G/H.

(d) It only remains to find the five-forms that generate $H^5(G/H; \mathbf{R})$. These are all known where G/H is itself a simple group. If $G/H = SU(N)$ with $N \geq 3$ then $H^5(G/H; \mathbf{R})$ has a single generator:

$$\Omega_5 = \frac{i}{240\pi^2} \text{Tr} \left\{ U^{-1}\, dU \wedge U^{-1}\, dU \wedge U^{-1} dU \right.$$
$$\left. \wedge U^{-1}\, dU \wedge U^{-1}\, dU \right\} . \tag{14}$$

For the QCD case of $SU(3) \times SU(3)$ spontaneously broken to $SU(3)$, we have $G/H = SU(3)$, and the unique generator (15) is the Wess-Zumino-Witten five dimensional Lagrangian density. So at least as far as the strong interactions at low energy are concerned, we have not been missing anything. Where G/H is any simple Lie group other than $SU(N)$ with $N \geq 3$, the cohomology is trivial, and so the four-dimensional Lagrangian density must be G-invariant. This includes the original case of $SU(2) \times SU(2)$ spontaneously broken to $SU(2)$, where $G/H = SU(2)$.

Of course, a great deal is known about the fifth cohomology groups $H^5(G/H; \mathbf{R})$ even where G/H is not a simple Lie group. One interesting result is that if G itself is one of the simple Lie groups other than $SU(N)$ with $N \geq 3$, then G/H has trivial fifth cohomology group for any subgroup $H \subset G$, and so the Lagrangian density must be G-invariant. These groups G also have vanishing triangle anomalies for fermions in any representation of G, because in all representations the generators t_α satisfy

$$\text{Tr} \left\{ t_\alpha (t_\beta t_\gamma + t_\gamma t_\beta) \right\} = 0$$

so here we would not have expected a Wess-Zumino-Witten term anyway. D'Hoker is continuing with the study of general coset spaces G/H, to map out the detailed relation between the possibility of ABJ anomalies and non-G-invariant terms in the four-dimensional Lagrangian density. It is remarkable that in all the cases we have studied, the possibility of anomalies could have been discovered (or ruled out) within the effective field theory of soft Goldstone bosons, without ever looking at a fermion loop.

3 The Nonrelativistic Quark Model: Sum Rules *vs* Large N_c

The derivation of results of the non-relativistic quark model from quantum chromodynamics has long remained problematical. Recently, the large N_c approximation[13] has been used[14-18] to derive some of the quark model results for baryons. Specifically, it is found that the nucleon doublet is connected by one-pion transitions to a 'tower' of narrow baryon states, with spin and isospin $J = T = 1/2, 3/2, \cdots N_c/2$, and positive parity. According to this picture, the amplitudes for pion transitions between the tower states together with the spin and isospin operators form a contracted $SU(4)$ algebra[15-17], under which the baryon tower transforms irreducibly. The one-pion transition amplitudes derived in this way are just those of the non- relativistic quark model.

I want to point out that strikingly similar results can be derived in a very different way. In this approach no use is made of the large N_c approximation or dynamical models like the non-relativistic quark model or Skyrme model, beyond the qualitative assumption that there is a tower of narrow baryon states, with spin and isospin $J = T = 1/2, 3/2, \cdots N_c/2$, and positive parity, that are connected only to each other by one-pion transitions. By the use of well-known sum rules saturated with narrow tower states, we will be able to show (1) that the tower states are degenerate[3], and (2) that the pion transition amplitudes are part of an *uncontracted* $SU(4) \times O(3)$ Lie algebra, under which the baryonic tower transforms as a symmetric rank-N_c tensor[4] of $SU(4)$ and a singlet under $SO(3)$. Here N_c is any integer, including $N_c = 3$; we keep N_c a free parameter to facilitate comparison with work based on the large N_c limit. By relying hardly at all here on the large N_c approximation, this approach offers a prospect of a more convincing derivation of the main results of the non-relativistic quark model.

To derive these results, lets first recall the relevant sum rules and their algebraic consequences. We usually think of spontaneously broken symmetries like $SU(2) \times SU(2)$ as being manifested entirely in low energy theorems for the interaction of Goldstone bosons. There are similar low-energy theorems for the interactions of soft photons. But these low energy theorems when married to dispersion relations yield sum rules, like the celebrated Adler-Weisberger, Drell-Hearn, and Cabibbo-Radicati sum rules. Other sum rules known as superconver-

---

[3] This is a somewhat surprising result. In general for large N_c it is only the lower tower states with $T = J = O(1)$ that become degenerate when $N_c \to \infty$. The work described here shows that in order to understand the splittings of the baryon tower masses, it will be necessary to take into account single-pion transitions from tower to non-tower states, rather than $1/N_c$ corrections to matrix elements between tower states.

[4] Of course, if one also assumes that N_c is large, the leading terms in the matrix elements of the $SU(4)$ generators between low tower states in this representation will grow as N_c. Since the spin and isospin matrices are for these states are of order N_c^0, they and the leading terms in the pion transition amplitudes for $N_c \to \infty$ will furnish a contracted $SU(4)$ algebra, as found in references [15] - [17].

gence relations are provided by assumptions that limit the asymptotic behaviour of scattering amplitudes at high energy. When we assume that these sum rules are saturated by any number of particles and narrow resonances, and write down all the sum rules for scattering of Goldstone bosons and/or photons not only on stable targets but also on all the resonances, we find that they take remarkably similar algebraic forms[19-21]. This is a very old story, going back more years than I care to remember. The new thing I want to discuss here is the solution of these sum rules under a specific assumption about the menu of baryonic spins and isospins for general N_c.

Consider pion scattering on an arbitrary hadronic target. The saturated sum rules can be expressed in terms of the matrix elements for pion transitions $\alpha \rightarrow \beta + \pi_a$ between stable or resonant states α, β with helicities λ and λ'. For any such transition, we can adopt a Lorentz frame in which the initial and final states have *collinear* momenta $\mathbf{p}$ and $\mathbf{p'}$, say, in the 3-direction. Using invariance under rotations around the 3-axis and boosts along the 3-axis, these matrix elements may be written as:

$$< \beta, \mathbf{p'}, \lambda'; \pi_a, q|S|\alpha, \mathbf{p}, \lambda > \equiv \frac{2(m_\alpha^2 - m_\beta^2)}{(2\pi)^{9/2}(8q^0 p^0 p'^0)^{1/2} F_\pi}$$
$$\times [X_a(\lambda)]_{\beta\alpha}\, \delta_{\lambda'\lambda}\delta^4(p' + q - p) \tag{15}$$

with a coefficient $[X_a(\lambda)]_{\beta\alpha}$ that is independent of $|\mathbf{p}|$ and $|\mathbf{p'}|$. (The axial coupling $g_A \simeq 1.25$ is just the helicity $+1/2$ proton-neutron element of the matrix $X_1 + iX_2$.) Parity conservation tells us that

$$[X_a(-\lambda)]_{\beta\alpha} = -\Pi_\alpha \Pi_\beta (-1)^{J_\alpha - J_\beta}[X_a(\lambda)]_{\beta\alpha} \tag{16}$$

where Π_α and J_α are the parity and spin of the baryonic state α. Isospin invariance tells us that

$$[T_a, X_b(\lambda)] = i\epsilon_{abc}X_c(\lambda) \quad \text{where} \quad [T_a, T_b] = i\epsilon_{abc}T_c .$$

In this language, when all of the Adler-Weisberger sum rules for scattering of a pion on all single-hadron states (either stable particles or narrow resonances) are saturated with single-hadron states, these sum rules read simply[19]

$$[X_a(\lambda), X_b(\lambda)] = i\epsilon_{abc} T_c . \tag{17}$$

Thus for each helicity the reduced pion matrix elements $X_a(\lambda)$ and the isospin matrix T_a together form an $SU(2) \times SU(2)$ algebra[5]. There are also two superconvergence relations that follow from the absence of $T = 2$ Regge trajectories with $\alpha(0) > 0$ in the cross channel. One takes the form[19]

$$[X_a(\lambda), [X_b(\lambda), m^2] \propto \delta_{ab} . \tag{18}$$

---

[5] This result can be generalized to arbitrary groups G broken to arbitrary subgroups H: the reduced amplitudes for Goldstone boson emission together with the unbroken symmetry generators furnish a representation of the algebra of G. This holds even where G/H is not a symmetric space, i. e., where terms linear in the broken as well as the unbroken generators appear in the commutators of the broken generators with each other.

The other is a spin-flip superconvergence relation, that connects different helicities[20]

$$X_a(\lambda \pm 1)X_b(\lambda \pm 1)\,m - X_a(\lambda \pm 1)\,m\,X_b(\lambda) - X_b(\lambda \pm 1)\,m\,X_a(\lambda)$$
$$+m\,X_b(\lambda)X_a(\lambda) \; \propto \; \delta_{ab} \tag{19}$$

where m is the hadronic mass matrix.

To derive the results of the quark model for baryon states, we shall make use of two lemmas, that may also have applications in other contexts.

LEMMA 1: Any set of hadronic states that furnish a representation of the commutation relations (17) and (18), in which for each helicity any given isospin appears at most once, must be degenerate.

Proof: Eq. (18) may be written as $[X_a(\lambda), [X_b(\lambda), m^2]] = m_4^2(\lambda)\delta_{ab}$. By taking the commutator of this with $X_c(\lambda)$ and using the Jacobi identity, it is easy to see that $[X_c(\lambda), m_4^2(\lambda)] = [X_c(\lambda), m^2]$. Hence the mass-squared matrix may be written $m^2 = m_4^2(\lambda) + m_0^2(\lambda)$, where $m_0^2(\lambda)$ is a chiral scalar, satisfying $[X_c(\lambda), m_0^2] = 0$. Also, $m_4^2(\lambda)$ and $m_b^2(\lambda) \equiv [X_b(\lambda), m_4^2(\lambda)]$ form a chiral four-vector, in the sense that $[X_a(\lambda), m_b^2(\lambda)] = m_4^2(\lambda)\delta_{ab}$. Now, since we assume that for a given helicity, each isospin occurs just once, each isospin for a given helicity can come from just one irreducible representation of $SU(2) \times SU(2)$. The mass m commutes with isospin, so m^2 can have matrix elements only between baryonic states with the same isospin, and hence belonging to the same representation of $SU(2) \times SU(2)$. But a $(1/2, 1/2)$ operator like $m_4^2(\lambda)$ can have no matrix elements between two states that belong to the same irreducible representation (A, B) of $SU(2) \times SU(2)$, so in this representation all matrix elements of m_4^2 must vanish. This leaves us with $m^2 = m_0^2$, and since this commutes with X_a all hadron states connected by one-pion transitions must have the same mass.

LEMMA 2: Any set of degenerate hadronic states of the same parity that furnish a representation of the commutation relations (17) and (19) also furnish a representation of an $SU(4) \times O(3)$ algebra with $SU(4)$ generators T_a, S_α, and D_{ai} and $O(3)$ generators $\tilde{S}_i \equiv J_i - S_i$, satisfying the commutation relations[6]

$$[T_a, T_b] = i\epsilon_{abc}T_c \qquad [S_i, S_j] = i\epsilon_{ijk}S_k \qquad [T_a, S_i] = 0 \tag{20}$$

$$[T_a, D_{bi}] = i\epsilon_{abc}D_{ci} \qquad [S_i, D_{aj}] = i\epsilon_{ijk}D_{ak} \tag{21}$$

$$[D_{ai}, D_{bj}] = i\delta_{ij}\epsilon_{abc}T_c + i\delta_{ab}\epsilon_{ijk}S_k \tag{22}$$

$$[\tilde{S}_i, \tilde{S}_j] = i\epsilon_{ijk}\tilde{S}_k \tag{23}$$

$$[\tilde{S}_i, T_a] = [\tilde{S}_i, S_j] = [\tilde{S}_i, D_{aj}] = 0 \tag{24}$$

where

$$[D_{a3}]_{\lambda'\beta,\lambda\alpha} = \delta_{\lambda'\lambda}[X_a(\lambda)]_{\beta\alpha} \tag{25}$$

and $J_i = \tilde{S}_i + S_i$ is the usual spin matrix acting on helicity indices, with

$$[J_3]_{\lambda'\beta,\lambda\alpha} = \delta_{\lambda'\lambda}\delta_{\beta\alpha}\lambda \,. \tag{26}$$

---

[6] We use a, b, c, etc. for isovector indices and i, j, k, etc. for spatial vector indices.

Proof: For a transition between equal parity states, in the rest frame of the initial particle the invariant pion transition amplitude $< \beta, \mathbf{p}', \lambda'; \pi_a, q|S|\alpha, \mathbf{p}, \lambda >$ $\times \sqrt{q^0 p'^0 p^0}$ must be an odd function of the momentum of the final particle, whose magnitude is proportional to $m_\alpha^2 - m_\beta^2$. Hence for $m_\alpha^2 - m_\beta^2 \to 0$, this invariant amplitude must be proportional to a linear combination of components of the final momentum vector. Rotational invariance requires that the coefficients must also form a three-vector. In particular, for $\mathbf{p}'$ in the 3- direction the coefficient of p'_3, which is proportional to X_a, must be the third component of a quantity that transforms like a spatial 3-vector and an isovector, in the sense that $\delta_{\lambda',\lambda}[X_a(\lambda)]_{\beta\alpha} = [D_{a3}]_{\lambda'\beta,\lambda\alpha}$ where $[T_a, D_{bi}] = i\epsilon_{abc}D_{ci}$ and $[J_i, D_{aj}] = i\epsilon_{ijk}D_{ak}$. Next, we must consider the commutation relations of the D_{ai} with each other. The commutators of two D's may be written as a sum of two terms, one symmetric in space indices and antisymmetric in isospin indices, and the other vice versa:

$$[D_{ai}, D_{bj}] = i\epsilon_{abc}A_{ij,c} + i\epsilon_{ijk}B_{ab,k}$$

with $A_{ij,c} = A_j$ and $B_{ab,k} = B_{ba,k}$. The commutation relation (17) now takes the form $A_{33,a} = T_a$. From rotational invariance (or formally, by taking repeated commutators with J_i), we easily see then that $A_{ij,a} = \delta_{ij}T_a$. To find $B_{ab,k}$, we must use the spin-flip superconvergence relation (19), which can be rewritten in the form $[D_{a3}, [D_{b3}, m(J_1 \pm iJ_2)]] \propto \delta_{ab}$, or, since the states connected by D_{ai} are degenerate, $[D_{a3}, D_{b1} \pm iD_{b2}] \propto \delta_{ab}$. It follows that $B_{ab,2}$ and $B_{ab,1}$ are proportional to δ_{ab}, and by rotational invariance the same is true of $B_{ab,3}$, so[7] $B_{ab,i} = \delta_{ab}S_i$, verifying Eq. (22). From (22), we have $S_i = -i\epsilon_{ijk}[D_{aj}, D_{ak}]/6$. Using (22) again to calculate the commutator of this with D_{bj}, we easily obtain the commutator $[S_i, D_{bj}] = i\epsilon_{ijk}D_{bk}$, and using this together with the above expression for S_j in terms of the D's, we also find $[S_i, S_j] = \epsilon_{ijk}S_k$, verifying the remainder of the commutators (20)-(21). We have already mentioned that D_{aj} is a 3-vector, in the sense that $[J_i, D_{aj}] = i\epsilon_{ijk}D_{ak}$, so the same is true of S_j. It follows then that $\tilde{S}_i \equiv J_i - S_i$ satisfies the commutation relations (23) and (24).

To apply Lemma 1 to the tower states, we note that for each helicity λ, the tower contains isospins $T = |\lambda|, |\lambda| + 1, \cdots N_c/2$. Since each isospin occurs just once, for a given helicity each isospin can come from just one irreducible representation of $SU(2) \times SU(2)$. Therefore according to Lemma 1, the tower states must be degenerate.

Lemma 2 then tells us that the pion transition amplitudes are part of an $SU(4) \times O(3)$ algebra. It is easy to see that the baryons transform under $SU(4) \times O(3)$ as a symmetric $SU(4)$ tensor[8] of rank N_c and an $SO(3)$ singlet, because this

---

[7] This result was obtained in reference [19] by a weak argument, that the algebra containing the pion transition amplitudes should not contain any $T = 2$ operators. The present argument, based on the superconvergence relation (19), avoids this handwaving.

[8] There are actually two of these representations, the contravariant and covariant tensors of rank N_c. They differ only in the sign of D_{ai}, so we can choose either of these representations by adjusting the sign of the one-pion state.

is the only representation of $SU(4) \times O(3)$ that contains just the spins and isospin states of the baryon tower. The matrix elements of D_{ai} in this representation may be calculated by representing it by $\sum \sigma_i t_a$, just as in the non-relativistic quark model. *Thus once we assume that general Adler-Weisberger and superconvergence sum rules are saturated by the tower states, the other consequences of the non-relativistic quark model for pion transitions and g_A follow immediately from these sum rules, with no further need for the large N_c approximation.*

All of the above results apply also for baryons that contain some number N_h of heavy quarks. Since pion transition amplitudes and baryon masses are independent of the spin 3-component of the heavy quarks, it is only necessary to replace N_c everywhere above with $N_c - N_h$.

I am grateful for helpful conversations about the large N_c approximation with Howard Georgi, Vadim Kaplunovsky, and Aneesh Manohar.

References

[1]   The terms in (1) involving pions alone or pions and a single nucleon bilinear were given by S. Weinberg, Phys. Rev. Lett. **18**, 1 88 (1967). It was realized later that the terms involving C_S and C_T make contributions of the same order in small energies: S. Weinberg, Phys. Lett. **B251**, 288 (1990); Nucl. Phys. **B363**, 3 (1991).

[2]   S. Weinberg, Physica **96A**, 327 (1979).

[3]   J. Gasser and H. Leutwyler, Phys. Lett. **125B**, 321, 325 (1985); Ann. Phys. **158**, 142 (1984).

[4]   V. Bernard, N. Kaiser, and U-G. Meissner, Nucl. Phys. **B 383**, 442 (1992); Ulf-G. Meissner, Lectures delivered at the XXXII. Internationale Universitätswochen für Kern- und Teilchenphsik, Schladming, February 24 - March 6, 1993, hep-ph 9303298.

[5]   C. Ordóñez and U. van Kolck, Phys. Lett. B **291**, 459 (1992); C. Ordóñez, L. Ray, and and U. van Kolck, Phys. Rev. Lett. **72**, 1982 (1994).

[6]   S. Weinberg, Transactions of the N. Y. Academy of Sciences **38**, 185 (1977).

[7]   After this talk was presented, H. Leutwyler informed me of a paper by A. Krause, Helv. Phys. Acta **63**, 3 (1990). This paper gave an $SU(3) \times SU(3)$ effective Lagrangian that includes a mass term from which it would be possible to derive the B term in (4), but did not consider the implications of this Lagrangian for pion nucleon interactions.

[8]   S. Weinberg, Phys. Rev. **166**, 1568 (1968).

[9]   S. Coleman, J. Wess and B. Zumino, Phys. Rev. **177**, 2239 (1969) ; C.G. Callan, S. Coleman, J. Wess and B. Zumino, Phys. Rev. **177**, 2247 (1969).

[10]   J. Wess and B. Zumino, Phys. Lett. **37B**, 95 (1971).

[11]   E. Witten, Nucl. Phys. **B223**, 422 (1983).

[12]   E. D'Hoker and S. Weinberg, UCLA-Texas preprint, to be published in Physical Review D.

[13]   G. 't Hooft, Nucl. Phys. **B 72**, 461 (1974); S. Coleman, in *Aspects of Symmetry* (Cambridge University Press, Cambridge, 1985).

[14]   E. Witten, Nucl. Phys. **B160**, 57 (1979).

[15]   J.-L. Gervais and B. Sakita, Phys. Rev. Lett. **52**, 87 (1984).

[16]   R. Dashen and A. V. Manohar, Phys. Lett. **B315**, 425, 438 (1993); E. Jenkins, Phys. Lett. **315**, 431, 447 (1993); R. Dashen, E. Jenkins, and A. V. Manohar, Phys. Rev. D **49**, 4713 (1994).

[17]   C. D. Carone, H. Georgi, and S. Osofsky, Phys. Lett. **F 322**, 227 (1994); C. D. Carone, H. Georgi, L. Kaplan, and D. Morin, Harvard preprint HUTP-94/A008 (1994).

[18]   A. Wirzba, M. Kirchbach, and D. O. Riska, Darmstadt-Helsinki preprint (1993), hep-ph/9311299.

[19]   S. Weinberg, Phys. Rev. **177**, 2604 (1969). Some algebraic errors in the Appendix to this paper are corrected here.

[20]   S. Weinberg, Phys. Rev. Lett. **22**, 1023 (1969).

[21]   S. Weinberg, in *Lectures on Elementary Particles and Quantum Field Theory, 1970 Brandeis University Summer Institute on Theoretical Physics* (M.I.T. Press, Cambridge, MA, 1970), pp. 285-393.

Foundations and scope of chiral perturbation theory

H. Leutwyler [1]

Universität Bern, Institut für theoretische Physik, Sidlerstr. 5, CH-3012 Bern, Switzerland

Abstract. The aim of this introductory lecture is to review the arguments, according to which the symmetry properties of the strong interaction reveal themselves at low energies. I first discuss the symmetries of QCD, then sketch the method used to work out their implications and finally take up a few specific issues, where new experimental results are of particular interest to test the predictions.

Chromodynamics is a gauge theory. The form of the interaction among the gluons and quarks is fully determined by gauge invariance. This implies, in particular, that the various different quark flavours, $u, d, \ldots$ interact with the gluons in precisely the same manner. As far as the strong interaction is concerned, the only distinction between, say, an s-quark and a c-quark is that the mass is different. In this respect, the situation is the same as in electrodynamics, where the interaction of the charged leptons with the photon is also universal, such that the only difference between e, μ and τ is the mass. As an immediate consequence, the properties of a bound state like the $\Lambda_s = (uds)$ are identical with those of the $\Lambda_c = (udc)$, except for the fact that m_c is larger than m_s.

1 Isospin symmetry

A striking property of the observed pattern of bound states is that they come in nearly degenerate *isospin* multiplets: (p, n), (π^+, π^0, π^-), (K^+, K^0), ... In fact, the splittings within these multiplets are so small that, for a long time, isospin was taken for an *exact* symmetry of the strong interaction; the observed small mass difference between neutron and proton or K^0 and K^+ was blamed on the electromagnetic interaction. We now know that this picture is incorrect: the bulk of isospin breaking does not originate in the electromagnetic fields, which surround the various particles, but is due to the fact that the d-quark is somewhat heavier than the u-quark.

From a theoretical point of view, the quark masses are free parameters — QCD makes sense for any value of $m_u, m_d, \ldots$ It is perfectly legitimate to compare the real situation with a theoretical one, where some of the quark masses are given values, which differ from those found in nature. In connection with isospin symmetry, the theoretical limiting case of interest is a fictitious world, with $m_u = m_d$. In this limit, the flavours u and d become indistinguishable. The Hamiltonian acquires an exact

---

[1] Work supported in part by Schweizerischer Nationalfonds

symmetry with respect to the transformation

$$u \; \to \; \alpha u + \beta d \qquad\qquad V = \begin{pmatrix} \alpha & \beta \\ \gamma & \delta \end{pmatrix} \; ,$$
$$d \; \to \; \gamma u + \delta d \quad ,$$

provided the 2×2 matrix V is unitary, $V \in U(2)$. Even for $m_u \neq m_d$, the Hamiltonian of QCD is invariant under a change of phase of the quark fields. The extra symmetry, occurring if the masses of u and d are taken to be the same, is contained in the subgroup SU(2), which results if the phase of the matrix V is subject to the condition $\det V = 1$. The above transformation law states that u and d form an isospin doublet, while the remaining flavours s, c, ... are singlets.

In reality, m_u differs from m_d. The isospin group SU(2) only represents an *approximate* symmetry. The piece of the QCD Hamiltonian, which breaks isospin symmetry, may be exhibited by rewriting the mass term of the u and d quarks in the form

$$m_u \, \bar{u} u + m_d \, \bar{d} d = \tfrac{1}{2}(m_u + m_d)(\bar{u} u + \bar{d} d) + \tfrac{1}{2}(m_d - m_u)(\bar{d} d - \bar{u} u) \; .$$

The remainder of the Hamiltonian is invariant under isospin transformations and the same is true of the operator $\bar{u} u + \bar{d} d$. The QCD Hamiltonian thus consists of an isospin invariant part $\bar{H}_0$ and a symmetry breaking term $\bar{H}_{sb}$, proportional to the mass difference $m_d - m_u$,

$$H_{\text{QCD}} = \bar{H}_0 + \bar{H}_{sb} \; , \qquad \bar{H}_{sb} = \tfrac{1}{2}(m_d - m_u)\int d^3x\,(\bar{d} d - \bar{u} u) \; . \tag{1}$$

The strength of isospin breaking is controlled by the quantity $m_d - m_u$, which plays the role of a *symmetry breaking parameter*. The fact that the multiplets are nearly degenerate implies that the operator $\bar{H}_{sb}$ only represents a small perturbation — the mass difference $m_d - m_u$ must be very small. QCD thus provides a remarkably simple explanation for the fact that the strong interaction is nearly invariant under isospin rotations: it so happens that the difference between m_u and m_d is small — this is all there is to it.

The symmetry breaking also shows up in the properties of the vector currents, e.g. in those of $\bar{u}\gamma^\mu d$. The integral of the corresponding charge density over space, $I^+ = \int d^3x \, u^\dagger d$, is the isospin raising operator, converting a d-quark into a u-quark. The divergence of the current is given by

$$\partial_\mu(\bar{u}\gamma^\mu d) = i\,(m_u - m_d)\bar{u} d \; , \tag{2}$$

and only vanishes for $m_u = m_d$, the condition for the charge I^+ to be conserved. In the symmetry limit, there are three such conserved charges, the three components of isospin, $\vec{I} = (I^1, I^2, I^3)$. The isospin raising operator considered above is the combination $I^+ = I^1 + i\,I^2$. Since $\bar{H}_0$ is invariant under isospin rotations, it conserves isospin,

$$[\vec{I}, \bar{H}_0] = 0 \; . \tag{3}$$

2 Eightfold way

On the basis of the few strange particles, which had been discovered in the course of the 1950's, Gell-Mann and Ne'eman [1] inferred that the strong interaction possesses a further approximate symmetry, of the same qualitative nature as isospin, but more strongly broken. The symmetry, termed the *eightfold way*, played a decisive role in unravelling the quark degrees of freedom. By now, it has become evident that the mesonic and baryonic levels are indeed grouped in multiplets of SU(3) — singlets, octets, decuplets — and there is also good phenomenological support for the corresponding symmetry relations among the various observable quantities.

In the framework of QCD, eightfold way symmetry occurs in the theoretical limit, where the three lightest quarks are given the same mass, $m_u = m_d = m_s$. The Hamiltonian then becomes invariant under the transformation

$$\begin{pmatrix} u \\ d \\ s \end{pmatrix} \rightarrow V \begin{pmatrix} u \\ d \\ s \end{pmatrix} \qquad V \in \text{SU}(3)$$

of the quarks fields and the spectrum of the theory consists of degenerate multiplets of this group. The degeneracy is lifted by the mass differences $m_s - m_d$ and $m_d - m_u$, which represent the symmetry breaking parameters in this case. Since the eightfold way does represent an approximate symmetry of the strong interaction, both of these mass differences must be small. Moreover, the observed level pattern requires $|m_d - m_u| \ll |m_s - m_d|$.

Formally, the above discussion may be extended to include additional flavours. One may even consider the theoretical limit, where all of the quarks are given the same mass. The extension, however, does not correspond to an approximate symmetry. The lightest pseudoscalar bound state with the quantum numbers of $\bar{d}c$, e.g., sits at $M_{D^+} \simeq 1.87\,\text{GeV}$. If the mass of the charmed quark is set equal to m_u, this state becomes degenerate with the π^+. Clearly, the mass difference $m_c - m_u$, which plays the role of a symmetry breaking parameter in this case, does not represent a small perturbation. We do not know why the quark masses follow the pattern observed in nature, nor do we understand the equally queer pattern of lepton masses. It so happens that the mass differences between u, d and s are small, such that the eightfold way represents a decent approximate symmetry.

3 Chiral symmetry

The approximate symmetries discussed above explain why the bound states of QCD exhibit a multiplet pattern, but they do not account for an observation which is equally striking and which plays a crucial role in strong interaction physics — the mass gap of the theory, M_π, is remarkably small. The approximate symmetry, hiding behind this observation, was discovered by Nambu [2]. It originates in a phenomenon, which is well-known from neutrino physics: right- and left-handed components of *massless* fermions do not communicate.

The symmetry, which forbids right-left-transitions, manifests itself in the properties of the axial vector currents, such as $\bar{u}\gamma^\mu\gamma_5 d$. The corresponding continuity equation reads

$$\partial_\mu(\bar{u}\gamma^\mu\gamma_5 d) = i\,(m_u + m_d)\,\bar{u}\gamma_5 d \ . \tag{4}$$

While the divergence of the vector current $\bar{u}\gamma^\mu d$ is proportional to the difference $m_u - m_d$, the one of the axial current is proportional to the sum $m_u + m_d$. If the two masses are set equal, the vector current is conserved and the Hamitonian becomes symmetric with respect to isospin rotations. If they are not only taken equal, but equal to zero, then the axial current is conserved, too, such that the corresponding charge $I_5^+ = \int d^3x\, d^\dagger \gamma_5 u$ also commutes with the Hamiltonian — QCD acquires an additional symmetry.

The isospin operator I^+ converts a d-quark into a u-quark, irrespective of the helicity. The operator I_5^+, however, acts differently on the right- and left-handed components. The sum $\frac{1}{2}(I^+ + I_5^+)$ takes a righthanded d-quark into a righthanded u-quark, but leaves left-handed ones alone. This implies that, for massless quarks, the Hamiltonian is invariant with respect to a set of *chiral* transformations: independent isospin rotations of the right- and left-handed components of u and d,

$$\begin{pmatrix} u_R \\ d_R \end{pmatrix} \rightarrow V_R \begin{pmatrix} u_R \\ d_R \end{pmatrix} \ , \qquad \begin{pmatrix} u_L \\ d_L \end{pmatrix} \rightarrow V_L \begin{pmatrix} u_L \\ d_L \end{pmatrix} \ , \qquad V_R,\ V_L \in \mathrm{SU}(2) \ .$$

The corresponding symmetry group is the direct product of two separate isospin groups, $\mathrm{SU}(2)_R \times \mathrm{SU}(2)_L$. The symmetry is generated two sets of isospin operators: ordinary isospin, $\vec{I}$ and chiral isospin, $\vec{I}_5$. The particular operator considered above is the linear combination $I_5^+ = I_5^1 + i\, I_5^2$.

In reality, chiral symmetry is broken, because m_u and m_d do not vanish. As above, the Hamiltonian may be split into a piece which is invariant under the symmetry group of interest and a piece which breaks the symmetry. In the present case, the symmetry breaking part is the full mass term of the u and d quarks,

$$H_{\mathrm{QCD}} = H_0' + H_{\mathrm{sb}}' \ , \quad H_{\mathrm{sb}}' = \int d^3x (m_u\, \bar{u}u + m_d\, \bar{d}d) \ . \tag{5}$$

The symmetric part conserves ordinary as well as chiral isospin,

$$[\vec{I}, H_0'] = 0 \ , \qquad [\vec{I}_5, H_0'] = 0 \ . \tag{6}$$

Note that the symmetry group exclusively acts on u and d — the remaining quarks $s, c, \ldots$ are singlets. The corresponding mass terms $m_s \bar{s}s + m_c\, \bar{c}c + \ldots$ do not break the symmetry and are included in H_0'.

4 Spontaneous symmetry breakdown

Much before QCD was discovered, Nambu pointed out that chiral symmetry breaks down spontaneously. The phenomenon plays a crucial role for the properties of the strong interaction at low energy. To discuss it, I return to the theoretical scenario, where m_u and m_d are set equal to zero.

In this framework, isospin is conserved. The isospin group $\mathrm{SU}(2)$ represents the prototype of a "manifest" symmetry, with all the consequences known from quantum mechanics: (i) The energy levels form degenerate multiplets. (ii) The operators $\vec{I}$ generate transitions within the multiplets, taking a neutron, e.g., into a proton, $I^+|n\rangle = |p\rangle$. (iii) The ground state is an isospin singlet,

$$\vec{I}\,|0\rangle = 0 \ . \tag{7}$$

If chiral symmetry was realized in the same manner, the energy levels would be grouped into degenerate multiplets of the group $\mathrm{SU}(2)_R \times \mathrm{SU}(2)_L$. Since the chiral isospin operators $\vec{I}_5$ carry negative parity, the multiplets would then necessarily contain members of opposite parity. The listings of the Particle Data Group, however, do not show any trace of such a pattern. A particle with the quantum numbers of $I_5^+|n\rangle$ and nearly the same mass as the neutron, e.g., is not observed in nature.

In fact, the symmetry of the Hamiltonian does not ensure that the corresponding eigenstates form multiplets of the symmetry group. In particular, the state with the lowest eigenvalue of the Hamiltonian need not be a singlet. In the case of a magnet, e.g., the Hamiltonian is invariant under rotations of the spin directions, but the ground state fails to be invariant, because the spins are aligned and thereby single out a direction. Whenever the state with the lowest eigenvalue is less symmetric than the Hamiltonian, the symmetry is called "spontaneously broken" or "hidden". Chiral symmetry belongs to this category. For dynamical reasons, the most important state — the vacuum — is symmetric only under ordinary isospin rotations, but does not remain invariant if a chiral rotation is applied,

$$\vec{I}_5 |0\rangle \neq 0 \ . \tag{8}$$

Since the Hamiltonian commutes with chiral isospin, the three states $\vec{I}_5 |0\rangle$ have the same energy as the vacuum, $E = 0$. The operators $\vec{I}_5$ do not carry momentum, either, so that the states $\vec{I}_5 |0\rangle$ have $\vec{P} = 0$. This indicates that the spectrum of physical states contains three massless particles. Indeed, the Goldstone theorem [3] rigorously shows that spontaneous symmetry breakdown gives rise to massless particles, "Goldstone bosons". Their quantum numbers are those of the states $\vec{I}_5 |0\rangle$: spin zero, negative parity and $I = 1$.

The three lightest mesons carry precisely these quantum numbers: π^+, π^0, π^-. The chiral isospin operators act like creation or annihilation operators for pions: Applied to the vacuum, they generate a state containing a pion, $I_5^+ |0\rangle = |\pi^+\rangle$. Applied to a neutron, they do not lead to a parity partner, but instead yield a state containing a neutron and a pion, $I_5^+|n\rangle = |n\pi^+\rangle$, etc.

5 Pion mass

The above discussion concerns the theoretical world, where u and d are assumed to be massless, such that the group $SU(2)_R \times SU(2)_L$ represents an exact symmetry. The Hamiltonian of QCD contains a quark mass term, which breaks chiral symmetry. To see how this affects the mass of the Goldstone bosons, consider the transition matrix element of the axial current $\bar{u}\gamma^\mu\gamma_5 d$, from the vacuum to a one-pion state. Lorentz invariance implies that this matrix element is determined by the pion momentum p^μ, up to a constant,

$$\langle \pi^+(p)| \bar{u}(x)\gamma^\mu \gamma_5 d(x) |0\rangle = -ip^\mu \sqrt{2}\, F_\pi\, e^{ipx} \ .$$

The value of the constant is measured in pion decay, $F_\pi \simeq 93$ MeV. For the divergence $\partial_\mu(\bar{u}\gamma^\mu\gamma_5 d)$, this yields an expression proportional to $p^2 = M_{\pi^+}^2$. Denoting the analogous matrix element of the pseudoscalar density by G_π,

$$\langle \pi^+(p)| \bar{u}(x)\gamma_5 d(x) |0\rangle = i\sqrt{2}\, G_\pi\, e^{ipx} \ ,$$

the conservation law (4) thus implies the exact relation

$$M_{\pi^+}^2 = (m_u + m_d)(G_\pi/F_\pi) \ . \tag{9}$$

The relation confirms that, when the symmetry breaking parameters m_u, m_d are put equal to zero, the pion mass vanishes, independently of the masses of the other quark flavours. The group $SU(2)_R \times SU(2)_L$ then represents a spontaneously broken, *exact* symmetry, with three strictly massless Goldstone bosons. When the quark masses are turned on, the Goldstone bosons pick up mass: M_{π^+} grows in proportion to $\sqrt{m_u + m_d}$. The pions remain light, provided m_u and m_d are small. The quark mass term of the Hamiltonian then amounts to a small perturbation, such that the group $SU(2)_R \times SU(2)_L$ still represents an *approximate* symmetry, with approximately massless Goldstone bosons.

Moreover, as noted in section 2, the observed level pattern also requires the differences between m_u, m_d and m_s to be small. Hence the strange quark must be light, too, such that the corresponding mass term may also be treated as a perturbation. The decomposition of the Hamiltonian then takes the form

$$H_{\text{QCD}} = H_0 + H_{\text{sb}} \ , \quad H_{\text{sb}} = \int d^3x (m_u \bar{u}u + m_d \bar{d}d + m_s \bar{s}s) \ . \tag{10}$$

The first term, H_0, describes three massless flavours (u, d, s) as well as three massive ones (c, b, t). It is symmetric with respect to independent rotations of the right- and left-handed components of u, d and s, i.e., with respect to the group $SU(3)_R \times SU(3)_L$. The perturbation series, which results if H_{sb} is treated as a perturbation, amounts to an expansion of the matrix elements and eigenvalues in powers of m_u, m_d and m_s. The inequality $|m_d - m_u| \ll |m_s - m_d|$, which follows from the fact that isospin breaking is much smaller than the breaking of eightfold way symmetry, implies that the s-quark is considerably heavier than the other two, $m_u, m_d \ll m_s$.

The above arguments rely on two phenomenological observations:

(a) The pion mass is small compared to the masses of all other hadrons. This indicates that the strong interaction possess an approximate, spontaneously broken symmetry, with the pions as the corresponding Goldstone bosons. Indeed, the Hamiltonian of QCD exhibits an approximate symmetry with the proper quantum numbers, provided both m_u and m_d are small.

(b) The multiplet structure seen in the particle data tables indicates that the eightfold way is an approximate symmetry of the strong interaction. For QCD to possess such a symmetry, the mass differences $m_d - m_u$ and $m_s - m_d$ must be small.

Combining the two observations, one concludes that the mass of the strange quark also amounts to a small perturbation: The two groups $SU(2)_R \times SU(2)_L$ and $SU(3)$ can be approximate symmetries of the QCD Hamiltonian only if $SU(3)_R \times SU(3)_L$ represents an approximate symmetry, too. The masses of the other quarks occurring in the Standard Model, on the other hand, cannot be treated as a perturbation. Since the corresponding fields $c(x)$, $b(x)$ and $t(x)$ are singlets with respect to $SU(3)_R \times SU(3)_L$, their contribution may be included in the symmetric part of the Hamiltonian, H_0. Their presence does not significantly affect the low energy structure of the theory.

The decomposition of the QCD Hamiltonian in eq. (10) may be compared with the standard perturbative splitting

$$H_{\text{QCD}} = H_{\text{free}} + H_{\text{int}} \ ,$$

where the first term describes free quarks and gluons, while the second accounts for their interaction. The corresponding expansion parameter is the coupling constant g. Since QCD is asymptotically free, the effective coupling becomes weak at large momentum transfers — processes which exclusively involve large momenta may indeed be analyzed by treating the interaction as a perturbation. Perturbation theory, however, fails in the low energy domain, where the effective coupling is strong, such that it is not meaningful to truncate the expansion in powers of H_{int} after the first few terms. In particular, the structure of the ground state cannot be analyzed in this way, while the above decomposition, which retains the interaction among the quarks and gluons in the "unperturbed" Hamiltonian H_0 and only treats m_u, m_d and m_s as perturbations, is perfectly suitable for that purpose. Note that the character of the perturbation series in powers of H_{sb} is quite different from the one in powers of H_{int}: while the eigenstates of H_{free} are known explicitly, this is not the case with H_0, which still describes a highly nontrivial, interacting system. H_0 differs from the full Hamiltonian only in one respect: it possesses an exact group of chiral symmetries.

6   Quark masses

There is an immediate experimental check of the above theoretical arguments: the spontaneous breakdown of the symmetry $SU(3)_R \times SU(3)_L$ to the subgroup $SU(3)_{R+L}$ generates eight Goldstone bosons. They are not massless, because the quark masses m_u, m_d and m_s break the symmetry, but since the breaking is supposed to be small, these levels should remain lowest. Indeed, the eight lightest bound states, π^+, π^0, π^-, $K^+, K^0, \bar{K}^0, K^-, \eta$, carry precisely the required quantum numbers, both with respect to spin/parity and to flavour.

As a further confirmation of the picture, one may compare the mass splittings within the pseudoscalar octet with those of the other multiplets. The mass *differences* are comparable: $M_\eta - M_\pi \simeq 410$ MeV, $M_\Xi - M_N \simeq 380$ MeV. The mass *ratios* of the Goldstone bosons, however, deviate much more strongly from unity than those of the other multiplets: while the various levels of the baryon octet differ from their mean mass by less than 20 %, the mass of the η is four times as large as the mass of the pion. The above symmetry considerations neatly explain why this is so. For ordinary multiplets, the eigenvalue of H_0 is different from zero; the perturbation H_{sb} only generates a correction, whose magnitude depends on the level in question, because H_{sb} breaks SU(3). In the case of the Goldstone bosons, however, the entire mass is due to the perturbation — the pattern of levels directly reveals the asymmetries of the operator H_{sb}. As discussed above, M_{π^+} is proportional to $\sqrt{m_u + m_d}$. The same analysis applies to the currents $\bar{s}\gamma^\mu\gamma_5 u$ and $\bar{s}\gamma^\mu\gamma_5 d$, which generate transitions from the vacuum to the states $|K^+\rangle$ and $|K^0\rangle$. Since the corresponding divergences are proportional to $(m_u + m_s)$ and $(m_d + m_s)$, one now obtains $M_{K^+} \propto \sqrt{m_u + m_s}$ and $M_{K^0} \propto \sqrt{m_d + m_s}$. The mass ratios of the Goldstone bosons strongly deviate from unity, because m_s happens to be large compared to m_u and m_d.

The level shifts generated by the symmetry breaking may be analyzed by treating the mass term in eq.(10) as a perturbation. To first order in the perturbation, the result obeys the Gell-Mann–Okubo formula. The calculation also applies to the pseudoscalar octet, where the unperturbed levels sit at $M = 0$, provided the shifts in the *square* of the mass are considered. Indeed, M_π^2, M_K^2 and M_η^2 obey the formula remarkably well, confirming that the mass pattern of the pseudoscalar octet is perfectly consistent with the claim that SU(3) is a decent approximate symmetry of the strong interaction.[2]

The first order mass formulae for the pseudoscalar octet may also be used to estimate the relative size of the three quark masses [4]. The most remarkable feature of the resulting pattern is that the quark masses strongly break isospin symmetry: m_u and m_d are quite different [5]. This may be verified as follows. Consider the mass difference between K^0 and K^+. If m_u and m_d where the same, the splitting would exclusively be due to the electromagnetic interaction. Since the main contribution from this interaction is the self energy of the electric field surrounding the K^+, this particle would have to be heavier than the K^0. The observed splitting, $M_{K^0} - M_{K^+} = 4$ MeV is of opposite sign. Hence the difference between m_d and m_u must make a significant contribution, opposite to the electromagnetic one, $m_d > m_u$ (the same conclusion also follows from the mass difference between neutron and proton). In first order perturbation theory, the mass ratio $(M_{K^0}^2 - M_{K^+}^2)/M_{\pi^+}^2$ is given by the relative size of isospin breaking in the quark masses, $r = (m_d - m_u)/(m_u + m_d)$. Using the observed meson masses, this gives $r \simeq 0.20$. If the electromagnetic self energy is taken into account, the result becomes even larger, because the two contributions are of opposite sign: $r \simeq 0.29$ [4].

The reason why, nevertheless, isopin is a nearly perfect symmetry of the strong interaction is essentially the same as for the case of SU(3) breaking, discussed above: The relative magnitude of isospin breaking in the quark masses does not represent an adequate estimate for the magnitude of the isospin breaking effects occurring in the bound states. What counts, instead, is the magnitude of the isospin breaking part of the Hamiltonian, $\bar{H}_{sb}$, compared to the isospin symmetric piece, $\bar{H}_0$ (see eq.(1)). This is particularly evident in the case of the nucleon, where the splitting is of the order of 1 MeV, while the isospin invariant part is responsible for the mean mass and is of order 1 GeV. In algebraic terms, the matrix elements of $\bar{H}_{sb}$ are of order $m_d - m_u$, while those of $\bar{H}_0$ are determined by the scale Λ_{QCD}, so that the magnitude of isospin breaking is determined by the ratio $(m_d - m_u)/\Lambda_{QCD}$, rather than $(m_d - m_u)/(m_u + m_d)$.

For the kaons, isospin breaking is enhanced, because these particles get their mass from m_s, not from the scale of QCD: the ratio $(M_{K^0} - M_{K^+})/(M_{K^0} + M_{K^+})$ is of order $(m_d - m_u)/m_s$. One might expect that the most important isospin breaking effects occur in the pion multiplet, where the matrix elements of $\bar{H}_0$ are suppressed even more strongly. It so happens, however, that the strong breaking of SU(3) symmetry seen in the pseudoscalar octet does not repeat itself here, because the matrix elements of the perturbation, $\langle \pi | \bar{H}_{sb} | \pi \rangle$ are suppressed, too: The mass splitting $M_{\pi^+} - M_{\pi^0}$ is of second order in the perturbation, proportional to $(m_d - m_u)^2$. Numerically, the effect is tiny, of order 0.2 MeV; the observed mass difference is due almost entirely to the electromagnetic interaction. The mathematical origin of this qualitative difference between the two cases is that, in contrast to SU(3), the group SU(2) does not have a d-symbol. For this reason, the pion mass is shielded from isospin breaking, so that the range of the forces generated by pion exchange is nearly charge independent.

---

[2] The experimental values of the decay constants F_π, F_K, which represent the bound state wave functions at the origin, also confirm the picture: The asymmetry seen there, $F_K/F_\pi = 1.22$ is quite typical of the SU(3) breaking effects observed in other multiplets.

7 Effective field theory

At low energies, the behaviour of scattering amplitudes or current matrix elements can be described in terms of a *Taylor series expansion* in powers of the momenta. The electromagnetic form factor of the pion, e.g., may be exanded in powers of the momentum transfer t. In this case, the first two Taylor coefficients are related to the total charge of the particle and to the mean square radius of the charge distribution, respectively,

$$f_{\pi^+}(t) = 1 + \tfrac{1}{6}\langle r^2 \rangle_{\pi^+} t + O(t^2) \ . \tag{11}$$

Scattering lengths and effective ranges are analogous low energy constants occurring in the Taylor series expansion of scattering amplitudes.

The occurrence of light particles gives rise to singularities in the low energy domain, which limit the range of validity of the Taylor series representation. The form factor $f_{\pi^+}(t)$, e.g., contains a branch cut at $t = 4M_\pi^2$, such that the formula (11) provides an adequate representation only for $|t| \ll 4M_\pi^2$. The problem becomes even more acute if m_u and m_d are set equal to zero. The pion mass then disappears, the branch cut sits at $t = 0$ and the Taylor series does not work at all. I first discuss the method used in the low energy analysis for this extreme case, returning to the physical situation with $m_u, m_d \neq 0$ below.

The reason why the spectrum of QCD with two massless quarks contains three massless bound states is understood: they are the Goldstone bosons of a hidden symmetry. The symmetry, which gives birth to these, at the same time also determines their low energy properties. This makes it possible to explicitly work out the poles and branch cuts generated by the exchange of Goldstone bosons. The remaining singularities are located comparatively far from the origin, the nearest one being due to the ρ-meson. The result is a modified Taylor series expansion in powers of the momenta, which works, despite the presence of massless particles. In the case of the $\pi\pi$ scattering amplitude, e.g., the radius of convergence of the modified series is given by $s = M_\rho^2$, where s is the square of the energy in the center of mass system (the first few terms of the series only yield a decent description of the amplitude if s is smaller than the radius of convergence, say $s < \tfrac{1}{2}M_\rho^2 \rightarrow \sqrt{s} < 540$ MeV).

As pointed out by Weinberg [6], the modified expansion may explicitly be constructed by means of an effective field theory, which is referred to as *chiral perturbation theory* and involves the following ingredients:

(i) The quark and gluon fields of QCD are replaced by a set of pion fields, describing the degrees of freedom of the Goldstone bosons. It is convenient to collect these in a 2×2 matrix $U(x) \in SU(2)$.

(ii) The Lagrangian of QCD is replaced by an effective Lagrangian, which only involves the field $U(x)$, and its derivatives

$$\mathcal{L}_{\mathrm{QCD}} \ \longrightarrow \ \mathcal{L}_{eff}(U, \partial U, \partial^2 U, \ldots) \ .$$

(iii) The low energy expansion corresponds to an expansion of the effective Lagrangian, ordered according to the number of the derivatives of the field $U(x)$. Lorentz invariance only permits terms with an even number of derivatives,

$$\mathcal{L}_{eff} = \mathcal{L}_{eff}^2 + \mathcal{L}_{eff}^4 + \mathcal{L}_{eff}^6 + \dots$$

Chiral symmetry very strongly constrains the form of the terms occurring in the series. In particular, it excludes momentum independent interaction vertices: Goldstone bosons can only interact if they carry momentum. This property is essential for the consistency of the low energy analysis, which treats the momenta as expansion parameters. The leading contribution involves two derivatives,

$$\mathcal{L}_{eff}^2 = \tfrac{1}{4}F_\pi^2 \text{tr}\{\partial_\mu U^+ \partial^\mu U\} \ , \tag{12}$$

and is fully determined by the pion decay constant. At order p^4, the symmetry permits two independent terms,[3]

$$\mathcal{L}_{eff}^4 = \tfrac{1}{4}l_1(\text{tr}\{\partial_\mu U^+ \partial^\mu U\})^2 + \tfrac{1}{4}l_2 \text{tr}\{\partial_\mu U^+ \partial_\nu U\}\text{tr}\{\partial^\mu U^+ \partial^\nu U\} \ , \tag{13}$$

etc. For most applications, the derivative expansion is needed only to this order.

The most remarkable property of the method is that it does not mutilate the theory under investigation: The effective field theory framework is no more than an efficient machinery, which allows one to work out the modified Taylor series, referred to above. If the effective Lagrangian includes all of the terms permitted by the symmetry, the effective theory is mathematically equivalent to QCD [6,7,8]. It exclusively exploits the symmetry properties of QCD and involves an infinite number of effective coupling constants, $F_\pi, l_1, l_2, \dots$, which represent the Taylor coefficients of the modified expansion.

In QCD, the symmetry, which controls the low energy properties of the Goldstone bosons, is only an approximate one. The constraints imposed by the hidden, approximate symmetry can still be worked out, at the price of expanding the quantities of physical interest in powers of the symmetry breaking parameters m_u and m_d. The low energy analysis then involves a combined expansion, which treats both, the momenta and the quark masses as small parameters. The effective Lagrangian picks up additional terms, proportional to powers of the quark mass matrix,

$$m = \begin{pmatrix} m_u & \\ & m_d \end{pmatrix}$$

It is convenient to count m like two powers of momentum, such that the expansion of the effective Lagrangian still starts at $O(p^2)$ and only contains even terms. The leading contribution picks up a term linear in m,

$$\mathcal{L}_{eff}^2 = \tfrac{1}{4}F_\pi^2 \text{tr}\{\partial_\mu U^+ \partial^\mu U\} + \tfrac{1}{2}F_\pi^2 B \,\text{tr}\{m(U + U^\dagger)\} \ . \tag{14}$$

Likewise, $\mathcal{L}_{eff}^4$ receives additional contributions, involving two further effective coupling constants, l_3, l_4, etc.

The expression (14) represents a compact summary of the soft pion theorems established in the 1960's: The leading terms in the low energy expansion of the

---

[3]In the framework of the effective theory, the anomalies of QCD manifest themselves through an extra contribution, the Wess-Zumino term, which is also of order p^4 and is proportional to the number of colours.

scattering amplitudes and current matrix elements are given by the tree graphs of this Lagrangian. The coupling constant B, needed to account for the symmetry breaking effects generated by the quark masses at leading order, represents the coefficient of the linear term in the expansion of the pion mass, $M_\pi^2 = (m_u + m_d)B + O(m^2)$. According to section 5, this constant also determines the vacuum-to-pion matrix element of the pseudoscalar density, $G_\pi = F_\pi B + O(m)$. Furthermore, the relation of Gell-Mann, Oakes and Renner, $F_\pi^2 M_\pi^2 = -(m_u + m_d)\langle 0|\bar{u}u|0\rangle + O(m^2)$, which immediately follows from the above expression for the effective Lagrangian, shows that the magnitude of the quark condensate is also related to the value of B.

The effective field theory represents an efficient and systematic framework, which allows one to work out the corrections to the soft pion predictions, those arising from the quark masses as well as those from the terms of higher order in the momenta. The evaluation is based on a perturbative expansion of the quantum fluctuations of the effective field. In addition to the tree graphs relevant for the soft pion results, graphs containing vertices from the higher order contributions $\mathcal{L}_{eff}^4, \mathcal{L}_{eff}^6 \ldots$ and loop graphs contribute. The leading term of the effective Lagrangian describes a nonrenormalizable theory, the "nonlinear σ-model". The higher order terms in the derivative expansion, however, automatically contain the relevant counter terms. The divergences occurring in the loop graphs merely renormalize the effective coupling constants. The effective theory is a perfectly renormalizable scheme, order by order in the low energy expansion, so that, in principle, the result of the calculation does not depend on who it is who did it.

8 Universality

The properties of the effective theory are governed by the hidden symmetry, which is responsible for the occurrence of Goldstone bosons. In particular, the form of the effective Lagrangian only depends on the symmetry group G of the Hamiltonian and on the subgroup H ⊂ G, under which the ground state is invariant. The Goldstone bosons live on the difference between the two groups, i.e., on the quotient G/H. The specific dynamical properties of the underlying theory do not play any role. To discuss the consequences of this observation, I again assume that G is an exact symmetry.

In the case of QCD with two massless quarks, G is the group $SU(2)_R \times SU(2)_L$ of chiral isospin rotations, while H = SU(2) is the ordinary isospin group. The Higgs model is another example of a theory with spontaneously broken symmetry. It plays a crucial role in the Standard Model, where it describes the generation of mass. The model involves a scalar field $\vec{\phi}$ with four components. The Hamiltonian is invariant under rotations of the vector $\vec{\phi}$, which form the group G = O(4). Since the field picks up a vacuum expectation value, the symmetry is spontaneously broken to the subgroup of those rotations, which leave the vector $\langle 0|\vec{\phi}|0\rangle$ alone, H = O(3). It so happens that these groups are the same as those above, relevant for QCD.[4] The fact that the symmetries are the same implies that the effective field theories are

---

[4] The structure of the effective Lagrangian rigorously follows from the Ward identities for the Green functions of the currents, which also reveal the occurrence of anomalies [7]. The form of the Ward identities is controlled by the structure of G and H in the infinitesimal neighbourhood of the neutral element. In this sense, the symmetry groups of the two models are the same: O(4) and O(3) are *locally* isomorphic to SU(2)×SU(2) and SU(2), respectively.

identical: (i) In either case, there are three Goldstone bosons, described by a matrix field $U(x) \in$ SU(2). (ii) The form of the effective Lagrangian is precisely the same. In particular, the expression

$$\mathcal{L}^2_{eff} = \tfrac{1}{4} F^2_\pi \text{tr}\{\partial_\mu U^+ \partial^\mu U\}$$

is valid in either case. At the level of the effective theory, the only difference between these two physically quite distinct models is that the numerical values of the effective coupling constants are different. In the case of QCD, the one occurring at leading order of the derivative expansion is the pion decay constant, $F_\pi \simeq 93\,\text{MeV}$, while in the Higgs model, this coupling constant is larger by more than three orders of magnitude, $F_\pi \simeq 250\,\text{GeV}$. At next-to-leading order, the effective coupling constants are also different; in particular, in QCD, the anomaly coefficient is equal to N_c, while in the Higgs model, it vanishes.

As an illustration, I compare the condensates of the two theories, which play a role analogous to the spontaneous magnetization $\langle \vec{M} \rangle$ of a ferromagnet (or the staggered magnetization of an antiferromagnet). At low temperatures, the magnetization singles out a direction — the ground state spontaneously breaks the symmetry of the Hamiltonian with respect to rotations. As the system is heated, the spontaneous magnetization decreases, because the thermal disorder acts against the alignment of the spins. If the temperature is high enough, disorder wins, the spontaneous magnetization disappears and rotational symmetry is restored. The temperature at which this happens is the Curie temperature. Quantities, which allow one to distinguish the ordered from the disordered phase are called *order parameters*. The magnetization is the prototype of such a parameter.

In QCD, the most important order parameter (the one of lowest dimension) is the quark condensate. At nonzero temperatures, the condensate is given by the thermal expectation value

$$\langle \bar{u}u \rangle_T = \frac{\text{Tr}\{\bar{u}u \exp(-H/kT)\}}{\text{Tr}\{\exp(-H/kT)\}} .$$

The condensate melts if the temperature is increased. At a critical temperature, somewhere in the range $140\,\text{MeV} < T_c < 180\,\text{MeV}$, the quark condensate disappears and chiral symmetry is restored. The same qualitative behaviour also occurs in the Higgs model, where the expectation value $\langle \vec{\phi} \rangle_T$ of the scalar field represents the most prominent order parameter.

At low temperatures, the thermal trace is dominated by states of low energy. Massless particles generate contributions which are proportional to powers of the temperature, while massive ones like the ρ-meson are suppressed by the corresponding Boltzmann factor, $\exp(-M_\rho/kT)$. In the case of a spontaneously broken symmetry, the massless particles are the Goldstone bosons and their contributions may be worked out by means of effective field theory. For the quark condensate, the calculation has been done [9], up to and including terms of order T^6:

$$\langle \bar{u}u \rangle_T = \langle 0 | \bar{u}u | 0 \rangle \left\{ 1 - \frac{T^2}{8F^2_\pi} - \frac{T^4}{384F^4_\pi} - \frac{T^6}{288F^6_\pi} \ln(T_1/T) + O(T^8) \right\} .$$

The formula is exact — for massless quarks, the temperature scale relevant at low T is the pion decay constant. The additional logarithmic scale T_1 occurring at order

T^6 is determined by the effective coupling constants l_1, l_2, which enter the expression (13) for the effective Lagrangian of order p^4. Since these are known from the phenomenology of $\pi\pi$ scattering, the value of T_1 is also known: $T_1 = 470 \pm 110$ MeV.

Now comes the point I wish to make. The effective Lagrangians relevant for QCD and for the Higgs model are the same. Since the operators of which we are considering the expectation values also transform in the same manner, their low temperature expansions are identical. The above formula thus holds, without any change whatsoever, also for the Higgs condensate,

$$\langle \vec{\phi} \rangle_T = \langle 0| \vec{\phi} |0\rangle \left\{ 1 - \frac{T^2}{8F_\pi^2} - \frac{T^4}{384F_\pi^4} - \frac{T^6}{288F_\pi^6} \ln(T_1/T) + O(T^8) \right\} \ .$$

In fact, the universal term of order T^2 was discovered in the framework of this model, in connection with work on the electroweak phase transition [10].

The effective Lagrangian of a Heisenberg antiferromagnet is also of the same structure,[5] so that the above formula even holds for the staggered magnetization, except for one modification: the Clebsch-Gordan coefficients, which accompany the various powers of T are different, because the symmetry groups differ: The Hamiltonian now is invariant under ordinary rotations, G = O(3), while the ground state spontaneously breaks the symmetry to the subgroup H = O(2) of the rotations around the direction singled out by the magnetization.

These examples illustrate the physical nature of effective theories: At long wavelength, the microscopic structure does not play any role. The behaviour only depends on those degrees of freedom, which require little excitation energy. The hidden symmetry, which is responsible for the absence of an energy gap and for the occurrence of Goldstone bosons, at the same time also determines their low energy properties. For this reason, the form of the effective Lagrangian is controlled by the symmetries of the system and is, therefore, universal. The microscopic structure of the underlying theory exclusively manifests itself in the numerical values of the effective coupling constants. The temperature expansion also clearly exhibits the limitations of the method. The truncated series can be trusted only at low temperatures, where the first term represents the dominant contribution. According to the above formula, the quark condensate drops to about half of the vacuum expectation value when the temperature reaches 160 MeV — the formula does not make much sense beyond this point. In particular, the behaviour of the quark condensate in the vicinity of the chiral phase transition is beyond the reach of the effective theory discussed here.

9   Experimental aspects

The DAFNE Handbook [12] provides an excellent overview over many of the processes, where new data will contribute to make progress in understanding the low energy

---

[5]Since the ground state of a magnet fails to be Lorentz invariant, the derivative expansion of the effective Lagrangian contains additional contributions. For a cubic lattice, however, the leading term is of the same form as in relativistically invariant theories, except that the velocity of light is to be replaced by the velocity of propagation for magnons of long wavelength. The low energy properties of a ferromagnet, on the other hand, are quite different. The corresponding effective Lagrangian is dominated by a topological term, related to the fact that the generators of the symmetry acquire nonzero expectation values in the ground state [11].

structure of QCD. I only add a few comments.

One of the issues, about which very little is known experimentally, is the explicit breaking of chiral and isospin symmetry, generated by m_u and m_d. Because the group $SU(2)_R \times SU(2)_L$ represents an almost exact symmetry of the strong interaction, the symmetry breaking part of the Hamiltonian only generates very small effects. An excellent place to check the theoretical ideas about the implications of symmetry breaking is $\pi\pi$ scattering. As shown by Weinberg [13], nearly 30 years ago, chiral symmetry leads to parameter free soft pion predictions for the corresponding S-wave scattering lengths a_0, a_2. There is a beautiful proposal [14] to accurately measure the combination $a_0 - a_2$, by producing $\pi^+\pi^-$ atoms and measuring the rate of their decay into $\pi^0\pi^0$. The corrections to the soft pion results have been worked out [15], so that a very accurate prediction is available for test. The S-wave scattering lengths are closely related to the σ-term matrix element $\sigma_{\pi\pi} = \langle \pi | m_u\bar{u}u + m_d\bar{d}d | \pi \rangle$ and are also proportional to $m_u + m_d$. The quantity $a_0 - a_2$ thus represents a direct measure of the asymmetries produced by the quark masses. The experiment, in particular, would provide a sensitive test of the standard hypothesis, according to which the expansion of the pion mass in powers of the quark masses,

$$M_\pi^2 = M^2 \left\{ 1 - \frac{M^2}{32\pi^2 F_\pi^2} \bar{l}_3 + O(M^4) \right\} \ , \qquad M^2 \equiv (m_u + m_d)B \ ,$$

is dominated by the first term. In the standard picture, the contribution of order $(m_u + m_d)^2$, which is proportional to the effective coupling constant $\bar{l}_3$, amounts to a small correction of order 2%; the corresponding contribution to $a_0 - a_2$ is three times smaller. As pointed out by Knecht et al. [16], the arguments which underly this estimate are theoretical: There is no direct experimental evidence, which would rule out an entirely different picture. A number like $\bar{l}_3 = -100$, e.g., would increase the result for $a_0 - a_2$ by about 25% and bring it into agreement with the central value of the currently available data. Conversely, if this value should be confirmed within narrow error bars, one would have to conclude that the "correction" in the expansion of M_π^2 is almost as large as the leading term. Needless to say that this would give rise to a major earthquake in the current understanding of QCD. The quark mass pattern discussed above is based on the standard picture, where it is assumed that the Gell-Mann-Oakes-Renner relation is not ruined by higher order terms. This is the only way I know of to understand the success of the Gell-Mann-Okubo formula for the pseudoscalar octet — if the symmetry breaking observed in $\pi\pi$ scattering should disagree with the theoretical predictions, the standard picture would require thorough revision, even at the qualitative level. Only few of us expect this to be the outcome of the investigation, but the earmark of an important experiment is the product of the likelihood for a discovery with the physical significance thereof, the likelihood as such may be quite small.

The analogous issue in pion-nucleon scattering is a dynosaur. It is notoriously difficult to accurately measure $\sigma_{\pi N}$. At the present time, the experimental uncertainties in this quantity amount to about 20%, comparable to those in the $\pi\pi$ S-wave scattering lengths. There are beautiful new data on the related πN scattering lengths, based on bound states of π^-p and π^-d [17], analogous to the $\pi^+\pi^-$ atoms of the proposal mentioned above. These data attain a precision, where even isospin breaking effects due to $m_d - m_u$ can be measured, provided the theoretical results [18], used to express the pion-deuteron scattering lengths in terms of those of proton and neutron, can be trusted at the accuracy needed here. The new data should give ample incentive for a careful reanalysis of the three-body problem, which arises if a pion of zero momentum encounters a deuteron. Evidently, the experimental discrepancies in low energy πN scattering should be resolved. For a measurement of small quantitities like $\sigma_{\pi N}$, the dominating contribution from the Born term, i.e., the value of the coupling constant $g_{\pi N}$, needs to be known to very high precision.

On the theoretical side, considerable progress in the chiral perturbation theory of the πN interaction is being made. The predictions are weaker here, because, in the hidden symmetry game, the nucleons are only spectators, not actors like the Goldstone bosons. Accordingly, the number of effective coupling constants, which need to be taken from phenomenology, is larger. In the case of the σ-term, e.g., the symmetry implies that the matrix element $\langle \pi | \bar{s} s | \pi \rangle$ vanishes if m_u, m_d are sent to zero, while this is not the case for the corresponding nucleon matrix element. Also, the $\pi \pi$ scattering matrix elements are shielded from the perturbations generated by $m_d - m_u$, but the πN scattering matrix elements are not — *small* quantities like the σ-term or the isospin even S-wave scattering length may pick up *comparatively* large charge asymmetries [4]. The fact that the excitation energy of the Δ is relatively small does not really present a problem; unless one attempts to use the effective theory in the vicinity of the resonance or beyond, the corresponding singularity may be expanded in the standard fashion, absorbing the Taylor coefficients in the relevant low energy constants [19]. The expansion of the πN scattering amplitude in powers of the momenta, however, contains odd as well as even powers — one needs to carry the expansion beyond the first two terms to achieve the same precision as the one available for $\pi \pi$ scattering [20]. Work on this problem is of interest, in particular, in connection with the ongoing experiments on pion photo- and electroproduction, whose significance as probes of the low energy structure is becoming increasingly evident and which were discussed in detail at this workshop.

Another topic, where the experimental situation needs to be clarified, is η decay. It is important to resolve the discrepancy between the older data, based on the Primakoff effect and the more recent ones, from photon-photon-collisions. The rate of the decay into three pions measures the ratio $(m_d^2 - m_u^2)/m_s^2$ of quark masses [21]. Also, the available information on the Dalitz plot distribution of the $\pi^+ \pi^- \pi^0$ final state and on the ratio $\Gamma_{\eta \to 3\pi^0}/\Gamma_{\eta \to \pi^0 \pi^+ \pi^-}$ leaves to be desired. Incidentally, the world average of the partly inconsistent data on these quantities is not in satisfactory agreement with the theoretical predictions.

There are many other items of interest, which are by no means less interesting — processes generated by the Wess-Zumino term, to only name one category — but I stop here, thanking Aron Bernstein, Barry Holstein and their coworkers for a very informative meeting.

References

[1] M. Gell-Mann, *The Eightfold Way: A Theory of Strong Interaction Symmetry*, California Insitute of Technology Report CTSL-20 (1961); Y. Ne'eman, *Nucl. Phys.* 26 (1961) 222.

[2] Y. Nambu, *Phys. Rev. Lett.* 4 (1960) 380.

[3] J. Goldstone, *Nuovo Cim.* 19 (1961) 154;
G. S. Guralnik, C. R. Hagen and T. W. B. Kibble, in *Advances in particle physics*, Vol. 2, p. 567, ed. R. L. Cool and R. E. Marshak (Wiley, New York, 1968);
S. Coleman, Erice Lectures 1973, in *Laws of hadronic matter*, Academic Press London and New York (1975), reprinted in
S. Coleman, *Aspects of symmetry*, Cambridge Univ. Press (1985).

[4] S. Weinberg, in *A Festschrift for I.I. Rabi*, ed. L. Motz (New York Acad. Sci, 1977), p. 185.

[5] J. Gasser and H. Leutwyler, *Nucl. Phys.* B94 (1975) 269.

[6] S. Weinberg, *Physica* A96 (1979) 327.

[7] H. Leutwyler, *Ann. Phys. (N.Y.)*, 235 (1994) 165.

[8] E. D'Hoker and S. Weinberg, *General effective actions*, UCLA-Texas preprint, to be published in *Phys. Rev. D*;
S. Weinberg, in these proceedings.

[9] P. Gerber and H. Leutwyler, *Nucl. Phys.* B321 (1989) 387.

[10] P. Binétruy and M.K. Gaillard, *Phys. Rev.* D32 (1985) 931.

[11] S. Randjbar-Daemi, A. Salam and J. Strathdee, *Phys. Rev.* B48 (1993) 3190;
H. Leutwyler, *Phys. Rev.* D49 (1994) 3033.

[12] *The DAFNE Physics Handbook*, eds. L. Maiani, G. Pancheri and N. Paver, INFN-Frascati (1992).

[13] S. Weinberg, *Phys. Rev. Lett.* 17 (1966) 616.

[14] G. Czapek et al., *Letter of intent*, CERN/SPSLC 92-44.

[15] J. Gasser and H. Leutwyler, *Phys. Lett.* B125 (1983) 325.

[16] M. Knecht, in these proceedings.

[17] H. J. Leisi et al., in these proceedings.

[18] A. W. Thomas and R. H. Landau, *Phys. Rep.* 58 (1980) 122;
T. Ericson and W. Weise, *Pions and nuclei*, Oxford Univ. Press (1988);
S. Weinberg, *Phys. Lett.* B295 (1992) 114.

[19] S. Mallik, *Massive states in chiral perturbation theory*, preprint Saha Inst. of Nucl. Phys., Calcutta, hep-ph 9410344.

[20] The model for the πN interaction developed by P. F. A. Goudsmit, H. J. Leisi and E. Matsinos, *Phys. Lett.* B271 (1991) 290; *Phys. Lett.* B299 (1993) shows that the low energy data may be understood in terms of effective fields. The expansion of the resonance denominators occurring in the tree graphs of this model should yield a decent approximation for the effective Lagrangian.

[21] J. Gasser and H. Leutwyler, *Nucl. Phys.* B250 (1985) 539;
J. Donoghue, B. Holstein and D. Wyler, *Phys. Rev.* D47 (1993) 2089; *Phys. Rev. Lett.* 69 (1992) 3444;
A. V. Anisovich, *Dispersion relation technique for three-pion system and the P-wave interaction in $\eta \rightarrow 3\pi$ decay*, preprint Petersburg Nuclear Physics Institute, Gatchina TH-62-1993/1931;
J. Kambor, C. Wiesendanger and D. Wyler, *Final state interactions and Khuri-Treiman equations in $\eta \rightarrow 3\pi$ decays*, preprint IPNO/TH 94-93, ZU-TH 41/94 in preparation;
A. V. Anisovich and H. Leutwyler, *Measuring the quark mass ratio $(m_d^2 - m_u^2)/m_s^2$ by means of $\eta \rightarrow 3\pi$*, preprint BUTP-95/1 in preparation.

Aspects of Nucleon Chiral Perturbation Theory

Ulf-G. Meißner

Centre de Recherches Nucléaire, Physique Théorique, BP 28 Cr, F–67037
Strasbourg Cedex 2, France

1 Introduction

This lecture is concerned with the structure of hadrons at low energies, where the
strong coupling constant is large. Most of the hadronic world discussed here will
be made up of the light u, d (and s) quarks since these are the constituents of the
low-lying hadrons. The best way to gain information about the strongly interacting
particles is the use of well-understood probes, such as the photon or the massive weak
gauge bosons. At very low energies, the dynamics of the strong interactions is governed
by constraints from chiral symmetry. This leads to the use of effective field theory
methods which in the present context is called baryon chiral perturbation theory. In
this lecture, I will briefly outline the basic framework of this effective field theory
and use nucleon Compton and pion–nucleon scattering to discuss the strengths and
limitations of it. The basic degrees of freedom are the pseudoscalar Goldstone bosons
chirally coupled to the matter fields like e.g. the nucleons. The very low-energy face
of the low-lying baryons is therefore of hadronic nature, essentially point-like Dirac
particles surrounded by a cloud of Goldstone bosons. Naturally, I can only cover
a small fraction of the many interesting phenomena related to low energy hadron
physics. I have chosen to mostly talk about the nucleon since after all it makes up
large chunks of the stable matter surrounding us and also is a good intermediary
between the nuclear and the high energy physicists present at this workshop. Most
of the methods presented here can easily be applied to other problems, and as it will
become obvious at many places, we still have a long way to go to understand all the
intriguing features of the nucleon in a systematic and controled fashion. Whenever
possible, I will avoid to talk about models, with the exception of some circumstances
where they can be used to estimate some of the low–energy constants entering the
chiral perturbation theory machinery. In fact, I will consider one of these constants
and discuss to what extent we can understand its numerical value from the so–called
resonance exchange saturation picture. Further aspects of nucleon structure related
to photo- and electropionproduction within the framework of CHPT are discussed in
V. Bernard's lecture [1].

Lecture delivered at the Workshop on Chiral Dynamics: Theory and Experiments,
Massachusetts Institute of Technology, Cambridge, USA, July 25 - 29, 1994, CRN
94-44

2 Chiral Perturbation Theory with Nucleons

The interactions of the strongly interacting particles at low energies are severely.constrained by the approximate chiral symmetry of the QCD Lagrangian. This is particularly evident for the pseudoscalar Goldstone bosons which are directly related to the spontaneous symmetry violation. In this section I will be concerned with the inclusion the low-lying spin-1/2 baryons (the nucleons) to the effective field theory. I will consider the two flavor case and mostly work in the isospin limit $m_u = m_d = \hat{m}$. For a more detailed account, I refer to A. Manohar's lecture [2]. The inclusion of such matter fields is less straightforward since these particles are not related to the symmetry violation. However, their interactions with the Goldstone bosons is dictated by chiral symmetry. Let us denote by Ψ the isospinor doublet including the neutron and the proton. It is most convenient to choose a non-linear realization of the chiral symmetry so that Ψ transforms as $\Psi \to K\Psi$ under $SU(2)_L \times SU(2)_R$, where K is a complicated function that does not only depend on the group elements $g_{L,R}$ of the $SU(2)_{L,R}$ but also on the Goldstone boson fields collected in $U(x)$, i.e. $K(x) = K(g_L, g_R, U(x))$ defines a local transformation. Expanding K in powers of the Goldstone boson fields, one realizes that a chiral transformation is linked to absorption or emission of pions (which was the theme in the days of "current algebra" techniques). Let us restrict the discussion to processes with one incoming and one outgoing baryon, such as πN scattering, pion photo- and electroproduction or nucleon Compton scattering (otherwise, we would have to add contact n-fermion terms with $n \geq 4$). In that case, the underlying effective Lagrangian formulated in terms of the asymptotically observed fields takes the form

$$\mathcal{L}_{\pi N}^{(1)} = \bar{\Psi}\left(i\gamma^\mu D_\mu - m + \frac{1}{2}g_A\gamma^\mu\gamma_5 u_\mu\right)\Psi \tag{2.1}$$

with m the nucleon mass (in the chiral limit), $u_\mu = iu^\dagger \nabla_\mu U u^\dagger$, $u = \sqrt{U}$ and D_μ (∇_μ) the chiral covariant derivative acting on the nucleons (pions). Finally, g_A is the axial-vector coupling constant measured in neutron β-deacy, $g_A = 1.26$. Notice that the lowest order effective Lagrangian contains one derivative and therefore is of dimension one as indicated by the superscript '(1)'. In contrast to the meson sector, $\mathcal{L}_{\pi\pi}^{(2,4,\cdots)}$, odd powers of the small momentum q are allowed (thus, to leading order, no quark mass insertion appears since $\hat{m} \sim q^2$). It is instructive to expand (2.1) in powers of the Goldstone and external fields. From the vectorial term, one gets the minimal photon-baryon coupling, the two-Goldstone seagull (Weinberg term) and many others. Expansion of the axial-vectors leads to the pseudovector meson-baryon coupling, the celebrated Kroll-Rudermann term and much more. Calculating tree diagrams based on (2.1) leads to the current algebra results. This is, however, not sufficient. First, tree diagrams are always real (i.e. unitarity is violated) and second, the Goldstone nature of the pions can lead to large (non-analytic) corrections. Therefore, one has to include loop diagrams making use of the chiral power counting first spelled out by Weinberg [3] for the meson sector. In the presence of baryons, the loop expansion is more complicated. First, since odd powers in q are allowed, a one-loop calculation of order q^3 involves contact terms of dimension two and three, i.e. combinations of zero or one quark mass insertions with zero to three derivatives. These terms are collected in $\mathcal{L}_{\pi N}^{(2,3)}$ and a complete list of them can be found in Krause's paper [4] (for the case of SU(3)). Second, the finiteness of the nucleon mass in the chiral limit and the fact that its value is comparable to the chiral symmetry breaking scale $\Lambda \sim M_\rho$ complicates the low energy structure. This has been discussed in detail by Gasser et al. [5]. Let me just give one illustrative example. The one loop contribution to the nucleon mass not only gives the celebrated non-analytic contribution proportional to $M_\pi^3 \sim \hat{m}^{3/2}$ but also an infinite shift of m which has to be compensated by a counterterm of dimension zero. It is a general feature that loops produce analytic contributions at orders below what one would naively expect (e.g. below q^3 from one loop diagrams). Therefore, in a CHPT calculation involving baryons one has to worry more about higher order contributions than it is the case in the meson sector. There is one way of curing this problem, namely to go into the extreme non-relativistic limit [6] and consider the baryons as very heavy (static) sources. Then, by a clever definition of velocity-dependent fields, one can eliminate the baryon mass term from the lowest order effective Lagrangian and expand all interaction vertices and baryon propagators in increasing powers of $1/m$. To be specific, one writes (I follow here ref.[7])

$$\Psi(x) = \exp[-imv \cdot x](H(x) + h(x)) \qquad (2.2)$$

where $H(x)$ and $h(x)$ are velocity–eigenstates (remember that a non–relativistic nucleon has a good four–velocity v_μ) and then eliminates the "small" component $h(x)$. This is similar to a Foldy-Wouthuysen transformation known from QED. The lowest order effective Lagrangian takes the form

$$\mathcal{L}_{\pi N}^{(1)} = \bar{H}(iv^\mu D_\mu + g_A S^\mu u_\mu)H \qquad (2.3)$$

with S_μ the covariant spin–vector (à la Pauli-Lubanski). In this limit one recovers a consistent derivative expansion since the troublesome mass term has been shifted into a string of interaction vertices. A lucid discussion of the chiral counting rules in the presence of heavy baryons can be found in ref.[8]. For example, the one loop contribution of the Goldstone bosons to the baryon self-energy is nothing but the non-analytic M_ϕ^3 ($\phi = \pi, K, \eta$) terms together with three contact terms from $\mathcal{L}_{MB}^{(2)}$ (in SU(3)). However, one has to be somewhat careful still. The essence of the heavy mass formalism is that one works with old-fashioned time-ordered perturbation theory. So one has to watch out for the appearance of possible small energy denominators (infrared singularities). This problem has been addressed by Weinberg [9] in his discussion about the nature of the nuclear forces. The dangerous diagrams are the ones where cutting one pion line (this only concerns pions which are not in the asymptotic in- or out-states) separates the diagram into two disconnected pieces (one therefore speaks of reducible diagrams). These diagrams should be inserted in a Schrödinger equation or a relativistic generalization thereof with the irreducible ones entering as a potential. So the full CHPT machinery is applied to the irreducible diagrams. This should be kept in mind. For the purposes I am discussing, we do not need to worry about these complications. Being aware of them, it is then straightforward to apply baryon CHPT to many nuclear and particle physics problems [10-14]. I will illustrate this on two particular examples in the next sections. Before doing that, however, I would like to stress that most calculations are only in their infancy. It is believed that for a good quantitative description one has to perform systematic calculations to order q^4, i.e. beyond next-to-leading order, as I will discuss in the context of nucleon Compton scattering. A systematic analysis to this order in the chiral expansion is not yet available. In Manohar's lecture [2,15], an alternative approach of including the low–lying spin-3/2 decuplet in the effective field theory is discussed (based on phenomenological considerations supplemented with some arguments from the large N_c world). In that fashion, one sums up a certain subset of graphs starting at order q^4. A critical discussion of this approach can e.g. be found in ref.[16].

3 Nucleon Compton Scattering

Consider low-energy (real) photons scattering off a proton, $\gamma(k)+p(p_1) \rightarrow \gamma(k')+p(p_2)$ in the gauge $\epsilon_0 = \epsilon'_0 = 0$ (with ϵ_μ denoting the polarization vector of the incoming photon). In the cm-system we have $k_0 = k'_0 = \omega$ and the invariant momentum transfer squared is $t = (k - k')^2 = -2\omega^2(1 - \cos\theta)$. The T–matrix takes the form

$$T = e^2 \sum_{i=1}^{6} A_i(\omega, t)\, \mathcal{O}_i \tag{2.4}$$

in the operator basis of ref.[17]. Under crossing ($\omega \rightarrow -\omega$) the $A_{1,2}$ are even and the $A_{3,4,5,6}$ are odd. Furthermore, below the single pion production threshold, $\omega_{\rm thr} = M_\pi$, the A_i are real. Clearly, the nucleon structure is encoded in these invariant functions. With them at hand, one can readily calculate the differential cross section and polarisation observables like the parallell asymmetry $\mathcal{A}_\parallel$ (polarized photons scatter on polarized protons with the proton spin (anti)parallel to the photon direction) or the perpendicular asymmetry $\mathcal{A}_\perp$ (with the proton spin perpendicular to the photon direction) (explicit formulae are given in ref.[14]). In forward direction, the scattering amplitude takes the form

$$\frac{1}{4\pi}T(\omega) = f_1(\omega^2)\, \vec{\epsilon}'^* \cdot \vec{\epsilon} + i\omega\, f_2(\omega^2)\, \vec{\sigma} \cdot (\vec{\epsilon}'^* \times \vec{\epsilon}) \quad . \tag{2.5}$$

The energy expansion of the spin-independent amplitude $f_1(\omega^2)$ reads

$$f_1(\omega^2) = -\frac{e^2 Z^2}{4\pi m} + (\bar{\alpha} + \bar{\beta})\omega^2 + \mathcal{O}(\omega^4) \tag{2.6}$$

where the first energy-independent term is nothing but the Thomson amplitude mandated by gauge invariance. Therefore, to leading order, the photon only probes some global properties like the mass or electric charge of the spin-1/2 target. At next-to-leading order, the non-perturbative structure is parametrized by two constants, the so-called electric and magnetic polarizabilities. To lowest order, q^3, these are given by a few loop diagrams, i.e. they belong to the rare class of observables free of low–energy constants. The lowest order results [18]

$$\bar{\alpha}_p = \bar{\alpha}_n = 10\bar{\beta}_p = 10\bar{\beta}_n = \frac{5e^2 g_A^2}{384\pi^2 F_\pi^2} \frac{1}{M_\pi} = 13.6 \cdot 10^{-4}\,{\rm fm}^3 \tag{2.7}$$

already describe the two main features of the data, namely that (a) the neutron and the proton behave essentially as (induced) electric dipoles and that (b) $(\bar{\alpha} + \bar{\beta})_p \simeq$

$(\bar{\alpha} + \bar{\beta})_n$ (see e.g. the contributions by Nathan and Bergstrom [19]). A few remarks concerning the results (2.7) are in order. In the chiral limit of vanishing pion mass, $\bar{\alpha}_{p,n}$ and $\bar{\beta}_{p,n}$ diverge as $1/M_\pi$. This is expected since the two photons probe the long-ranged pion cloud, i.e. there is no more Yukawa suppression as in the case for a finite pion mass. Furthermore, a well-known dispersion sum rule relates $(\bar{\alpha} + \bar{\beta})$ to the total nucleon photoabsorption cross section. The latter is, of course, also well-behaved in the chiral limit which at first sight seems to be at variance with the behaviour of the expansion of the scattering amplitude. But be aware that the general form of (2.6) has been derived under the assumption that there is a well defined low-energy limit. However, as has been pointed out by many [20], the strong magnetic (M1) $N\Delta\gamma$ transition leads to a potentially large Δ conntribution, $\bar{\beta}_{p,n}^\Delta \simeq 10 \cdot 10^{-4}$ fm^3. From the CHPT point of view, such contributions start at order q^4 since they are $\sim F_{\mu\nu}F^{\mu\nu}$ (with $F_{\mu\nu}$ the canonical photon field strength tensor which counts as q^2). This problem was addressed in a systematic fashion in refs.[21], where *all* terms of $\mathcal{O}(q^4)$ were considered (not only some as in previous works). These new terms fall into two categories. The first one consists of one loop diagrams with exactly one insertion from $\mathcal{L}_{\pi N}^{(2)}$. The corresponding low-energy constants $c_{1,2,3}$ can be estimated from resonance exchange or determined from data on elastic πN scattering (as discussed in section 4 and 5). The second class are genuine new counter terms from $\mathcal{L}_{\pi N}^{(4)}$, their coefficients could only be estimated making use of the resonance saturation principle (which works well in the meson sector [22]). I will come back to this in section 5. The pertinent results for the electromagnetic polarizabilities take the generic form [21]

$$(\bar{\alpha}, \bar{\beta})_{p,n} = \frac{C_1}{M_\pi} + C_2 \ln M_\pi + C_3 \tag{2.8}$$

where the constant C_1 can be read off from eq.(2.7). The loops of order q^4 contribute to the second and third term whereas the large local Δ contribution enters prominently in C_3. Including the theoretical uncertainties in estimating the corresponding low-energy constants and also the possible contributions from loops involving strangeness, one arrives at the following theoretical predictions:

$$\bar{\alpha}_p = 10.5 \pm 2.0, \quad \bar{\alpha}_n = 13.4 \pm 1.5, \quad \bar{\beta}_p = 3.5 \pm 3.6, \quad \bar{\beta}_n = 7.8 \pm 3.6, \tag{2.9}$$

in units of 10^{-4} fm^3. These agree (with the exception of $\bar{\beta}_n$) very well with the data. The two main lessons learned from this improved calculation are: (1) The chiral expansion for electric polarizabilities converges quickly and (2) in the case of $\bar{\beta}_p$, the coefficient C_2 is large so that the $\ln M_\pi$ term cancels most of the large and positive Δ contribution. This is a novel effect which goes in the right direction and shows once more that one has to include all terms at a given order. However, since there are large cancellations in the predictions for the magnetic polarizabilities, one would like to see the result of a q^5 calculation. On the experimental side, it would be of importance to perform independent measurements of the electric and magnetic polarizabilities to (a) test the dispersion sum rule and (b) to lower the uncertainties in the individual polarizabilities (these are considerably larger than the usually quoted ones if one does not impose the constraint from the sum rule).

The spin-dependent amplitude $f_2(\omega^2)$ has an expansion analogous to (2.6),

$$f_2(\omega^2) = f_2(0) + \gamma \omega^2 + \mathcal{O}(\omega^4) \tag{2.10}$$

with the Taylor coefficient $f_2(0)$ given by celebrated LET due to Low, Gell-Mann and Goldberger [23], $f_2(0) = -(e^2\kappa^2)/(8\pi m^2)$, with κ denoting the anomalous magnetic moment of the particle the photon scatters off. In CHPT, κ does not appear in the lowest order effective Lagrangian but is given by loops and counter terms from $\mathcal{L}_{\pi N}^{(2,3)}$ (this is frequently overlooked). The physics of the so–called "spin–dependent" polarizability γ is discussed in some detail in refs.[7,24]. Here, I just want to point out that the LEGS collaboration at Brookhaven intends to measure this interesting nucleon structure constant [25]. Also, in ref.[26] the interesting observation was made that the multipole predictions for the nucleon spin–polarizability and for the so–called Drell–Hearn–Gerasimov sum rule are incompatible. This again points towards the importance of independent experimental determinations of these quantities.

4 Topics in Pion–Nucleon Scattering

In this section, I will mostly discuss the chiral corrections to the S–wave πN scattering lengths and give some necessary definitions for the following section. Consider first the S–wave scattering of pions off a nucleon at rest in forward direction,

$$T^{ba} = T^+(\omega)\delta^{ba} + T^-(\omega)i\epsilon^{bac}\tau^c \qquad (2.11)$$

with $a(b)$ the isospin index of the incoming (outgoing) pion and $\omega = v \cdot q = q_0$ denotes the pion cms energy. Under crossing, the functions $T^\pm$ behave as $T^\pm(\omega) = \pm T^\pm(-\omega)$. At threshold, $\omega_{\text{thr}} = M_\pi$ (remember that I work to lowest order in the $1/m$ expansion), the amplitude is given by its scattering lenghts,

$$a^\pm = \frac{1}{4\pi}\frac{1}{1+\mu}T^\pm(\omega_{\text{thr}}) \quad . \qquad (2.12)$$

These are related to the also often used $a_{1/2}$ and $a_{3/2}$ via $a_{1/2} = a^+ + 2a^-$ and $a_{3/2} = a^+ - a^-$, respectively. For the later discussion, we also need the so–called axial polarizability. For that, consider T^+ not longer in forward direction and subtract the nucleon born terms (as indicated by the overbar),

$$\bar{T}^+(\omega, \vec{q}, \vec{q}\,') = t_0(\omega) + t_1(\omega)\vec{q}\,' \cdot \vec{q} + \ldots \qquad (2.13)$$

with the kinematics $\omega = \nu = v \cdot q = v \cdot q'$ and $t = (q - q')^2 = 2(M_\pi^2 - \omega^2 + \vec{q}\,' \cdot \vec{q})$. The axial polarizability is then defined via

$$\alpha_A = 2c_{01}^+ \equiv t_1(0) \qquad (2.14)$$

where for completeness I have also given the relation to the low–energy expansion parameter c_{01}^+ commonly used in the πN community.

One of the most splendid successes of current algebra in the sixties was Weinberg's prediction [27] of the S–wave πN scattering lengths,

$$a^- = \frac{M_\pi}{8\pi F_\pi^2} = 8.8 \cdot 10^{-2}/M_\pi, \quad a^+ = 0 \qquad (2.15)$$

in good agreement with the data, $a^- = 9.2 \pm 0.2$ and $a^+ = -0.8 \pm 0.4$ (in units of $10^{-2}/M_\pi$) [28]. I should stress that in view of the confused situation about low–energy πN scattering, these scattering lenghts certainly should be assigned much larger uncertainties [29]. For the sake of the argument, I will however stick to the Karlsruhe–Helsinki values [28]. Of course, one has to worry whether the chiral corrections will spoil this remarkable agreement. In ref.[30], this question was addressed. Besides the canonical one–loop diagrams, one has to include three finite contact terms from $\mathcal{L}_{\pi N}^{(2)}$, which due to crossing contribute to T^+,

$$\mathcal{L}_{\pi N}^{(2)} = c_1 \bar{H} H \, \text{Tr}\,(\chi_+) + \left(c_2 - \frac{g_A^2}{8m}\right)\bar{H}(v \cdot u)^2 H + c_3 \bar{H} u \cdot u H \qquad (2.16)$$

but in fact, only the combination $C \equiv c_2 + c_3 - 2c_1 + g_A^2/8m$ enters the result for a^+. The isospin–odd amplitude T^- has to be renormalized via a combination of four scale-dependent counter terms, $b^r(\lambda) = b_1^r(\lambda) + b_2^r(\lambda) + b_3^r(\lambda) - 2b_4^r(\lambda)$ (here, λ is the scale of dimensional regularization) with the corresponding contact terms from $\mathcal{L}_{\pi N}^{(3)}$ given in [30] together with $b_4 \bar{H}[\chi_-, v \cdot u]H$ (which was omitted in that paper, the conclusions and numbers, however, remain unchanged). Due to crossing, $\mathcal{L}_{\pi N}^{(3)}$ terms contribute to a^-. Defining $L \equiv M_\pi/8\pi F_\pi^2$, one arrives at

$$a^- = L\left[1 - \mu - \mu^2(1 + \frac{g_A^2}{4})\right] + \frac{L^2 M_\pi}{\pi}\left(1 - 2\ln\frac{M_\pi}{\lambda}\right) - 64\pi L^2 M_\pi F_\pi b^r(\lambda)$$

$$a^+ = 32\pi F_\pi^2 L^2 \, C\,(1+\mu) + \frac{3}{4}g_A^2 L^2 M_\pi \qquad (2.17)$$

which shows that the exact knowledge of the low–energy constants is much more important for a^+ than for a^- because in the latter case their contribution is suppressed with respect to the leading term by two powers of M_π. To get a handle on the numerical values for $c_{1,2,3}$ and $b_{1,2,3,4}$, the following procedure was used in ref.[30]. While c_1 is uniquely fixed form the pion–nucleon σ–term [7], the other low–energy constants were estimated from resonance exchange. In this case, one has contributions from the Δ, the Roper and also from scalar exchange. The quality of this procedure will be discussed in section 5.

Let me first consider the result for the isospin–odd scattering length. One finds

$$a^- = (8.76 + 0.40) \cdot 10^{-2}/M_\pi = 9.16 \cdot 10^{-2}/M_\pi \qquad (2.18)$$

which shows that the chiral corrections of order M_π^2 and M_π^3 are small (approximately 5% of the lowest order result) and move the prediction closer to the empirical value. Furthermore, the dependence of this result on the actual value of $b^r(\lambda)$ is very weak. Matters are different for a^+. Here, the contact terms play a prominent role and the chiral prediction is very sensitive to the choice of certain resonance parameters, one related to the scalar exchange and the other to the Δ contribution (for a more detailed discussion, see ref.[30]). Therefore, in the absence of more stringent bounds on these parameters one can only draw the conclusion that the chiral prediction for a^+ is within the empirical bounds for reasonable values of the resonance parameters. Also, while for a^- the convergence in $\mu = M_\pi/m$ is rapid, it is much slower in the case of a^+. This indicates that one should perform a q^4 calculation as it was done in case for the nucleon polarizabilities discussed in section 3.

5 Anatomy of a Low–Energy Constant

To get an idea about the quality of the resonance saturation principle used in the previous sections, I will consider here the low–energy constant c_3 defined in eq.(2.16). First, however, let me briefly review the underlying idea of estimating low–energy constants from resonance exchange [22]. As the starting point, consider meson resonances (M = V,A,S,P) chirally coupled to the Goldstone fields collected in U and the matter fields (N) plus baryonic excitations (N^*) and integrate out the meson and nucleon resonances

$$\exp i \int dx \, \mathcal{L}_{\text{eff}}[U, N] = \int [dM][dN^*] \exp i \int dx \, \tilde{\mathcal{L}}_{\text{eff}}[U, M, N, N^*] \qquad (2.19)$$

so that one is left with a string of higher dimensional operators contributing to $\mathcal{L}_{\text{eff}}[U, N]$ in a manifestly chirally invariant manner and with coefficients given entirely in terms of resonance masses and coupling constants of the resonance fields to the Goldstone bosons. In the meson sector, i.e. considering neither baryons nor their exciations, this scheme works remarkably well,

$$L_i = \sum_{M=V,A,S,P} L_i^M + L_i(\lambda) \qquad (2.20)$$

where the scale–dependent remainder $L_i(\lambda)$ can be neglected if one choses λ to take a value in the resoance region (say between 500 MeV and 1 GeV in the meson sector). But what about the baryon sector? To get an idea, I calculate the axial polarizability defined in eq.(2.14). This amounts to the evaluation of 5 finite one loop diagrams (and their crossed partners) plus the contact term contribution proportional to c_3,

$$\alpha_A = -\frac{2c_3}{F_\pi^2} - \frac{g_A^2 M_\pi}{8\pi F_\pi^4}\left(\frac{77}{48} + g_A^2\right) = 2.28 \pm 0.10 \, M_\pi^{-3} \qquad (2.21)$$

using the central value given in [31] but enlarging the uncertainty by a factor 2.5 (as was suggested to me by M. Sainio [32]). This amounts to

$$c_3 = -5.2 \pm 0.2 \, \text{GeV}^{-1} \quad . \qquad (2.22)$$

In the resonance exchange picture, the dominant contributions to c_3 stem from intermediate Δ's and scalar exchange with a small correction from excitation of the Roper resonance. Varying the corresponding couplings within their allowed values leads to

$$
\begin{aligned}
c_3^{\text{res}} &= c_3^\Delta + c_3^{N^*} + c_3^S \\
&= (-2.5\ldots-3.2) + (-0.1\ldots-0.2) + (-1.0\ldots-1.6)\,\text{GeV}^{-1} \qquad (2.23) \\
&= -3.6\ldots-5.0\,\text{GeV}^{-1}
\end{aligned}
$$

at the scale $\lambda = m_\Delta$. Comparison of (2.23) with (2.22) reveals that at least for this particular low–energy constant, the resonance exchange saturation principle seems to work. Clearly, a more systematic analysis has to be performed to draw a final conclusion. However, it is also mandatory to get more high precision low–energy data, at present there are just too few of these to determine all low–energy constants up-to-and–including order q^3 (or higher) and compare with predictions from resonance exchange.

6 An Amazingly Accurate QCD Prediction

The structure of the nucleon as probed by the weak charged currents is encoded in two form factors, the axial and pseudoscalar ones,

$$< N(p')|A_\mu^a|N(p) >= \bar{u}(p')\left[\gamma_\mu G_A(t) + \frac{(p'-p)_\mu}{2m}G_P(t)\right]\gamma_5 \frac{\tau^a}{2}u(p) \qquad (2.24)$$

with $t = (p'-p)^2$ and $A_\mu^a = \bar{q}\gamma_\mu\gamma_5(\tau^a/2)q$ the isovector axial current. The axial form factor $G_A(t)$ is discussed in V. Bernard's lecture [1] so I will here concentrate on the induced pseudoscalar form factor $G_P(t)$ as measured e.g. in muon capture, $\mu^- + p \rightarrow \nu_\mu + n$, i.e. at $t = -0.88M_\mu^2 \simeq -0.5M_\pi^2$. One defines the induced pseudoscalar coupling constant g_P via

$$g_P = \frac{M_\mu}{2m}G_P(t = -0.88M_\mu^2) \quad . \qquad (2.25)$$

The best empirical determination of $g_P = 8.7 \pm 1.6$ [33] is consistent with the PCAC (lowest order) prediction $g_P^{\text{PCAC}} = 8.9$. It is therefore believed that a measurement of g_P can test pion pole dominance but not more. However, one can do better in baryon chiral perturbation theory. For that, one simply uses the chiral Ward identity

$$\partial^\mu[\bar{q}\gamma_\mu\gamma_5 \frac{\tau^a}{2}q] = \hat{m}\bar{q}i\gamma_5\tau^a q \qquad (2.26)$$

and sandwiches it between nucleon states. One arrives at [34]

$$g_P = \frac{2M_\mu g_{\pi N} F_\pi}{M_\pi^2 + 0.88M_\mu^2} - \frac{1}{3}g_A M_\mu m r_A^2 + \mathcal{O}(q^2) \qquad (2.27)$$

with $g_{\pi N} = 13.31 \pm 0.34$ the strong pion–nucleon coupling constant and $r_A = 0.65 \pm 0.03$ fm the nucleon axial radius. The relation (2.27) is known since long [35] but its derivation solely based on the chiral Ward identity of QCD is new. The resulting prediction is [34]

$$g_P = (8.89 \pm 0.23) - (0.45 \pm 0.04) = 8.44 \pm 0.23 \qquad (2.28)$$

if one adds the uncertainties in quadrature. In fact, the largest uncertainty stems from the much debated value of the pion–nucleon coupling constant. Consequently, if one could measure g_P within an accuracy of 2% (as it seems to be feasible within present day technology [36]), one could cleanly test the QCD versus the lowest order (PCAC) prediction. In fact, one could turn the argument around and use such an accurate measurement to pin down the allowed range for the strong pion–nucleon coupling constant. Finally, let me make a remark on the form factor at other values of t. The recently published data on $G_p(t)$ for $t = -0.07, -0.139$ and -0.179 GeV^2 [37] are not accurate enough to cleanly distinguish between the lowest order and the one–loop QCD prediction.

7 Concluding Remarks

The standard model of the strong and electroweak interactions enjoys a spectacular success, particularly at high energies. At low energies, the symmetries help to formulate an effective field theory (EFT) which can be used to perform precise calculations. The relevant degrees of freedom of this EFT called chiral perturbation theory are the pseudo–Goldstone bosons and other hadrons, but not the quarks and gluons. The pions, kaons and etas play a special role in that they are linked directly to the spontaneous symmetry violation QCD is believed to undergo. The exact mechanism of this phase transition which generates the almost massless degrees of freedom is not yet understood. In the effective Lagrangian, a whole string of terms with increasing energy dimension is present, rendering the theory non–renormalizable. This is, howewer, of no relevance since the EFT is not supposed to be of use at all scales, but in the case at hand for energies below the typical resonance masses (say 1 GeV). The power of chiral perturbation theory stems from the observation that it is a systematic and simultaneous expansion in small momenta (energies) and quark (pion) masses. It is of utmost importance that to a given order one includes *all* terms demanded by the symmetry requirements. This means that beyond leading order the so–called low–energy constants enter, which are not given by the symmetries. The finite parts of these constants have to be determined from experiment or can be estimated from some principles like resonance exchange saturation. While in the meson sector the machinery exists and is fully operative, calculations in the baryon sector are hampered by the fact that not sufficiently many accurate low–energy data exist to pin down all appearing low–energy constants. However, with the new CW machines and the renewed interest in low–energy domain, we will eventually leave this transitional stage and will achieve a more satisfactory description of the effective pion–nucleon field theory [38]. The extension to the case of three flavors is also only in its infancy since the small parameter $M_K/4\pi F_\pi \simeq 0.4$ is not that small whereas in the two–flavor sector we deal with $M_\pi/4\pi F_\pi \simeq 0.1$. Furthermore, the closeness of the spin-3/2 decuplet has triggered some speculations that one should include these degrees of freedom in the EFT. Again, a systematic investigation of this approach is not yet available, so that at present one can not draw a final conclusion on it. To summarize, let me emphasis that *low–energy hadron physics is as interesting as any other field in physics and that exciting times are ahead of us.* Many challenging problems, both theoretical and experimental, remain to be solved.

8 Acknowledgements

I would like to thank the organizers for their invitation and making such an interesting workshop possible. The work reported here has been done in collaboration with V. Bernard, N. Kaiser and A. Schmidt, to whom I express my gratitude.

References

[1] V. Bernard, these proceedings

[2] A.V. Manohar, these proceedings

[3] S. Weinberg, *Physica* **96A** (1979) 327

[4] A. Krause, *Helv. Phys. Acta* **63** (1990) 3

[5] J. Gasser, M.E. Sainio and A. Švarc, *Nucl. Phys.* **B307** (1988) 779

[6] E. Jenkins and A.V. Manohar, *Phys. Lett.* **B255** (1991) 558

[7] V. Bernard, N. Kaiser, J. Kambor and Ulf-G. Meißner, *Nucl. Phys.* **B388** (1992) 315

[8] G. Ecker, *Czech. J. Phys.* **44** (1994) 405

[9] S. Weinberg, *Nucl. Phys.* **B363** (1991) 3

[10] E. Jenkins and A.V. Manohar, in "Effective Field Theories of the Standard Model", ed. Ulf–G. Meißner, World Scientific, Singapore, 1992

[11] Ulf-G. Meißner, *Int. J. Mod. Phys.* **E1** (1992) 561

[12] Ulf-G. Meißner, *Rep. Prog. Phys.* **56** (1993) 903

[13] T.-S. Park, D.-P. Min and M. Rho, *Phys. Rep.* **233** (1993) 341

[14] V. Bernard, Ulf-G. Meißner, N. Kaiser and A. Wirzba, "Chiral Symmetry in Nuclear Physics", in preparation

[15] E. Jenkins and A.V. Manohar, *Phys. Lett.* **B259** (1991) 353

[16] V. Bernard, N. Kaiser and Ulf-G. Meißner, *Z. Phys.* **C60** (1993) 111

[17] A.C. Hearn and E. Leader, *Phys. Rev.* **126** (1962) 789

[18] V. Bernard, N. Kaiser and Ulf-G. Meißner, *Phys. Rev. Lett.* **67** (1991) 1515; *Nucl. Phys.* **373** (1992) 364

[19] J. Bergstrom, A. Nathan, these proceedings

[20] M.N. Butler and M.J. Savage, *Phys. Lett.* **B294** (1992) 369; A. L'vov, *Phys. Lett.* **B304** (1993) 29

[21] V. Bernard, N. Kaiser, Ulf-G. Meißner and A. Schmidt, *Phys. Lett.* **B319** (1993) 269; *Z. Phys.* **A348** (1994) 317

[22] G. Ecker, J. Gasser, A. Pich and E. de Rafael, *Nucl. Phys.* **B321** (1989) 311; J. F. Donoghue, C. Ramirez and G. Valencia, *Phys. Rev.* **D39** (1989) 1947

[23] F.E. Low, *Phys. Rev.* **96** (1954) 1428; M. Gell-Mann and M.L. Goldberger, *Phys. Rev.* **96** (1954) 1433

[24] Ulf-G. Meißner, in "Substructures of Matter as Revealed with Electroweak Probes", eds. L. Mathelitsch and W. Plessas, Springer, Berlin, 1994

[25] A.M. Sandorfi, private communication

[26] A.M. Sandorfi et al., Brookhaven preprint BNL–60616, 1994

[27] S. Weinberg, *Phys. Rev. Lett.* **17** (1966) 616

[28] R. Koch, *Nucl. Phys.* **A448** (1986) 707

[29] G. Höhler, M.E. Sainio, these proceedings

[30] V. Bernard, N. Kaiser and Ulf-G. Meißner, *Phys. Lett.* **B319** (1993) 269

[31] G. Höhler, in Landolt–Börnstein, vol.9 b2, ed. H. Schopper, Springer, Berlin, 1983

[32] M.E. Sainio, private communication

[33] G. Bardin et al., *Phys. Lett.* **B104** (1981) 320

[34] V. Bernard, N. Kaiser and Ulf-G. Meißner, *Phys. Rev.* **D50** (1994), in print

[35] S.L. Adler and Y. Dothan, *Phys. Rev.* **151** (1966) 1267; L. Wolfenstein, in: "High–Energy Physics and Nuclear Structure", ed. S. Devons, Plenum, New York, 1970

[36] D. Taqqu, talk presented at the International Workshop on "Large Experiments at Low Energy Hadron Machines", PSI, Switzerland, April 1994

[37] S. Choi et al., *Phys. Rev. Lett.* **71** (1993) 3927

[38] G. Ecker, *Phys. Lett.* **B** (1994), in print

Estimates of Low–Energy Parameters[1]

G. Ecker

Institut für Theoretische Physik, Universität Wien
Boltzmanngasse 5, A–1090 Wien, Austria

Abstract: The phenomenological and theoretical status of the low–energy parameters of effective chiral Lagrangians is reviewed.

1 Low–Energy Constants

Chiral perturbation theory (CHPT) is the effective field theory of the Standard Model (SM) at low energies. The effective chiral Lagrangians contain *all* terms compatible with the symmetries of the SM [1]. The associated coupling constants are called low–energy constants (LECs) [2, 3].

As a quantum field theory, CHPT guarantees the usual axiomatic properties like (perturbative) unitarity, analyticity, cluster decomposition, etc. The LECs parametrize the most general solutions of the Ward identities, but they are themselves not constrained by the symmetries. They can be interpreted as describing the influence of all degrees of freedom not explicitly contained in the chiral Lagrangians. The great number of LECs reflect the loss of information in the transition from the SM to its effective realization.

CHPT is a quantum field theory that must be renormalized order by order in the loop expansion. Most of the LECs encountered in non–leading orders are needed for renormalization. The analysis of LECs in CHPT proceeds in two steps:

(i) Regularization and renormalization

In order to absorb the divergences occurring in the loop expansion, the LECs (denoted generically as C_i) are decomposed in two parts:

$$C_i = C_i^r(\mu) + D_i \lambda(\mu) . \tag{1}$$

The divergence is contained in the function $\lambda(\mu)$ depending on a scheme dependent renormalization scale μ. Consequently, the measurable renormalized LECs $C_i^r(\mu)$ are in general scale dependent. The constants D_i govern the scale dependence of the $C_i^r(\mu)$ and they also determine the so–called "chiral logs". In any physical amplitude, the scale dependence always cancels between the loop and the counterterm contributions containing the renormalized LECs.

(ii) Determination of the renormalized LECs

In principle, the LECs are calculable quantities in the SM. In practice, they

---

[1]Work supported in part by FWF, Project No. P09505–PHY and by HCM, EEC–Contract No. CHRX–CT920026 (EURODAΦNE).

are extracted from experimental input or estimated with additional model dependent assumptions.

As a guide–line for this talk, I will restrict the discussion of LECs to the chiral Lagrangians listed in Table 1 where the renormalization is completely known to the order indicated. Mainly for lack of time (and space), I will not discuss the odd–intrinsic parity sector of $\mathcal{O}(p^6)$ ("anomalous" sector; see Ref. [4] for a review). Since there were two plenary talks at this Workshop on general aspects of the meson–baryon system [5, 6], I will be very brief on this topic.

Table 1: Chiral Lagrangians with completely known renormalization properties to the order indicated. The term "strong" stands for strong, electromagnetic and semileptonic weak interactions (external gauge fields only).

interaction	chiral order
mesons	
strong	$\mathcal{O}(p^4)$
odd–intrinsic parity	$\mathcal{O}(p^6)$
non–leptonic weak	$\mathcal{O}(G_F p^4)$
virtual photons	$\mathcal{O}(e^2 p^2)$
meson–baryon system	
strong $[SU(2)]$	$\mathcal{O}(p^3)$

2   Strong interactions of mesons

Adopting the standard chiral counting where quark masses are counted as $\mathcal{O}(p^2)$, the chiral Lagrangian of lowest order, $\mathcal{O}(p^2)$, contains two LECs: F and B [2, 3]. They are related to the pion decay constant and to the quark condensate in the chiral limit:

$$F_\pi = F[1 + \mathcal{O}(m_{quark})] = 93.2 MeV \tag{2}$$

$$\langle 0|\bar{u}u|0\rangle = -F^2 B[1 + \mathcal{O}(m_{quark})] \ .$$

Restricting the discussion to chiral $SU(3)$, there are ten measurable LECs at $\mathcal{O}(p^4)$: $L_1,\ldots, L_{10}$ [3]. Chiral dimensional analysis [1, 7] suggests as an approximate upper bound

$$|L_i| \lesssim \frac{N_f}{4(4\pi)^2} \simeq 5 \cdot 10^{-3} \tag{3}$$

for $N_f = 3$ light flavours. The present values for the L_i^r at $\mu = M_\rho$ displayed in Table 2 are still remarkably similar to the original values [3]. Chiral dimensional analysis is quite successful here: only L_3, L_9 and L_{10} are close to the upper bound (3) [cf. also Eq. (8)].

The main new input in Table 2 comes from a recent analysis of K_{e4} decays [8]. In particular, incorporation of the K_{e4} data allows for a test of the Zweig rule classification of the L_i [3]:

$$2L_1 - L_2, L_4, L_6 \qquad \mathcal{O}(1) \tag{4}$$

$$L_1, L_2, L_3, L_5, L_8, L_9, L_{10} \qquad \mathcal{O}(N_c) \tag{5}$$

$$L_7 \qquad \mathcal{O}(N_c^2). \tag{6}$$

The recent analysis of Ref. [8] finds

$$\frac{L_2 - 2L_1}{L_3} = -0.17 \begin{array}{c} +0.12 \\ -0.22 \end{array} \tag{7}$$

in support of the Zweig rule.

Table 2: Current phenomenological values and source for the renormalized coupling constants $L_i^r(M_\rho)$, taken from Ref. [9]. The quantities Γ_i in the fourth column determine the scale dependence of the $L_i^r(\mu)$.

i	$L_i^r(M_\rho) \times 10^3$	source	Γ_i
1	0.4 ± 0.3	$K_{e4}, \pi\pi \to \pi\pi$	$3/32$
2	1.35 ± 0.3	$K_{e4}, \pi\pi \to \pi\pi$	$3/16$
3	-3.5 ± 1.1	$K_{e4}, \pi\pi \to \pi\pi$	0
4	-0.3 ± 0.5	Zweig rule	$1/8$
5	1.4 ± 0.5	$F_K : F_\pi$	$3/8$
6	-0.2 ± 0.3	Zweig rule	$11/144$
7	-0.4 ± 0.2	Gell-Mann–Okubo,L_5, L_8	0
8	0.9 ± 0.3	$M_{K^0} - M_{K^+}, L_5,$	$5/48$
		$(2m_s - m_u - m_d) : (m_d - m_u)$	
9	6.9 ± 0.7	$\langle r^2 \rangle_V^\pi$	$1/4$
10	-5.5 ± 0.7	$\pi \to e\nu\gamma$	$-1/4$

2.1 Lattice

A very promising first attempt to calculate the L_i "directly" from QCD has been undertaken by Myint and Rebbi [10]. Their work documents the feasibility of extracting LECs from the lattice (see also Ref. [11]). The first analytic results [10] obtained in the strong–coupling and large–N_c limits are not too realistic, but the announced numerical analysis is expected to come closer to the real world.

2.2 Resonance exchange

The actual values of the L_i can be understood in terms of meson resonance exchange [12, 13]. The situation can be summarized as follows:

Chiral duality:

The $L_i^r(M_\rho)$ are practically saturated by resonance exchange.

Chiral VMD:

Whenever spin-1 resonances can contribute at all ($i = 1, 2, 3, 9, 10$), the $L_i^r(M_\rho)$ are almost completely dominated by V and A exchange.

With additional QCD–inspired assumptions, all V, A couplings can be expressed in terms of F_π and $M_V \simeq M_\rho$ only [14]:

$$8L_1^V = 4L_2^V = -\frac{4}{3}L_3^V = L_9^V = -\frac{4}{3}L_{10}^{V+A} = \frac{F_\pi^2}{2M_V^2} \,. \tag{8}$$

These relations are in very good agreement with the phenomenological values of the $L_i^r(M_\rho)$.

There are two objections sometimes raised against the interpretation of resonance dominated L_i.

Objection # 1 : Resonance exchange produces scale independent L_i.

A more precise formulation of resonance dominance [12] clarifies the situation: decomposing the $L_i^r(\mu)$ into resonance contributions L_i^R and remainders $\widehat{L}_i(\mu)$ carrying the scale dependence,

$$L_i^r(\mu) = \sum_R L_i^R + \widehat{L}_i(\mu) \,, \tag{9}$$

there is a range in μ (depending on the renormalization scheme) with

$$|\widehat{L}_i(\mu)| \ll |L_i^r(\mu)| \quad \forall\, i \,. \tag{10}$$

Objection # 2 : The resonance parameters are determined at $p^2 = M_R^2$, but the L_i describe physics for $p^2 \ll M_R^2$.

To resolve this puzzle, let us consider the $VV - AA$ two–point functions as a specific example:

$$i \int d^4x \, e^{ip\cdot x}\langle 0|T\{V_\mu^i(x)V_\nu^j(0) - A_\mu^i(x)A_\nu^j(0)\}|0\rangle =$$

$$= (p_\mu p_\nu - g_{\mu\nu}p^2)\Pi_{LR,ij}^{(1)}(p^2) + p_\mu p_\nu \,\Pi_{LR,ij}^{(0)}(p^2) \,. \tag{11}$$

Any given model for the spin–1 resonances will produce a spectral function $\Pi_{LR}^{(1)}$ of the following structure in the narrow–width approximation (V exchange only):

$$\Pi_{LR}^{(1)}(p^2) = \frac{P_V(p^2)^2}{M_V^2 - p^2} + P_c(p^2) \,, \tag{12}$$

where $P_V(p^2)$, $P_c(p^2)$ are model dependent polynomials characterizing the off-shell behaviour of V exchange. From resonance decays (e.g., $\rho \rightarrow e^+e^-$ in this case) one can fix $F_V = P_V(M_V^2)$, but unless $P_V(p^2)$ is a constant, the decay width tells us nothing about $P_V(p^2)$ for $p^2 \neq M_V^2$.

The solution of the puzzle is provided by the connection between low and high energies characteristic for a quantum field theory like QCD. In fact, QCD requires a relation between the polynomials $P_V(p^2)$ and $P_c(p^2)$ such that the off–shell behaviour of $P_V(p^2)$ is actually irrelevant. The key to the solution is the unsubtracted dispersion relation satisfied by $\Pi_{LR}^{(1)}$:

$$\Pi_{LR}^{(1)}(p^2) = \int_0^\infty \frac{ds}{s-p^2}[\rho_V^{(1)}(s) - \rho_A^{(1)}(s)] . \tag{13}$$

Again in the narrow–width approximation, V exchange yields

$$\rho_V^{(1)}(s) = F_V^2 \, \delta(s - M_V^2) \tag{14}$$

and therefore [recall $F_V = P_V(M_V^2)$]

$$\Pi_{LR}^{(1)}(p^2) = \frac{P_V(M_V^2)^2}{M_V^2 - p^2} = \frac{P_V(p^2)^2}{M_V^2 - p^2} + P_c(p^2) , \tag{15}$$

fixing the counterterm polynomial $P_c(p^2)$ uniquely:

$$P_c(p^2) = (M_V^2 - p^2)^{-1}[P_V(M_V^2)^2 - P_V(p^2)^2] . \tag{16}$$

The general conclusion is that Green functions at small p^2 are determined by the on–shell resonance parameters, *independently* of higher–order couplings [14, 15]. It is this universality that implies in particular the equivalence of all realistic models for spin–1 meson resonances to $\mathcal{O}(p^4)$ [14].

The symmetry breaking sector (involving the quark masses) is characterized at $\mathcal{O}(p^4)$ by the LECs L_4, L_5, L_6, L_7, L_8. These LECs are only sensitive to scalar (octet S and singlet S_1) and pseudoscalar (η') exchange. In the case of L_5, only the scalar octet can contribute. Saturating the unsubtracted dispersion relation for the scalar form factor $\langle\pi|\bar{u}s|K\rangle$ with S exchange, one finds [16]

$$L_5 \simeq \frac{F_\pi^2}{4M_S^2} . \tag{17}$$

The LEC L_5 governs $SU(3)$ breaking in the meson decay constants [3]:

$$\begin{aligned}
\frac{F_K}{F_\pi} &= 1 + \frac{4L_5^r}{F^2}(M_K^2 - M_\pi^2) + \text{chiral logs} \\
&\simeq 1 + \frac{M_K^2 - M_\pi^2}{M_S^2} + \text{chiral logs} \\
&\simeq 1.22 \quad [17] .
\end{aligned} \tag{18}$$

With $M_S \simeq M_{\eta'} \simeq 1$ GeV, one understands why $SU(3)$ breaking in the meson sector is generally of $\mathcal{O}(25\%)$, except for the pseudoscalar masses.

2.3 Sum rules

An alternative approach to use phenomenological input for the determination of LECs is provided by sum rules. Consider again the two–point functions (11) as an example. To $\mathcal{O}(p^4)$, the spin–1 spectral function is of the form [2, 3]

$$\Pi_{LR}^{(1)}(p^2) = F_4(p^2, M^2; \mu^2) - 4L_{10}^r(\mu) \tag{19}$$

where F_4 contains the scale dependent one–loop contribution $(M = M_\pi, M_K)$. For $p^2 = 0$ one obtains the sum rule [18, 19]

$$\int \frac{ds}{s}[\rho_V^{(1)}(s) - \rho_A^{(1)}(s)] = F_4(0, M^2; \mu^2) - 4L_{10}^r(\mu) =: -4\overline{L_{10}} \tag{20}$$

$$\overline{L_{10}} = L_{10}^r(\mu) + \frac{1}{384\pi^2}\left[\ln\frac{M_K^2}{\mu^2} + 2\ln\frac{M_\pi^2}{\mu^2} + 3\right].$$

Saturating the integral with resonance exchange leads to the scale independent version of the V, A exchange contributions to L_{10}:

$$\overline{L_{10}} = -\frac{F_V^2}{4M_V^2} + \frac{F_A^2}{4M_A^2} = L_{10}^r(M_\rho) - 1.2 \cdot 10^{-3} \equiv L_{10}^r(350 \text{ MeV}) . \tag{21}$$

The spectral function $\Pi_{LR}^{(1)}$ has recently been calculated to $\mathcal{O}(p^6)$ [20]:

$$\Pi_{LR}^{(1)}(p^2) = F_6(p^2, M^2; \mu^2) - 4L_{10}^r(\mu) + P_1(p^2, M^2; \mu^2) , \tag{22}$$

where F_6 contains both one– and two–loop contributions and P_1 is a first–order polynomial in p^2, M^2 containing the appropriate renormalized LECs of $\mathcal{O}(p^6)$. The first derivative of (22) gives rise to a sum rule for a combination of $\mathcal{O}(p^6)$ LECs relevant for $\gamma\gamma \to \pi^0\pi^0$ [21]

$$\left.\frac{d\Pi_{LR}^{(1)}(p^2)}{dp^2}\right|_{p^2=0} = \int \frac{ds}{s^2}[\rho_V^{(1)}(s) - \rho_A^{(1)}(s)] = F_6'(0, M^2; \mu^2) + P_1'(0, 0; \mu^2) . \tag{23}$$

A similar sum rule was derived in Ref. [22] within the framework of generalized CHPT.

The two–loop calculation can be used for still more sum rules:

$$(n-1)! \int \frac{ds}{s^n}[\rho_V^{(1)}(s) - \rho_A^{(1)}(s)] = F_6^{(n-1)}(0, M^2) \qquad n \geq 3 . \tag{24}$$

Although the LECs of $\mathcal{O}(p^6)$ drop out for $n \geq 3$, these sum rules can test combinations of the L_i entering at $\mathcal{O}(p^6)$ via one–loop diagrams [20].

2.4 Chiral models

There has been a lot of activity in recent years to obtain the LECs from different models for hadronic interactions at low energies. Since it was not my task to cover those developments, I refer to two recent reviews [23] for an up–to–date summary.

3 Non–leptonic weak interactions

The effective chiral Lagrangian describing the $\Delta S = 1$ non–leptonic weak interactions of mesons starts at $\mathcal{O}(G_F p^2)$:

$$\mathcal{L}_2^{\Delta S=1} = G_8 \langle \lambda L_\mu L^\mu \rangle + G_{27} \left(L_{\mu 23} L_{11}^\mu + \frac{2}{3} L_{\mu 21} L_{13}^\mu \right) + \text{h.c.} \tag{25}$$

$$\lambda = (\lambda_6 - i\lambda_7)/2 \qquad L_\mu = iF^2 U^\dagger D_\mu U .$$

The matrix field $U(\phi)$ incorporating the pseudoscalar Goldstone fields transforms linearly under the chiral group $SU(3)_L \times SU(3)_R$ and $\langle \ldots \rangle$ denotes the trace in flavour space. The octet and 27–plet coupling constants G_8, G_{27} can be extracted from $K \to \pi\pi$ decay rates:

$$|G_8| \simeq 9 \cdot 10^{-6} \text{ GeV}^{-2} \qquad G_{27}/G_8 \simeq 1/18 . \tag{26}$$

At $\mathcal{O}(G_F p^4)$, there are 22 measurable LECs in the octet part alone [24]. The renormalization of the one–loop functional and the resulting scale dependence of the LECs are completely known [25, 26]. Only a few combinations of the LECs have been determined phenomenologically (see Refs. [27, 28] for details). The resonance saturation of the strong LECs of $\mathcal{O}(p^4)$ served as a motivation to investigate the relevance of resonance exchange for the non–leptonic weak LECs [24]. However, in contrast to the strong case the weak resonance couplings are essentially unknown. Moreover, it is clear that resonance exchange cannot saturate all LECs in this case. There exist direct short–distance contributions to the LECs, e.g., due to electromagnetic penguin operators [29, 30]. In addition, the chiral anomaly contributes to and probably dominates the $\mathcal{O}(G_F p^4)$ terms involving an ε tensor [31].

A more predictive framework is based on the idea of factorization [32]. To first order in $1/N_c$ and α_s, the chiral representation of the 4–quark operators in the non–leptonic weak Hamiltonian factorizes into a product of currents, e.g.,

$$\bar{q}_L \gamma^\mu q_L \bar{q}_L \gamma_\mu q_L \to J^\mu J_\mu \tag{27}$$

$$J_\mu = \frac{\delta S_{\text{strong}}}{\delta \ell^\mu} = J_\mu^{(1)} + J_\mu^{(3)} + \cdots ,$$

where S_{strong} is the chiral action for the strong interactions and ℓ^μ is the octet of left–chiral external gauge fields. The left–chiral current J_μ is decomposed into pieces with chiral dimension $D = 1, 3, \ldots$:

$$J_\mu^{(1)} = -\frac{i}{2} F^2 U^\dagger D_\mu U \tag{28}$$

$$J_\mu^{(3)} = J_{\mu,\text{normal}}^{(3)}(L_i) + J_{\mu,\text{anom}}^{(3)} . \tag{29}$$

The current of $\mathcal{O}(p^3)$ has both a normal part depending on the L_i and a parameter–free anomalous part

$$J_{\mu,\text{anom}}^{(3)} = \frac{\delta S_{WZW}}{\delta \ell^\mu} \tag{30}$$

due to the Wess–Zumino–Witten functional [33].

The factorization model [24, 31] is defined by the following Lagrangian of $\mathcal{O}(G_F p^4)$:

$$\mathcal{L}_{4,FM}^{\Delta S=1} = 4k_f G_8 \langle \lambda_6 \{ J_\mu^{(1)}, J^{(3)\mu} \} \rangle , \tag{31}$$

with a fudge factor $k_f = \mathcal{O}(1)$ to allow for a different overall scale compared to $\mathcal{L}_2^{\Delta S=1}$. The factorization model can accommodate the available experimental information [24, 28], but it has not really been put to a decisive test yet. It can be improved systematically [30] using the method of the effective action [34].

4 Virtual photon corrections

The inclusion of virtual photons modifies the chiral counting. The chiral Lagrangian of lowest order is now of $\mathcal{O}(e^2 p^0)$ and consists of a single term [35, 12]:

$$\mathcal{L}_{(e^2 p^0)} = e^2 C \langle QUQU^\dagger \rangle , \tag{32}$$

where Q is the quark charge matrix. This Lagrangian yields both Dashen's theorem [36]

$$M_{\pi^+}^2 = M_{K^+}^2 = \frac{2e^2 C}{F^2} + \mathcal{O}(m_q) \tag{33}$$

and Sutherland's theorem [37]

$$A(\eta \to 3\pi) = \mathcal{O}(m_u - m_d, e^2 m_q) . \tag{34}$$

The complete list of terms in the chiral Lagrangian of $\mathcal{O}(e^2 p^2)$ has recently been given by Urech [38], together with the renormalization of the corresponding LECs. The main area of applications are isospin violating transitions, especially those that are suppressed at $\mathcal{O}(p^4)$. Due to the lack of information about the LECs of $\mathcal{O}(e^2 p^2)$, the field is still in the stage of simple chiral dimensional analysis. In a recent application to the rare decay $\eta \to \pi \ell \nu$, the corrections of $\mathcal{O}(e^2 p^2)$ were found to be small compared to the contributions with $m_u \neq m_d$ [39].

Another interesting application concerns the electromagnetic corrections of $\mathcal{O}(e^2 p^2) \equiv \mathcal{O}(e^2 m_q)$ to Dashen's theorem (33). The relevant quantity Δ_{EM} is defined for $m_u = m_d$ as

$$\Delta_{EM} := M_{K^+}^2 - M_{K^0}^2 - M_{\pi^+}^2 + M_{\pi^0}^2 . \tag{35}$$

The result of the one–loop calculation at $\mathcal{O}(e^2 m_q)$ is [38]

$$\Delta_{EM} = (-0.26; 0.49; 0.99 \pm 0.13) \cdot 10^{-3} \text{ GeV}^2 + \Delta M_{ct}^2(\mu) , \tag{36}$$

where the entries in brackets contain both the loop contributions and the strong LEC $L_5^r(\mu)$ for $\mu = 0.5$, 0.77 and 1 GeV, respectively. The μ dependence of the first term is of course cancelled by the scale dependence of $\Delta M_{ct}^2(\mu)$ containing the LECs of $\mathcal{O}(e^2 m_q)$. To appreciate the strong scale dependence of the two terms, one should recall

$$M_{\pi^+}^2 - M_{\pi^0}^2 \big|_{\text{expt}} = 1.26 \cdot 10^{-3} \text{ GeV}^2. \tag{37}$$

Turning to chiral dimensional analysis for an estimate of ΔM_{ct}^2, one finds

$$|\Delta M_{ct}^2| \lesssim 2.6 \cdot 10^{-3} \text{ GeV}^2 \qquad [38] \qquad (38)$$

$$-0.6 \lesssim \Delta_{EM} \cdot 10^3 \text{ GeV}^{-2} \lesssim 2.0 \qquad [39] \qquad (39)$$

The conclusion is that corrections to Dashen's theorem *could* be sizable.

With stronger, model dependent assumptions, the following estimates have been obtained:

$$\Delta_{EM} = \left\{ \begin{array}{c} 1.2 \\ 1.3 \pm 0.4 \end{array} \cdot 10^{-3} \text{ GeV}^2 \quad \begin{array}{c} [40] \\ [41] \end{array} \right. . \qquad (40)$$

5   Meson–baryon system

CHPT with baryons has become a very active field. It was reviewed at this Workshop by A.V. Manohar [5] and U.-G. Meißner [6].

In the heavy baryon formulation of CHPT (HBCHPT) [42], there is a correspondence between the loop and the derivative expansion similar to the meson sector. Unlike in the original relativistic framework [43], diagrams with L loops are at least $\mathcal{O}(p^{2L+1})$ in HBCHPT. Different from the purely mesonic case, the chiral expansion in the presence of fermions has terms of $\mathcal{O}(p^n)$ for all integer $n \geq 1$. An immediate consequence of the loop structure is that the LECs of both $\mathcal{O}(p)$ and $\mathcal{O}(p^2)$ are not renormalized in a mass independent renormalization scheme such as dimensional regularization (corresponding to the non-renormalization of F, B in the lowest–order mesonic Lagrangian).

Restricting myself to chiral $SU(2)$, there is a single LEC of $\mathcal{O}(p)$, the nucleon axial–vector coupling constant g_A in the chiral limit. The chiral pion–nucleon Lagrangian of $\mathcal{O}(p^2)$ has two types of terms: the terms coming from the expansion in $1/m$, where m is the nucleon mass in the chiral limit, and 7 additional chiral invariant terms with a priori undetermined LECs. The natural order of magnitude of those LECs is $\mathcal{O}(1/m)$. What is known phenomenologically at this time (see [6] and references therein), is in agreement with the naive estimate. Among other experimental input, the nucleon magnetic moments, the πN σ term and the πN scattering lengths can be used to pin down those LECs.

At $\mathcal{O}(p^3)$, the one–loop diagrams start to contribute. The complete renormalization of the one–loop functional has recently been carried out [44]. There are only partial results for the phenomenological values of the LECs of $\mathcal{O}(p^3)$ [6]. A major issue is how to understand the phenomenological values of the LECs, starting at $\mathcal{O}(p^2)$. It is clear that resonances will again play an important role. This concerns especially the Δ resonance. In fact, it has been an open question for some time whether one should incorporate the Δ resonance (and higher baryon resonances in the case of chiral $SU(3)$) in the LECs only [6] or as a dynamical field in the chiral Lagrangian [5]. Both approaches have their pros and cons and more experience is needed to decide either way.

6 Conclusions and outlook

CHPT predicts the structure of Green functions and amplitudes at low energies in terms of LECs which cannot be fixed from symmetry arguments alone. We can draw some lessons from the strong, electromagnetic and semileptonic weak interactions of mesons, the most developed area in this field:

- In order to derive reliable predictions from the SM, there is no alternative to a systematic analysis [2, 3].

- There are no simple generally applicable recipes like using "chiral logs" instead of the actual LECs.

- The LECs should and can be understood at various levels of theoretical sophistication (resonance exchange, sum rules, chiral models, lattice gauge theories, ...).

The status of the strongly interacting meson sector can serve as a guide–line for future work in the low–energy effective field theory of the SM. Among the topics to be considered in the near future, I only list some of the most obvious ones:

(i) The complete renormalization remains to be carried out at $\mathcal{O}(p^3)$ [chiral $SU(3)$] and $\mathcal{O}(p^4)$ in the meson–baryon system, to $\mathcal{O}(p^6)$ for mesons (strong interactions) and to $\mathcal{O}(G_F p^3)$ for the non–leptonic weak interactions of baryons.

(ii) A lot of phenomenology remains to be done, especially in the meson–baryon system.

(iii) Once the LECs have been determined phenomenologically in the different sectors, interpreting the LECs in a similar way as for mesons at $\mathcal{O}(p^4)$ will be important for understanding the structure of the SM at low energies.

Acknowledgements

I want to thank Aron Bernstein and Barry Holstein for the invitation to a most successful workshop. I am also grateful to Joachim Kambor for informing me about their work with E. Golowich [20] and to Helmut Neufeld for a critical reading of the manuscript.

References

[1] S. Weinberg, Physica **96A** (1979) 327.

[2] J. Gasser and H. Leutwyler, Ann. Phys. **158** (1984) 142.

[3] J. Gasser and H. Leutwyler, Nucl. Phys. **B250** (1985) 465.

[4] J. Bijnens, Int. J. Mod. Phys. **A8** (1993) 3045.

[5] A.V. Manohar, these Proceedings.

[6] U.-G. Meißner, these Proceedings.

[7] A.V. Manohar and H. Georgi, Nucl. Phys. **B234** (1984) 189;
M. Soldate and R. Sundrum, Nucl. Phys. **B340** (1990) 1;
R.S. Chivukula, M.J. Dugan and M. Golden, Phys. Rev **D47** (1993) 2930.

[8] J. Bijnens, G. Colangelo and J. Gasser, K_{l4} decays beyond one loop, Univ. Bern
preprint BUTP-94/4, hep-ph/9403390, to appear in Nucl. Phys. **B**.

[9] J. Bijnens, G. Ecker and J. Gasser, Chiral perturbation theory, contribution to The
Second DAΦNE Physics Handbook, Eds. L. Maiani, G. Pancheri and N. Paver, in
preparation.

[10] S. Myint and C. Rebbi, Nucl. Phys. **B421** (1994) 241.

[11] J.W. Negele, these Proceedings.

[12] G. Ecker, J. Gasser, A. Pich and E. de Rafael, Nucl. Phys. **B321** (1989) 311.

[13] J.F. Donoghue, C. Ramirez and G. Valencia, Phys. Rev. **D39** (1989) 1947.

[14] G. Ecker, J. Gasser, H. Leutwyler, A. Pich and E. de Rafael, Phys. Lett. **B223**
(1989) 425.

[15] G. Ecker, in Proc. of the 12th Warsaw Symposium on Elementary Particle Physics,
Kazimierz, Poland, June 1988, Eds. Z. Ajduk et al., World Scient. Publ. Co. (Singapore, 1990).

[16] H. Leutwyler, Nucl. Phys. **B337** (1990) 108.

[17] Review of Particle Properties, Phys. Rev. **D50** (1994), part 1.

[18] T. Das, V. Mathur and S. Okubo, Phys. Rev. Lett. **19** (1967) 859.

[19] J.F. Donoghue and B.R. Holstein, Phys. Rev. **D46** (1992) 4076.

[20] E. Golowich and J. Kambor, work in progress.

[21] S. Bellucci, J. Gasser and M.E. Sainio, Nucl. Phys. **B423** (1994) 80.

[22] M. Knecht, B. Moussallam and J. Stern, The reaction $\gamma\gamma \to \pi^0\pi^0$ in generalized
chiral perturbation theory, Orsay preprint IPNO/TH 94-08, hep-ph/9402318.

[23] J. Bijnens, Introduction to extended Nambu–Jona-Lasinio models,
preprint NORDITA-94/27 N,P, hep-ph/9406425;
E. de Rafael, Chiral Lagrangians and kaon CP violation, TASI Lectures 1994,
Boulder, Colorado, in preparation.

[24] G. Ecker, J. Kambor and D. Wyler, Nucl. Phys. **B394** (1993) 101.

[25] J. Kambor, J. Missimer and D. Wyler, Nucl. Phys. **B346** (1990) 17.

[26] G. Ecker, in Proc. of the IX. Int. Conference on the Problems of Quantum Field Theory, Dubna, April 1990, Ed. M.K. Volkov (Dubna, 1990);
G. Esposito-Farèse, Z. Phys. **C50** (1991) 255.

[27] J. Kambor, J. Missimer and D. Wyler, Phys. Lett. **B261** (1991) 496;
J. Kambor et al., Phys. Rev. Lett. **68** (1992) 1818.

[28] G. D'Ambrosio, G. Ecker, G. Isidori and H. Neufeld, Radiative non–leptonic kaon decays, contribution to The Second DAΦNE Physics Handbook, Eds. L. Maiani, G. Pancheri and N. Paver, in preparation;
G. Ecker, E. de Rafael and A. Pich, Chiral perturbation theory and rare kaon decays, in preparation.

[29] G. Ecker, A. Pich and E. de Rafael, Nucl. Phys. **B291** (1987) 692.

[30] C. Bruno and J. Prades, Z. Phys. **C57** (1993) 585.

[31] H.-Y. Cheng, Phys. Rev. **D42** (1990) 72;
J. Bijnens, G. Ecker and A. Pich, Phys. Lett. **B286** (1992) 341.

[32] A. Pich and E. de Rafael, Nucl. Phys. **B358** (1991) 311.

[33] J. Wess and B. Zumino, Phys. Lett. **37B** (1971) 95;
E. Witten, Nucl. Phys. **B223** (1983) 422.

[34] D. Espriu, E. de Rafael and J. Taron, Nucl. Phys. **B345** (1990) 22; Err. ibid. **B355** (1991) 278.

[35] J. Gasser and H. Leutwyler, unpublished work.

[36] R. Dashen, Phys. Rev. **183** (1969) 1245.

[37] D.G. Sutherland, Phys. Lett. **23** (1966) 384.

[38] R. Urech, Virtual photons in chiral perturbation theory, Univ. Bern preprint BUTP-94/9, hep-ph-9405341.

[39] H. Neufeld and H. Rupertsberger, Isospin breaking in chiral perturbation theory and the decays $\eta \to \pi \ell \nu$ and $\tau \to \eta \pi \nu$, Univ. Wien preprint UWThPh-1994-15, to appear in Nucl. Phys. **B**.

[40] J.F. Donoghue, B.R. Holstein and D. Wyler, Phys. Rev. **D47** (1993) 2089.

[41] J. Bijnens, Phys. Lett. **B306** (1993) 343.

[42] E. Jenkins and A.V. Manohar, Phys. Lett. **B255** (1991) 558.

[43] J. Gasser, M.E. Sainio and A. Švarc, Nucl. Phys. **B307** (1988) 779.

[44] G. Ecker, Chiral invariant renormalization of the pion–nucleon interaction, hep-ph/9402337, Phys. Lett. **B** (in print).

Part II: Other Approaches to QCD

Thoughts on Large N_c QCD*

Howard Georgi **

Lyman Laboratory of Physics
Harvard University
Cambridge, MA 02138

In the nonrelativistic quark model (NRQM) there is an SU(6) spin-flavor symmetry for light quark states, with the ground-state baryons in a 56, the $l = 1$ orbitally excited in a 70, etc. Dashen, Jenkins, and Manohar [1, 2] (DJM) argued that if the number of colors, N_c, is large, the NRQM spin-flavor structure of the $\ell = 0$ baryons and in particular the SU(6) symmetry is a rigorous consequence of QCD for baryons with small total spin.

One thing I want to do today is to give a simple explanation of what DJM are doing, to make it easy to understand what the assumptions are. [5, 6] But in addition, my main reason for giving this talk is that I have problems with DJM which I can encapsulate in the following phrases:

1. top versus bottom;
2. sloppy language;
3. 3 versus 2, 1 or 0;

I'll explain what I mean by these later.

Let me begin with what I believe is the simplest explanation of the DJM idea. [5] First, let's review the classic paper by Witten on baryons in the large N_c approximation. He shows that low-lying large N_c baryons can be described by a Hartree wave-function with all (or almost all, for low-lying excited states) quarks in the same ground-state wave function, bound in a potential produced by all the other $\mathcal{O}(N_c)$ quarks. [7] This gives a baryon mass that grows, it is $\mathcal{O}(N_c)$ as $N_c \to \infty$. However, the N_c counting here is not trivial. The leading $\mathcal{O}(N_c)$ effect comes from interactions involving any number of quarks!

- There are 1-quark contributions with planar dressings of the quark lines. Each of these is at most of order 1, but there are N_c of them, so they can give an $\mathcal{O}(N_c)$ effect.
- There are 2-quark contributions with planar gluon exchanges between two quark lines. These are at most of order $1/N_c$, but there are $N_c(N_c - 1)/2$ of them.

* Talk presented to the 1994 Chiral Dynamics Workshop, at MIT. #HUTP- 94/A024.
** Research supported in part by the National Science Foundation under Grant #PHY-9218167.

- There are 3-quark contributions with planar gluon exchanges among three quark lines. These are order $1/N_c^2$, but there are $N_c(N_c - 1)(N_c - 2)/6$ of them.
- etc.

One sentence in [7] hints that the spin structure of large N_c baryonic bound states of light quarks may be an interesting thing to study. Witten notes that while spin-orbit coupling will seriously deform the high-spin baryons away from an s-wave ground state, the low-spin, ground state baryons may not be deformed. It turns out that this notion can be made precise. The point is that the N_c counting above assumes coherent addition over all the quarks. For the spin dependent interaction, the counting is not so simple.

To see how this goes in detail, I believe that one must start with an assumption: **that we can describe the low- spin, large-N_c baryon states in a tensor product space of the spin-flavor indices of the N_c valence quarks, as in the NRQM** — in other words that the low-spin baryons have the spin-flavor and angular momentum structure of representations of nonrelativistic SU(6)×O(3). This does not, I think, actually follow from large N_c alone, however it follows from a very reasonable smoothness hypothesis and I would be astonished if it were wrong.

The assumption is certainly true for $m_q \gg \Lambda_{QCD}$ because NRQM can be derived from QCD. In this limit, the baryon states are in approximately degenerate multiplets for each spatial wavefunction. Different spatial wave-functions correspond to different SU(6)×O(3) representations. The ground-state wavefunction is the completely symmetric spin-flavor combination described by the Young tableaux

$$N_c$$

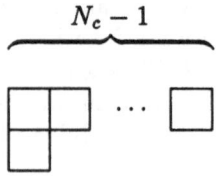

The first excited $\ell = 1$ baryons correspond to the Young tableaux

$$N_c - 1$$

combined with one unit of orbital angular momentum to get spin-flavor symmetry. And so on.

Now what happens as quarks become light? The important point is that for states with small total spin, the splittings between neighboring states are not only suppressed by powers of $1/m_q$, but also by powers of $1/N_c$. This is a rigorous result to all orders in QCD, true for all interactions, because there is no coherent addition of spin-dependent interactions for low spin states.

Thus if is no discontinuity in the physics as $m_q \to 0$, the low spin states at the bottom of each spin-flavor multiplet will be approximately degenerate and well-described in the same tensor product space that works at large m_q. In other words, the NRQM states should be an appropriate description of the states with low spin.

To my mind, the restriction to **low spin** is the hard and subtle part of the whole large N_c argument and the one that is left out is most discussions – – both in quark model language and in the Skyrme language. It is very important to remember that it is **not true at the top of the spin-flavor multiplets, where the baryon spin is of order N_c and the splittings between neighboring spin states are of order Λ_{QCD} for small quark mass.**

Thus in large N_c, there are not simply infinite multiplets of degenerate baryons. There are subtle interchanges of limits involved in any symmetry argument on these objects. Instead of spin-flavor symmetry, what emerges is a partial spin-flavor symmetry — not an approximate symmetry in the usual sense because symmetry breaking effects cannot be ignored on any multiplet.

The way I would explain what is going on is this: because the dimensions of the multiplets go to infinity as the small parameter $(1/N_c)$ goes to zero, we can derive reliable predictions at one end (for small spin) of the multiplets even though the symmetry is badly broken at the other Nevertheless, I believe that this argument justifies the use of the NRQM tensor product states to describe the low-spin baryon states for large-N_c.

However, there are obvious ambiguities in applying large-N_c arguments to $N_c = 3$ This is the first of the problems I mentioned with the DJM work. How do we identify states "near" the top and bottom of the multiplets for $N_c = 3$?

I have no idea how to deal with this problem — I will ignore it and use the expressions we derive for the entire baryon multiplets. But I will not be surprised if our results become less reliable as the baryon spin increases.

Let me now come to the second problem I have with DJM — what I call "sloppy language." I do not pretend to have all the answers to the question of how to use large N_c to make contact with physics, but I do know one thing that you should NOT do. This has to do with another ambiguity in comparing $N_c = \infty$ and $N_c = 3$ — the flavor states are very different.

Steve Weinberg mentioned in his talk that in $N_c = \infty \Rightarrow$ infinite tower of states with

$$(I, J) = \left(\tfrac{1}{2}, \tfrac{1}{2}\right), \left(\tfrac{3}{2}, \tfrac{3}{2}\right), \left(\tfrac{5}{2}, \tfrac{5}{2}\right), \cdots$$

Even Steve fell into the trap of identifying this tower with

$$\sim N, \Delta, \cdots \qquad ??$$

This is nonsense, for many reasons — for example

- it only makes sense only for two flavors
- and for N_c odd
- and it has the wrong structure of valence quarks!

These states, though they have the right spin and isospin, are unlike the nucleon and delta in all other respects. The $\left(\tfrac{1}{2}, \tfrac{1}{2}\right)$ state has a large number of u-d pairs in $(I, J) = (0, 0)$ states. Anyone who talks about these states as if they are analogs of the nucleon is guilty of talking about large N_c in a sloppy way. That doesn't necessarily mean that they are also guilty of sloppy thinking, but it should at least cause you to worry!

What to do instead? I think that the best plan is to try to extract general principles from large N_c — independent of # of flavors, etc. In particular, the only scheme I can find that makes sense is to express the large N_c predictions as a sum over valence quark states — and count powers of N_c.

In particular, what about the beautiful results of Dashen, Jenkins and Manohar? One problem here is that I have not checked everything they have done. There are very complicated expressions here, and unlike Steve Weinberg, I don't find algebra with Racah coefficients very beautiful. It looks to me as if the way they do their calculations could get them into trouble, but those that I can check easily give results that are equivalent to what I would get.

Let me describe the process of counting powers of N_c for matrix elements of operators between baryon states. The operators could be interpolating fields for mesons, for this is quite general. I will ignore flavor symmetry breaking for simplicity. The procedure is simple:

1. In the spin-flavor space of the NRQM for the baryon states of interest, write down the most general flavor- conserving expression for the matrix element. This essentially just amounts to figuring out which quarks contribute.
2. Assign each term in the expression a power of N_c given by the largest possible power that can appear *on the low spin states*. This is most conveniently determined by looking at Feynman diagrams contributing to the matrix element, making appropriate assumptions about the N_c dependence of individual quark matrix elements.

Among the Feynman graphs that contribute to the matrix element is a sum over all quarks of single-quark matrix elements. This has the spin-flavor structure of the NRQM. In all examples we know of, this gives a contribution to the leading N_c dependence. The reason that the suggestion above is nontrivial is that while multiquark diagrams are suppressed by powers of $1/N_c$, their effects can be enhanced by coherent contributions from the sum over the N_c quarks. This can give additional contributions of the same order in N_c as the NRQM but with a different spin- flavor structure.

The possible different spin-flavor structures on quark lines are:

1. Constant terms — these always sum coherently over the quarks, but the result has no spin- flavor structure and therefore is not interesting.
2. Spin terms —

$$\propto \sum_{\text{quarks } x} \sigma_x^j. \tag{1}$$

This never adds coherently on low spin states, so these contributions are down by $1/N_c$.
3. Flavor terms —

$$\propto \sum_{\text{quarks } x} \lambda_x^a. \tag{2}$$

This sometimes adds coherently, for — example

$$\sum_{\text{quarks } x} \lambda_x^8 \qquad (3)$$

acting on a low-spin state of u and d quarks is $N_c/\sqrt{12}$.

4. Spin-Flavor terms — these are proportional to

$$\propto \sum_{\text{quarks } x} \sigma_x^j \lambda_x^a. \qquad (4)$$

This can also add coherently; in fact, I will show in an Appendix that the SU(6) quadratic Casimir operator,

$$C_2 \equiv \left[\frac{1}{2f} \sum_j \left(\sum_{\text{quarks } x} \sigma_x^j \right)^2 \right.$$
$$+ \frac{1}{4} \sum_a \left(\sum_{\text{quarks } x} \lambda_x^a \right)^2$$
$$\left. + \frac{1}{4} \sum_{j,a} \left(\sum_{\text{quarks } x} \sigma_x^j \lambda_x^a \right)^2 \right] \qquad (5)$$
$$= \frac{2f-1}{2f} N_c^2 (1 + \mathcal{O}(1/N_c))$$

on any finitely excited large-N_c baryon state. Thus, generically, some spin-flavor matrix elements grow like N_c.

Here is a familiar and important example of one-quark contribution — the couplings of the vector mesons, ρ and ω, to the nucleon states. Both couplings grow with N_c, but they are dominated by different contributions. The leading contribution to the ω matrix element is the flavor coupling (in relativistic notation),

$$\omega_\mu \overline{N} \left(\sum_x \lambda_x^8 \right) \gamma^\mu N, \qquad (6)$$

while the leading contribution to the ρ coupling is the spin-flavor coupling,

$$(\partial_\mu \rho_\nu^a - \partial_\nu \rho_\mu^a) \overline{N} \left(\sum_x \sigma_x^{\mu\nu} \lambda_x^a \right) N. \qquad (7)$$

The spin-flavor coupling dominates for the ρ coupling because the isospin matrix element is small for low-spin states, and thus the flavor coupling does not grow with N_c. This is an example of what, in the Skyrme literature, is called the $I_t = J_t$ rule [4, 10]. Both of these leading terms are dominated by one-quark operators. Examples in which multiquark operators contribute at leading order in N_c will appear soon.

Another example of large N_c counting is the linearity of various quantities in the number of s quarks found by DJM. The assumption that leads to this result is that the number of s quarks is finite as $N_c \to \infty$ while the number or u and d quarks is of $\mathcal{O}(N_c)$. In this limit, no more than one s quark line can contribute **to anything** in leading nontrivial order, simply because no counting factor to compensate for extra s quark line. As Manohar showed you in his talk, this seems to work. **But this is crazy! Surely this is not the reason for the equal spacing in decuplet, for example!** I say that because in this application, the large N_c limit is as far away as it could possibly be — Manohar has assumed that the Ω^- contains a very large number of u and d quarks! Maybe 3 is large, but surely 0 is not!!!! This is the last of my general problems with DJM.

In the last few minutes of my talk, I want to talk briefly about the decays of $\ell = 1$ baryons, work I have done with my former students, Chris Carone and Sam Osofsky (who participated in the beginning of the work), and with current students Lev Kaplan and David Morin.

We were interested in studying the one-pion decays of 70- plet baryons to baryons in the 56. The reason is, as I will show you, that this can in principle distinguish between the kind of large N_c argument given above and other possibilities. While there are also $70 \to 70$ decays, we did not consider them. The decays to states in the 56 are generally favored by the kinematics, and indeed few $70 \to 70$ modes have been observed in experiment. In the Hartree language, the part of the interaction Hamiltonian that is of interest can be written

$$H_{\text{int}} = \sum_{n=1}^{N_c} \sum_{\substack{\{x_1, \ldots, x_n\} \\ \subset \{1, \ldots, N_c\}}}$$

$$\left.\int d^3 r_{x_1} \ldots d^3 r_{x_n} \Phi(r_{x_1})^\dagger_{x_1} \otimes \ldots \otimes \Phi(r_{x_n})^\dagger_{x_n} \right\} 56$$

$$\left. \times \mathcal{O}(r_{x_1}, \ldots, r_{x_n}) \times \right\} \pi$$

$$\left. \Psi(r_{x_1})_{x_1} \otimes \ldots \otimes \Psi(r_{x_{n-1}})_{x_{n-1}} \otimes \underbrace{\Psi_*(r_{x_n})_{x_n}}_{\substack{\text{excited} \\ \text{quark}}} \right\} 70$$

where $\mathcal{O}$ is the pion coupling to the axial-vector quark current.

This requires some explanation. The Φs are quark wavefunctions for 56 — the self-consistent solutions to the Hartree equation. The Ψs are quark wavefunctions for 70. The sum over n indicates that we have broken up the interaction into parts involving different numbers of quark lines; the second sum accounts for the possible quark interactions with fixed n that connect the initial to the final baryon state. This separation allows us to classify interactions by their order in the $1/N_c$ expansion. By Witten's counting arguments, a general n-body interaction is of order $1/N_c^{n-1}$. For example, the term in $\mathcal{O}(1/N_c^2)$ interaction involving three quark lines might come from an interaction like this:

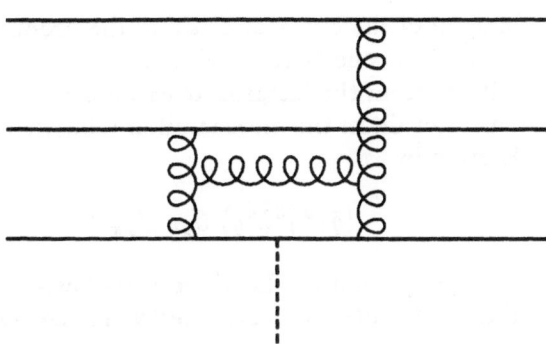

Finally, note that in the 70 one of the quarks is orbitally excited — this is indicates by the subscript ∗ Each term involves the wavefunction Ψ_*, regardless of the number of quark lines involved. This follows because we are only interested in interactions that contribute to the 70→56 decays, which necessarily involve the "de-excitation" of the orbitally excited quark.

It is much too difficult for us to compute the Hartree potential in a baryon composed of light quarks, we still can learn a great deal by studying the symmetry structure of (). As we argued earlier, it is plausible to represent the small-spin baryon states made from light quarks in the same space, and by the same representations, as the baryon states of the naive quark model. Thus, we work in a $(2f)^{N_c}$-dimensional tensor product space, where f is the number of quark flavors. The quark wavefunctions Φ and Ψ can be thought of as — $2f \times 2f$ matrices acting on the spin-flavor space of a single quark;

H as a whole can be thought of as a — $(2f)^{N_c} \times (2f)^{N_c}$ matrix acting on the $(2f)^{N_c}$-dimensional spin-flavor space in which we represent the baryon states. The Φ and Ψ are the solutions to the zeroth order Hartree equation, and therefore are spherically symmetric and spin-flavor independent.

Thus, we can replace these matrix wavefunctions by c-numbers

$$\Phi(\vec{r}) \rightarrow \phi(r) \quad , \quad \Psi(\vec{r}) \rightarrow \psi(r) = \phi(r) \tag{8}$$

where $r = |\vec{r}|$. Note that the Hartree potential is a collective phenomenon and to leading order is unaffected by the excitation of a single quark. This accounts for the equality shown in (8). In the 70 state, the excited quark has one unit of orbital angular momentum, so we know the form of its spatial wavefunction:

$$\Psi_*(\vec{r}) = f(r)\, Y_{l=1,m}(\theta, \varphi) = f(r)\, (\vec{r} \cdot \vec{\varepsilon}_m) \tag{9}$$

where $f(r)$ is a spin-flavor independent c-number. In (9) we have chosen to express the $l = 1$ spherical harmonics in terms of the vectors $\vec{\varepsilon}_m$, which are given by

$$\varepsilon_1 = \begin{pmatrix} \frac{-1}{\sqrt{2}} \\ \frac{-i}{\sqrt{2}} \\ 0 \end{pmatrix} \quad \varepsilon_0 = \begin{pmatrix} 0 \\ 0 \\ 1 \end{pmatrix} \quad \varepsilon_{-1} = \begin{pmatrix} \frac{1}{\sqrt{2}} \\ \frac{-i}{\sqrt{2}} \\ 0 \end{pmatrix} \tag{10}$$

Thus, () is the integral of the operator $\mathcal{O}$ times the product of $2N_c$ spherically symmetric functions, times one factor of $\vec{r} \cdot \vec{\varepsilon}_m$.

We can formally perform the integrals once we have specified the symmetry structure of the operator $\mathcal{O}$. In the more familiar relativistic notation, the pion-quark coupling is given by

$$(\bar{q}\gamma^\mu\gamma^5\lambda^a q)\,\partial_\mu\pi^a/f_\pi \tag{11}$$

where the λ^a are SU(3) generators. In the Hartree basis, the piece of the pion-quark coupling that contributes to baryon decays in the s-wave has the form

$$\mathcal{O} \sim \lambda^a(\vec{\sigma}\cdot\vec{r})\,\partial^0\pi^a/f_\pi \tag{12}$$

which, after integration, gives us a one-body interaction that is leading in $1/N_c$

$$a\,\lambda_*^a(\vec{\sigma}_*\cdot\vec{\varepsilon}_m)\,\partial^0\pi^a/f_\pi \tag{13}$$

where a is an unknown coefficient. The $*$ under the spin and flavor matrices indicates that each acts only in the subspace of the orbitally excited quark. Recall that a purely one-body interaction must act on the excited quark line, or there would be no way to change its orbital angular momentum. The spin-flavor structure of the operator in (13) is consistent with the predictions of the NRQM.

We can also write down a number of operators that are subleading in $1/N_c$ that involve two quark lines. However, we will only include two of these in our subsequent numerical analysis:

$$i\,b\,(\vec{\sigma}_* \times \vec{\varepsilon}_m) \cdot \left(\sum_{x\neq *}\lambda_x^a\vec{\sigma}_x\right)\,\partial^0\pi^a/f_\pi \tag{14}$$

$$c\left(\sum_{x\neq *}\lambda_x^a\right)(\vec{\sigma}_*\cdot\vec{\varepsilon}_m)\,\partial^0\pi^a/f_\pi \tag{15}$$

Our motivation for retaining these operators is that the sum over $\lambda^a\sigma$ in the case of (14) and the sum over λ^a in the case of (15) can both be coherent on low-spin states, and thus the matrix elements can be of order 1, rather than order $1/N_c$. This follows from the argument in the Appendix. Thus we will take our leading s- wave operators to be those given in (13), (14), and (15), which we will call operators A, B, and C respectively.

Analogous arguments to those that we have used to arrive at the operators responsible for the s-wave decays can also be used to determine the give operators responsible for decays through the d-wave. (Note that the decay channels in which the pion has odd orbital angular momentum are forbidden by parity.) The leading one-body operator is given by

$$d\,\lambda_*^a\left(\sigma_*^i\varepsilon_m^j + \sigma_*^j\varepsilon_m^i - \tfrac{2}{3}\delta^{ij}\vec{\sigma}_*\cdot\vec{\varepsilon}_m\right)\,\partial^i\partial^j\pi^a/f_\pi^2 \tag{16}$$

We also have two-body operators in the d-wave channel with the same kind of sum that we encountered in (14)

$$
\begin{aligned}
i\,e \sum_{x \neq *} & [(\vec{\sigma}_x \times \vec{\sigma}_*)^i \varepsilon_m^j \lambda_x^a + (\vec{\sigma}_x \times \vec{\sigma}_*)^j \varepsilon_m^i \lambda_x^a \\
& - \tfrac{2}{3}(\vec{\sigma}_x \times \vec{\sigma}_*) \cdot \vec{\varepsilon}_m \lambda_x^a \delta^{ij}]\ \partial^i \partial^j \pi^a / f_\pi^2
\end{aligned}
\tag{17}
$$

$$
\begin{aligned}
i\,f \sum_{x \neq *} & [(\vec{\sigma}_* \times \vec{\varepsilon}_m)^i \sigma_x^j \lambda_x^a + (\vec{\sigma}_* \times \vec{\varepsilon}_m)^j \sigma_x^i \lambda_x^a \\
& - \tfrac{2}{3}(\vec{\sigma}_* \times \vec{\varepsilon}_m) \cdot \sigma_x \lambda_x^a \delta^{ij}]\ \partial^i \partial^j \pi^a / f_\pi^2
\end{aligned}
\tag{18}
$$

There is also a third two-body operator involving the cross- product $(\vec{\sigma}_x \times \vec{\varepsilon}_m)$ which is not linearly independent of the two operators that we show above. Finally, there is also a d-wave operator analogous to (15)

$$
g \sum_{x \neq *} \lambda_x^a \left(\sigma_*^i \varepsilon_m^j + \sigma_*^j \varepsilon_m^i - \tfrac{2}{3}\delta^{ij}\, \vec{\sigma}_* \cdot \vec{\varepsilon}_m \right)\ \partial^i \partial^j \pi^a / f_\pi^2
\tag{19}
$$

Thus, we will retain (16), (17), (18), and (19) as our set of leading operators in considering the d-wave decays, and refer to them as operators D, E, F, and G.

We must now decide precisely which physical quantities we will fit, and select the corresponding experimental data. In addition, we must arrive at estimates of both the experimental and theoretical uncertainties. The experimental results we will use are the masses, total decay widths, and branching fractions given in the 1992 Review of Particle Properties (RPP) [11]. We use the experimentally measured masses, rather than large-N_c predictions, in computing partial decay widths. The masses are affected by large logarithmic corrections proportional to m_π^3 / f_π^2 which we would have to include if we were to do the calculation properly. For baryons in the 56, these one-loop corrections are relatively straightforward to compute, because we know the mass eigenstates. For baryons in the 70, however, we can determine the mass eigenstates only after including the one- loop corrections. This makes the problem of computing the masses nonlinear and thus, far more difficult. For this reason, the problem of predicting 70-plet masses in the Hartree picture is best treated separately.

Theoretical errors also have to be considered. Sources of these errors include subleading operators in the $1/N_c$ expansion, which we have ignored, as well as flavor SU(3) breaking operators. (The only explicit SU(3) breaking effect that we include is the difference between f_π and f_K.) As a rough estimate, we have assumed a 20% theoretical uncertainty for each partial width prediction, and have combined this uncertainty in quadrature with the experimental uncertainty. The primary effect of this addition is that the fit is not completely dominated by a few decays which have been measured extremely well experimentally, in particular, the $\Lambda(1520)$ d-wave decays. For the vast majority of decays, the theoretical error is not very important, but for consistency we have used the same value throughout. The choice of a precise value for the theoretical error does not substantially affect the final results.

I'm not going to discuss the details of the fit — see the paper — I'll just flash the results (I couldn't think of a good graphical way of displaying it).

Decay	f_s	f_d	f_{exp}
$N(1520) \rightarrow N\pi$	-	65.5	55.0 ± 12.1
$\rightarrow N\eta$	-	0.07	0.1 ± 0.1
$N(1535) \rightarrow N\pi$	52.6	-	45.0 ± 13.4
$\rightarrow N\eta$	30.0	-	40.0 ± 12.8
$\rightarrow \Delta\pi$	-	0.4	5.0 ± 5.0
$N(1650) \rightarrow N\pi$	78.4	-	70.0 ± 17.2
$\rightarrow N\eta$	0.9	-	1.0 ± 1.0
$\rightarrow \Lambda K$	3.2	-	7.0 ± 7.0
$\rightarrow \Delta\pi$	-	9.0	5.0 ± 5.0
$N(1675) \rightarrow N\pi$	-	38.3	45.0 ± 10.3
$\rightarrow N\eta$	-	2.1	1.0 ± 1.0
$\rightarrow \Lambda K$	-	0.005	0.1 ± 0.1
$\rightarrow \Delta\pi$	-	53.7	55.0 ± 12.1
$N(1700) \rightarrow N\pi$	-	13.2	10.0 ± 5.4
$\rightarrow \Lambda K$	-	0.09	0.2 ± 0.1
$\Delta(1620) \rightarrow N\pi$	18.7	-	25.0 ± 7.1
$\rightarrow \Delta\pi$	-	41.8	50.0 ± 14.1
$\Delta(1700) \rightarrow N\pi$	-	12.0	15.0 ± 5.8
$\Lambda(1520) \rightarrow N\overline{K}$	-	17.9	45.0 ± 9.1
$\rightarrow \Sigma\pi$	-	41.5	42.0 ± 8.5

Decay	f_s	f_d	f_{exp}
$\Lambda(1670) \rightarrow N\overline{K}$	20.2	-	20.0 ± 6.4
$\rightarrow \Sigma\pi$	40.2	-	40.0 ± 21.5
$\rightarrow \Lambda\eta$	25.1	-	25.0 ± 11.2
$\Lambda(1690) \rightarrow N\overline{K}$	-	21.7	25.0 ± 7.1
$\rightarrow \Sigma\pi$	-	30.3	30.0 ± 11.7
$\Lambda(1800) \rightarrow N\overline{K}$	32.6	-	32.5 ± 9.9
$\Lambda(1830) \rightarrow N\overline{K}$	-	1.3	6.5 ± 3.7
$\rightarrow \Sigma\pi$	-	83.2	55.0 ± 22.8
$\Sigma(1670) \rightarrow N\overline{K}$	-	4.0	10.0 ± 3.6
$\rightarrow \Lambda\pi$	-	11.6	10.0 ± 6.4
$\rightarrow \Sigma\pi$	-	44.4	45.0 ± 17.5
$\Sigma(1750) \rightarrow N\overline{K}$	28.1	-	25.0 ± 15.8
$\rightarrow \Sigma\pi$	4.2	-	4.0 ± 4.0
$\rightarrow \Sigma\eta$	6.5	-	35.0 ± 21.2
$\Sigma(1775) \rightarrow N\overline{K}$	-	17.3	40.0 ± 8.5
$\rightarrow \Lambda\pi$	-	25.6	17.0 ± 4.5
$\rightarrow \Sigma\pi$	-	3.4	3.5 ± 1.7
$\rightarrow \Sigma^*\pi$	-	6.7	10.0 ± 2.8

The quantities f_s and f_d are s-wave and d-wave partial widths.

The predicted branching fractions correspond to the parameter set $a = 0.536$, $b = -0.028$, $c = 0.101$, $d = 0.203$, $e = -0.015$, $f = -0.029$, $g = -0.002$, and the mixing angles $\theta_{N1} = 0.61$, $\theta_{N3} = 3.04$, $\theta_{A11} = 1.78$, $\theta_{A12} = 2.79$, $\theta_{A13} = 1.53$, $\theta_{A31} = 0.32$, $\theta_{A32} = 0.14$, $\theta_{A33} = 2.63$, $\theta_{\Sigma11} = 2.00$, $\theta_{\Sigma12} = 1.16$, $\theta_{\Sigma31} = 2.14$, $\theta_{\Sigma32} = 0.48$

With a few exceptions (notable ones being the $\Lambda(1520) \rightarrow N\overline{K}$ and $\Sigma(1775) \rightarrow N\overline{K}$ decays), the predictions are within the range of uncertainty given by the combined experimental and theoretical errors. The most interesting feature of the fit is the smallness of parameters b and c relative to a and of parameters e, f, and g relative to d.

Firstly, this fit is not any better than others using different methods, although our methods follow more directly from the underlying physics in a well-defined limit of QCD. But why are the two-body operators, B, C, E, F, and G so suppressed???? We expected a relative suppression in their coefficients, compensated by an enhancement in the matrix elements — but we found suppression much greater than a factor of $N_c = 3$. The parameters b, c, e, f, and g in fit are so small that the matrix elements of the two-body operators are suppressed even when the sums over quark lines are coherent.

One possible conclusion is that the large N_c argument works but enhancement the of coherent terms in excited baryons not big enough. After all, here it is really $N_c - 1$ that is assumed large, because the orbitally excited quark is distinguished.

More likely, in my view, is that there is something more to the success of the NRQM for baryons than large-N_c. Perhaps somehow, in spite of the fact that the quarks are not really heavy, they act in the process of $\ell = 1$ baryon decay as if they were.

Appendix on the Casimir operator

In this section we derive (5). The Casimir operator can be written

$$C_2 \equiv \sum_\alpha T_\alpha^2 \tag{20}$$

where the T_α are the SU(2f) generators, normalized so that

$$\operatorname{tr} T_\alpha T_\beta = \delta_{\alpha\beta} \tag{21}$$

in the defining, $2f$ dimensional representation. Rather than computing the Casimir operator directly in other representations, R, it is easier to compute the quantity $T(R)$, defined by

$$\operatorname{tr}_R T_\alpha T_\beta = T(R)\,\delta_{\alpha\beta}. \tag{22}$$

Then C_2 can be obtained as follows:

$$C_2 = (4f^2 - 1)\frac{T(R)}{D(R)}, \tag{23}$$

where $D(R)$ is the dimension of the representation, R. Thus, for example — in the defining representation, the Casimir operator is

$$C_2 = \frac{4f^2 - 1}{2f}. \tag{24}$$

The crucial step in obtaining (5) is to calculate $T(R)$ for the completely symmetric representation with N_c boxes. Let us call this representation $\{N_c\}$. We will calculate the trace of the square of a generator that is the analogue of λ^8 for SU(2f),

$$T_{2f-1} \equiv \frac{1}{\sqrt{2f(2f-1)}} \begin{pmatrix} 1 & 0 & \cdots & & 0 \\ 0 & 1 & \cdots & & 0 \\ \vdots & \vdots & \ddots & & \vdots \\ 0 & 0 & \cdots & & 1 - 2f \end{pmatrix} \tag{25}$$

Then we can compute the trace by noting that in $\{N_c\}$, there are

$$\begin{pmatrix} N_c + 2f - k - 2 \\ N_c - k \end{pmatrix} \tag{26}$$

states with k indices having value $2f$, on each of which the value of T_{2f-1}^2 is

$$\frac{1}{2f(2f-1)}[k(1 - 2f) + (N_c - k)]^2,$$

thus

$$T(\{N_c\}) = \text{tr}_{\{N_c\}} T_{2f-1}^2$$

$$= \frac{1}{2f(2f-1)} \sum_{k=0}^{N_c} [k(2f-1) - (N_c - k)]^2$$

$$\begin{pmatrix} N_c + 2f - k - 2 \\ N_c - k \end{pmatrix} = \begin{pmatrix} N_c + 2f \\ N_c - 1 \end{pmatrix}$$

This gives

$$C_2(\{N_c\}) = \frac{2f - 1}{2f} N_c(N_c + 2f) \tag{27}$$

in agreement with (5).

The reason that (5) is correct for any finitely excited baryon state is that the order N_c^2 term comes from the horizontal string of boxes in the Young tableaux with length of order N_c, a feature shared by all the finitely excited large-N_c baryon states. More precisely, note that (27) implies

$$C_2(\{N_c - \ell\}) = \frac{2f - 1}{2f} N_c^2 + \mathcal{O}(N_c) \tag{28}$$

for any fixed ℓ as $N_c \to \infty$. Note further that we can determine $T(R)$ for any finitely excited baryon state by starting with the representations, $\{N_c - \ell\}$, and using the recursion relations

$$T(r \otimes R) = D(r)T(R) + D(R)T(r)$$

$$T(r \oplus R) = T(r) + T(R)$$

The point is that the Clebsch-Gordon decomposition does not change C_2 to leading order in N_c because $T(\{N_c - \ell\})$ is higher order in N_c than $D(\{N_c - \ell\})$. Thus

$$\frac{T(r \otimes \{N_c - \ell\})}{D(r \otimes \{N_c - \ell\})}$$

$$= \frac{D(r)T(\{N_c - \ell\}) + D(\{N_c - \ell\})T(r)}{D(r)D(\{N_c - \ell\})}$$

$$= \frac{T(\{N_c - \ell\})}{D(\{N_c - \ell\})} + \mathcal{O}(1) \tag{29}$$

for any fixed r. Then the standard rules of Clebsch-Gordon decomposition can be used to establish (5) for any representation obtained from $\{N_c - \ell\}$ by adding a finite number of boxes.

Acknowledgments

This research was supported in part by the National Science Foundation, under grant PHY-9218167.

References

[1] R. Dashen and A. Manohar, Phys. Lett. **B315** 425 (1993); **B315** 438 (1993); E. Jenkins, Phys. Lett. **B315** 431 (1993); **B315** 441 (1993); **B315** 447 (1993); R. Dashen, E. Jenkins, A. Manohar, UCSD-PTH-93-21, Oct. 1993 (hep-ph 9310379).

[2] Results that are closely related to those of Dashen, Manohar and Jenkins have been derived previously by various groups [3, 4]. We focus on the Dashen *et. al* work because the results are expressed in a language that can be simply related to the NRQM.

[3] J.-L. Gervais and B. Sakita, Phys. Rev. Lett. **52** 527 (1984); Phys. Rev. **30** 1795 (1984).

[4] M.P. Mattis and M. Mukerjee, Phys. Rev. Lett. **61** 1344 (1988); M.P. Mattis and E. Braaten, Phys. Rev. **D39** 994, 2737 (1989).

[5] C. Carone, H. Georgi and S. Osofsky, Phys. Lett. **B322** 227 (1994).

[6] M. Luty and J. March-Russell, LBL-34778, Oct. 1993 (hep-ph 9310369), to be published in Nucl. Phys. **B**.

[7] E. Witten, Nucl. Phys. **B160** 57 (1979). See also G. Adkins, C. Nappi and E. Witten, Nucl. Phys. **B228** 552 (1983); A. Manohar, Nucl. Phys. **B248** 19 (1984).

[8] D. Faiman and D. Plane, Nucl. Phys. **B50** 379 (1972); A. Hey, P. Litchfield, and R. Cashmore, Nucl. Phys. **B95** 516 (1975).

[9] F. J. Gilman, M. Kugler, and S. Meshkov, Phys. Lett. **45B** 481 (1973); Phys. Rev. **D9** 715 (1974).

[10] J.T. Donohue, Phys. Rev. Lett. **58** 3 (1987); Phys. Rev. **D37** 631 (1988).

[11] Review of Particle Properties, M. Aguilar- Benitez *et al.*, Phys. Rev. **D45** Part II, 1 (1992).

[12] C. Carone, H. Georgi, L. Kaplan and D. Morin, HUTP-94/A008, to be published.

This article was processed using the LaTeX macro package with LMAMULT style

Pion Observables and QCD

Craig D. Roberts

Physics Division, Bldg. 203,
Argonne National Laboratory, Argonne, IL 60439-4843, USA

1 Introduction

The Dyson-Schwinger equations (DSEs) are a tower of coupled integral equations that relate the Green functions of QCD to one another. Solving these equations provides the solution of QCD. This tower of equations includes the equation for the quark self-energy, which is the analogue of the *gap equation* in superconductivity, and the Bethe-Salpeter equation, the solution of which is the quark-antiquark bound state amplitude in QCD. The application of this approach to solving Abelian and non-Abelian gauge theories is reviewed in Ref. [1].

The nonperturbative DSE approach is being developed as both: 1) a computationally less intensive alternative and; 2) a complement to numerical simulations of the lattice action of QCD. In recent years, significant progress has been made with the DSE approach so that it is now possible to make sensible and direct comparisons between quantities calculated using this approach and the results of numerical simulations of Abelian gauge theories.[2]

Herein the application of the DSE approach to the calculation of pion observables is described[3] using: the π-π scattering lengths (a_0^0, a_0^2, a_1^1, a_2^0, a_2^2) and associated partial wave amplitudes; the $\pi^0 \to \gamma\gamma$ decay width; and the charged pion form factor, $F_\pi(q^2)$, as illustrative examples. Since this approach provides a straightforward, microscopic description of dynamical chiral symmetry breaking (DχSB) and confinement, the calculation of pion observables is a simple and elegant illustrative example of its power and efficacy. The relevant DSEs are discussed in Sec. 2, the calculation of pion observables in Sec. 3 and concluding remarks are presented in Sec. 4.

2 Dyson-Schwinger Equations in QCD

2.1 Quark Propagator

In Euclidean space, with metric $\delta_{\mu\nu} = \text{diag}(1,1,1,1)$ and $\gamma_\mu = \gamma_\mu^\dagger$, and in a general covariant gauge, the inverse of the dressed quark propagator can be written as

$$S^{-1}(p) = i\gamma \cdot p + m + \Sigma(p) \equiv Z^{-1}(p^2)\left(i\gamma \cdot p + M(p^2)\right) \equiv i\gamma \cdot p\, A(p^2) + m + B(p^2), \tag{2.1}$$

with: m the renormalised, explicit chiral symmetry breaking mass (if present); $\Sigma(p)$ the self-energy; $M(p^2) = [m + B(p^2)]/A(p^2)$ the dynamical quark mass function; and

$Z(p^2) = A^{-1}(p^2)$ the momentum-dependent renormalisation of the quark wavefunction. The DSE for the inverse propagator is

$$S^{-1}(p) = i\gamma \cdot p + m + \frac{4}{3}g^2 \int \frac{d^4k}{(2\pi)^4} \gamma_\mu S(k) \Gamma_\nu(k, p) D_{\mu\nu}((p-k)^2), \qquad (2.2)$$

where $D_{\mu\nu}(q^2)$ is the dressed gluon propagator and Γ_ν is the proper quark-gluon vertex.

When the current-quark mass is zero, the solution of this equation determines whether or not chiral symmetry is dynamically broken in QCD. The quark condensate, $\langle \bar{q}q \rangle \propto \text{tr}[S(x = 0)]$, is a chiral symmetry order parameter. If, with $m = 0$ in Eq. (2.2), there is a solution with $B \neq 0$ then the quark has generated a mass via interaction with its own gluon field and the chiral symmetry is therefore dynamically broken.

The solution also provides information about quark confinement. The presence or absence of quark production thresholds in the S-matrix amplitudes that contribute to physical observables is determined by the analytic structure of the quark propagator, which one obtains by solving this equation.[4]

2.2 Gluon Propagator

In a general covariant gauge the dressed gluon propagator can be written:

$$D^{\mu\nu}(q^2) = \left[\left(\delta^{\mu\nu} - \frac{q^\mu q^\nu}{q^2} \right) \frac{1}{1 - \Pi(q^2)} + \xi \frac{q^\mu q^\nu}{q^2} \right] \frac{1}{q^2}, \qquad (2.3)$$

where $\Pi(q^2)$ is the gluon vacuum polarisation and ξ is the gauge parameter.

The Dyson-Schwinger equation for the gluon propagator is given diagrammatically in Fig. 1. The symmetrisation factors of 1/2 and 1/6 arise from the usual Feynman

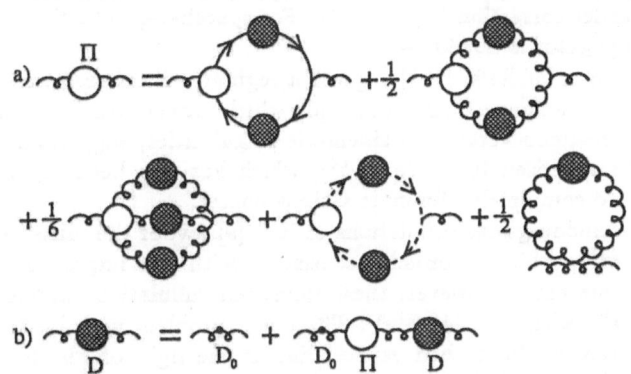

Fig. 1. The Dyson-Schwinger equation for the gluon propagator. [Here and below the broken line represents the propagator for the ghost field.]

rules, which also require a negative sign [unshown] to be included for every fermion and ghost loop. This equation has been studied extensively.[5-8] There have also been attempts to determine the gluon propagator from numerical simulations of lattice-QCD.[9,10]

The results of the DSE and lattice studies are summarised in Sec. 5.1 of Ref. [1] and are represented in Fig. 2. This figure illustrates that for spacelike-$q^2 > 1$ GeV^2

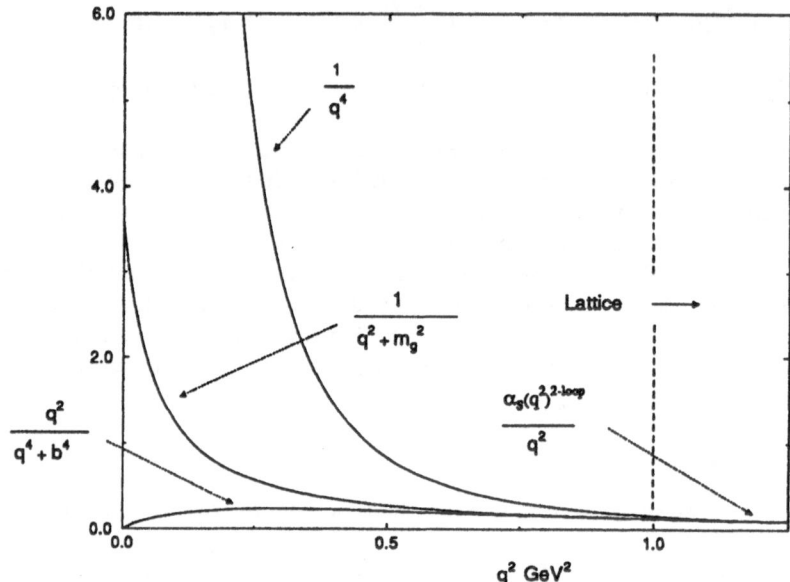

Fig. 2. Results of studies of the gluon propagator in QCD. Typical values of $m_g = 0.5$ GeV and $b = 0.7$ GeV have been used.

the gluon propagator is given by the two-loop, QCD renormalisation group result, with the next order correction being <10%. For spacelike-$q^2 < 1$ GeV^2, however, the form of the propagator is not known.

The DSE studies of Refs. [5–7] suggest a regularised infrared singularity, represented by $1/q^4$ in the figure. That of Ref. [8], which differs mainly in that the Ansatz used for the triple-gluon vertex has kinematic singularities, suggests an infrared vanishing form, characterised by $q^2/(q^4 + b^4)$, which has also been argued[11] to be the form necessary to completely eliminate Gribov copies.

The lattice Landau-gauge simulations of Ref. [9] favour the massive vector boson form, $1/(q^2 + m_g^2)$, which is broadly consistent with the improved simulations of Ref. [10]. In some cases, however, these improved simulations allowed a fit of the form $q^2/(q^4 + b^4)$, with $b \sim 340$ MeV. There is a problem with lattice simulations, however, indicated by the dashed vertical line at the right of Fig. 2. With present technology, the domain of $q^2 < 1$ GeV^2 is actually inaccessible in lattice studies and, since all forms of the propagator are the same outside this domain, it is clear that these results are both qualitatively and quantitatively unreliable.

2.3   Bound States: Bethe-Salpeter Equation

In quantum field theory, the Bethe-Salpeter amplitude for a two-body, quark-anti-quark bound state is obtained as the solution of the homogeneous Bethe-Salpeter equation:

$$\Gamma_M^{rs}(p; P) = -\int \frac{d^4 k}{(2\pi)^4}\, K_{ab;cd}^{rs;tu}(k, p; P)\, \left(S(k - \tfrac{1}{2}P)\Gamma_M(k; P)S(k + \tfrac{1}{2}P)\right)_{cd}^{tu}, \quad (2.4)$$

where Γ_M is the proper meson-quark vertex and K is the kernel. A commonly used approximation for the kernel, K, is *generalised ladder approximation*, in which

$$K = g^2\, D_{\mu\nu}(p - q) \left(\frac{\lambda^a}{2}\right)_{ac} \left(\frac{\lambda^a}{2}\right)_{db} (\gamma_\mu)_{rt}\, (\gamma_\mu)_{us} \qquad (2.5)$$

and $S(k)$ is obtained from the *rainbow approximation* to the quark DSE; i.e., Eq. (2.2) with $\Gamma_\mu = \gamma_\mu$. Solving for Γ_M in this approximation yields what is meant by the *dressed-quark core* of the bound state.

In the DSE approach the dichotomy of the pion as both a Goldstone boson and a $q\text{-}\bar{q}$ bound state is beautifully and easily understood. One has DχSB when, with $m = 0$ in Eq. (2.2), one obtains $B_{m=0}(p^2) \neq 0$ as a solution. In this case ($m = 0$) the Bethe-Salpeter equation in the pseudoscalar channel reduces to the quark DSE as $P^2 \to 0$, where P_μ is the centre-of-mass momentum of the bound state.[13] It follows *without fine-tuning*, therefore, that if the quark DSE admits $B_{m=0}(p^2) \neq 0$ as a solution when $m = 0$; i.e., one has DχSB, then there exists a massless ($P^2 = 0$), pseudoscalar, $q\text{-}\bar{q}$ bound state with the proper meson-quark vertex dominated by its γ_5 component; i.e., $\Gamma_\pi(p^2; P^2 = 0) = B_{m=0}(p^2)/f_\pi$, where f_π is both the Bethe-Salpeter normalisation and pion decay constant, which is a *calculated* quantity in this approach. For $m \neq 0$ but $m/\Lambda_{\rm QCD} \ll 1$ one has

$$\Gamma_\pi(p^2; P^2 = -m_\pi^2) \;\approx\; \frac{1}{f_\pi} B_{m=0}(p^2) \qquad (2.6)$$

$$m_\pi^2 f_\pi^2 \;\approx\; 2m\langle\bar{q}q\rangle. \qquad (2.7)$$

2.4 Solution of the Quark Dyson-Schwinger Equation

The DSE for the quark propagator, Eq. (2.2), has been much studied (see Sec. 6 of Ref. [1]). The numerical solution obtained using a gluon propagator with an integrable singularity in the infrared region (upper curve in Fig. 2) and a quark-gluon vertex of the Ball-Chiu type[12] can be represented well by the following algebraic parametrisation:[3]

$$\bar{\sigma}_S(x) = Ce^{-2x} + \qquad\qquad (2.8)$$
$$\frac{1 - e^{-b_1 x}}{b_1 x}\frac{1 - e^{-b_3 x}}{b_3 x}\left(b_0 + b_2\frac{1 - e^{-\Lambda x}}{\Lambda x}\right) + \frac{\bar{m}}{x + \bar{m}^2}\left(1 - e^{-2(x + \bar{m}^2)}\right)$$

$$\bar{\sigma}_V(x) = \frac{2(x + \bar{m}^2) - 1 + e^{-2(x + \bar{m}^2)}}{2(x + \bar{m}^2)^2} - \bar{m}Ce^{-2x}, \qquad (2.9)$$

where $S(p) = -i\gamma \cdot p\,\sigma_V(p^2) + \sigma_S(p^2)$, $x = p^2/(2D)$, $\bar{\sigma}_S(x) = \sqrt{2D}\sigma_S(p^2)$, $\bar{\sigma}_V(x) = (2D)\sigma_V(p^2)$, $\bar{m} = m/\sqrt{2D}$, and D sets the mass scale. This parametrisation provides a representation of the quark propagator that is an entire function in the complex p^2

plane but for an essential singularity, as suggested by Ref. [15], which is sufficient to ensure confinement.

Equations (2.8) and (2.9) provide a six parameter algebraic approximation to the quark propagator in QCD: C, $\bar{m}$, $b_0, \ldots, b_3$. [$\Lambda(= 10^{-4})$ simply decouples b_2 from the quark condensate.] The parameters can be chosen so as to provide an accurate approximation to a numerical solution of Eq. (2.2). However, as illustrated in Fig. 2, the form of the gluon propagator for $q^2 < 1$ GeV^2 is unknown. This parametrisation is therefore also an implicit parametrisation of the gluon propagator in this region. Using this quark propagator to calculate experimental observables, and choosing the parameters so as to obtain the best possible fit to these observables, one has a direct connection between observables and the form of the effective quark-quark interaction in the infrared. This provides a means for using precise, low-energy experimental data to determine the effective quark-quark interaction in the infrared.

3 Calculating Pion Observables

In much the same way as an effective action can be used to formalise the constraints of chiral Ward identities in QCD, one can formalise the *Abelian approximation* to QCD (which decouples the ghost fields and reduces the Slavnov-Taylor identities to Ward identities; see Ref. [6] and Sec. 5 of Ref. [1]) via a model field theory described by the action:[14]

$$S[\bar{q}, q] = \int d^4x \, d^4y \left\{ \bar{q}(x) \left[\gamma \cdot \partial + m \right] \delta^4(x - y) q(y) + \frac{1}{2} j_\mu^a(x) g^2 D_{\mu\nu}(x - y) j_\nu^a(y) \right\}$$
(3.1)

with $j_\mu^a(x) = \bar{q}(x) \, (\lambda^a/2) \, \gamma_\mu q(x)$ and $D_{\mu\nu}$ the gluon propagator. This model field theory is called the Global Colour-symmetry Model (GCM).

3.1 π-π Scattering

The π-π scattering T-matrix is completely specified by one scalar function, $A(s, t, u)$:

$$T_{\alpha\beta;\gamma\delta} = A(s, t, u) \, \delta_{\alpha\beta}\delta_{\gamma\delta} + A(t, s, u) \, \delta_{\alpha\gamma}\delta_{\beta\delta} + A(u, t, s) \, \delta_{\alpha\delta}\delta_{\beta\gamma} \, .$$
(3.2)

In the DSE-GCM approach the *tree-level* contribution to $A(s, t, u)$ is obtained from a sum of *intrinsically-finite* quark-loop diagrams, which, using Eqs. (2.8) and (2.9), have no quark production thresholds. Although it is not necessary, one can perform a derivative expansion of this sum of quark-loop contributions to obtain:[16]

$$A(s, t, u) = \frac{m_\pi^2 + 2s - t - u}{3 f_\pi^2} + \frac{4N_c}{3 f_\pi^4} \times$$
(3.3)

$$\left[K_1 \left(-12 m_\pi^4 + 6 m_\pi^2 (s + t + u) + 2 s^2 - t^2 - u^2 - 2(st + su + tu) \right) \right.$$

$$\left. + K_2 \left(-2 m_\pi^2 + s \right) \left(-2 m_\pi^2 + t + u \right) + K_3 \left(-2 m_\pi^4 + m_\pi^2 (s + t + u) - tu \right) \right],$$

where $f_\pi^2 = \dfrac{N_c}{8\pi^2} \int_0^\infty ds \, s \, B_{m\neq 0}^2 \times$
(3.4)

$$\left(\sigma_V^2 - 2 \left[\sigma_S \sigma_S' + s\sigma_V \sigma_V' \right] - s \left[\sigma_S \sigma_S'' - (\sigma_S')^2 \right] - s^2 \left[\sigma_V \sigma_V'' - (\sigma_V')^2 \right] \right) ,$$

and m_π is obtained from Eq. (2.7) with

$$-\langle \bar{q}q \rangle_\mu = \tag{3.5}$$

$$\left(\ln \frac{\mu^2}{\Lambda_{QCD}^2} \right)^\alpha \lim_{\Lambda_{UV}^2 \to \infty} \left(\ln \frac{\Lambda_{UV}^2}{\Lambda_{QCD}^2} \right)^{-\alpha} \frac{3}{4\pi^2} \int_0^{\Lambda_{UV}^2} ds\, s \left(\sigma_S(s) - \frac{m}{s+m^2} \right) .$$

The constants K_i are also given by one dimensional integrals whose integrands involve only the quark propagator.[16] Clearly, and importantly, each of the quantities appearing in Eq. (3.3) is determined once the quark propagator is specified. The utility of the derivative expansion is only that it facilitates a comparison with other calculations of $A(s, t, u)$.

3.2 $\pi^0 \to \gamma\gamma$

In Euclidean metric the matrix element for the decay $\pi^0 \to \gamma\gamma$ can be written

$$\mathcal{M}(k_1, k_2) = -2\,i\, \frac{\alpha_{em}}{\pi f_\pi} \epsilon_{\mu\nu\rho\sigma} \epsilon_\mu(k_1)\, \epsilon_\nu(k_2)\, k_{1\rho} k_{2\sigma}\, G(k_1 \cdot k_2) , \tag{3.6}$$

where k_i are the photon momenta and $\epsilon(k_i)$ are their polarisation vectors. Here, the π^0 momentum is $P = (k_1 + k_2)$ and $P^2 = 2\,k_1 \cdot k_2$.

Using Eq. (3.6) one finds easily that $\Gamma_{\pi^0 \to \gamma\gamma} = [m_\pi^3 \alpha_{em}^2 G(-m_\pi^2)^2]/[16\pi^3 f_\pi^2]$. Experimentally one has $\Gamma_{\pi^0 \to \gamma\gamma} = (7.74 \pm 0.56)$ eV, which corresponds to

$$g_{\pi^0\gamma\gamma} \equiv G(-m_\pi^2) = 0.504 \pm 0.019 , \tag{3.7}$$

using $m_{\pi^0} = 135$ MeV and $f_\pi = 93.1$ MeV.

In generalised impulse approximation $g_{\pi^0\gamma\gamma}$ is obtained from the sum of two quark-loop diagrams and, in the chiral limit ($P^2 = 0$), one has:[3]

$$g_{\pi^0\gamma\gamma}^0 \equiv G(0) = \int_0^\infty ds\, s\, B_{m=0}\, A\, \sigma_V \times \tag{3.8}$$

$$\left\{ A\, [\sigma_V\, \sigma_S + s\, (\sigma_V'\, \sigma_S - \sigma_V\, \sigma_S')] + s\, \sigma_V\, (A'\, \sigma_S - B'\, \sigma_V) \right\} .$$

Defining $C(s) = B_{m=0}(s)^2/[s\, A(s)^2]$ one obtains a dramatic simplification and, because of DχSB; i.e., because $f_\pi \Gamma_\pi = B_{m=0}$,

$$g_{\pi^0\gamma\gamma}^0 = \int_0^\infty dC\, \frac{1}{(1+C)^3} = \frac{1}{2} , \tag{3.9}$$

since $C(s = 0) = \infty$ and $C(s = \infty) = 0$. Hence, the experimental value is reproduced, *independent* of the details of $S(p)$. This illustrates the manner in which the Abelian anomaly is incorporated in the DSE framework. This result will be violated in any approach that does not properly incorporate DχSB; i.e., in any approach which violates the chiral-limit identity: $f_\pi \Gamma_\pi = B_{m=0}$.

3.3 $F_\pi(q^2)$

In generalised impulse approximation, in Euclidean metric, with γ_μ hermitian, the π-π-γ vertex is[3]

$$\Lambda_\mu(P+q,-P) = \frac{N_c}{f_\pi^2} \int \frac{d^4k}{(2\pi)^4} \tag{3.10}$$

$$\mathrm{tr}_D \left[i\bar{\Gamma}_\pi(k;P+q)S(k_{++})i\Gamma_\mu^{BC}(k_{++},k_{-+})S(k_{-+})i\Gamma_\pi(k-\tfrac{1}{2}q;-P)S(k_{--}) \right],$$

where q is the photon momentum, P is the initial momentum of the pion, $k_{\alpha\beta} = k + \frac{\alpha}{2}q + \frac{\beta}{2}P$ and[12]

$$\Gamma_\mu^{BC}(p,q) = \Sigma_A(p,q)\,\gamma_\mu + (p+q)_\mu \left\{ \Delta_A(p,q)\tfrac{1}{2}\left[\gamma\cdot p + \gamma\cdot q\right] - i\Delta_B(p,q) \right\} \tag{3.11}$$

with $\Sigma_H(p,q) \equiv [H(p^2)+H(q^2)]/2$ and $\Delta_H(p,q) \equiv [H(p^2)-H(q^2)]/[p^2-q^2]$, for $H = A$ or B. Γ_μ^{BC} satisfies the Ward-Takahashi identity and ensures the conservation of the pion current[3] so that $\Lambda_\mu(P+q,-P) = (2P_\mu + q_\mu)F_\pi(q^2)$.

3.4 Fitting the Parameters and Calculated Results

The parameters in Eqs. (2.8) and (2.9) are fixed by requiring a global best-fit to:

$$\frac{f_\pi}{\langle\bar{q}q\rangle^{\frac{1}{3}}} = 0.423 \pm 0.024 , \qquad f_\pi\, r_\pi = 0.318 \pm 0.006 , \qquad \frac{m_\pi^2}{\langle\bar{q}q\rangle^{\frac{2}{3}}} = 0.396 \pm 0.036 ; \tag{3.12}$$

the dimensionless π-π scattering lengths

$$\begin{aligned} a_0^0 &= 0.21 \pm 0.02 , & a_0^2 &= -0.040 \pm 0.003 , \\ a_1^1 &= 0.038 \pm 0.002 , & a_2^0 &= 0.0017 \pm 0.0003 ; \end{aligned} \tag{3.13}$$

and a least-squares fit to $F_\pi(q^2)$ on the spacelike-q^2 domain: $[0,4]$ GeV^2. The fitting procedure used f_π from Eq. (3.4) and $\langle\bar{q}q\rangle$ from Eq. (3.5), with $\Lambda_{QCD} = 0.20$ GeV and $\alpha = 1$ [not the true anomalous dimension because $\ln[p^2]$-corrections have not been included in Eqs. (2.8) and (2.9)] and the expressions for a_0^0, a_0^2, a_1^1, a_2^0 and r_π given in Ref. [16]. It yielded

$$\begin{aligned} C &= 0.0406 , \quad \overline{m} = 0.0119 , \\ b_0 &= 0.118 , \quad b_1 = 2.51 , \quad b_2 = 0.525 , \quad b_3 = 0.169 . \end{aligned} \tag{3.14}$$

The mass scale is set by requiring equality between the percentage error in f_π and r_π, which yields $D = 0.133$ GeV^2.

The low-energy physical observables calculated with this parameter set are compared with their physical values in Table 1, where $m^{\text{ave}} = (m_u + m_d)/2$ and the "experimental" value of $\langle\bar{q}q\rangle^{\frac{1}{3}}$ is that typically used in QCD sum rules analysis. The calculated quantities were evaluated at the listed value of m_π; i.e., the chiral limit expressions were not used, but the corrections are $< 1\%$ in each case. The agreement is excellent. The π-π partial wave amplitudes associated with a_0^0 and a_0^2, calculated

	Calculated	Experiment
f_π	0.0839 GeV	0.0931 ± 0.001
$-\langle \bar{q}q \rangle^{\frac{1}{3}}_{1\,\mathrm{GeV}^2}$	0.211	0.220 ± 0.01
$m^{\mathrm{ave}}_{1\,\mathrm{GeV}^2}$	0.0061	0.0075 ± 0.004
m_π	0.127	0.138
r_π	0.596 fm	0.663 ± 0.006
$g_{\pi^0\gamma\gamma}$	0.497 (dimensionless)	0.504 ± 0.019
a^0_0	0.174	0.21 ± 0.02
a^2_0	-0.0496	-0.040 ± 0.003
a^1_1	0.0307	0.038 ± 0.002
a^0_2	0.00161	0.0017 ± 0.0003
a^2_2	-0.000251	

Table 1. Comparison of *tree-level* DSE calculations with experiment.

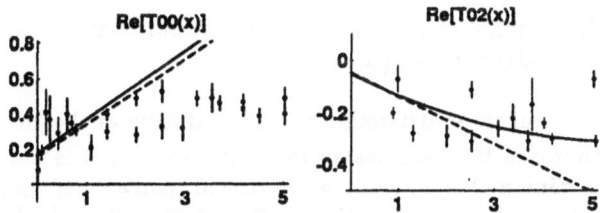

Fig. 3. Partial wave amplitudes in π-π scattering ($x = E^2/(4m_\pi^2) - 1$): solid line - this calculation; dashed line - current algebra.[17]

using the formulae in Ref. [16], are plotted in Fig. 3 and can be seen to be in agreement with the data up to $x \approx 3$, which corresponds to $E \approx 4m_\pi$. The same is true of the higher partial wave amplitudes.[3] The calculated form of $F_\pi(q^2)$ is presented in Fig. 4 and, given that the extraction of the "experimental" point at $q^2 = 6.3$ GeV^2, measured in pion electroproduction[20] is strongly model dependent, the agreement with the experimental data is again excellent.

4 Concluding Remarks

In principle, QCD can be solved using the Dyson-Schwinger equation (DSE) framework, however, in practice, obstacles to achieving this goal remain. The successes and remaining challenges are discussed in detail in Ref. [1].

Herein, the practical application of the DSE approach to the calculation of physical observables is illustrated. By its very nature, the approach directly incorporates all of the known large spacelike-q^2 behaviour of QCD. It has a phenomenological, model dependent aspect that is tied to the fact that the gluon propagator, $D_{\mu\nu}(q)$, is unknown for spacelike-$q^2 < 1$ GeV^2. This has the benefit that, in calculating experimental observables in this approach, one obtains a representation of these observables in terms of the infrared structure of $D_{\mu\nu}(q)$ and can thereby use precision

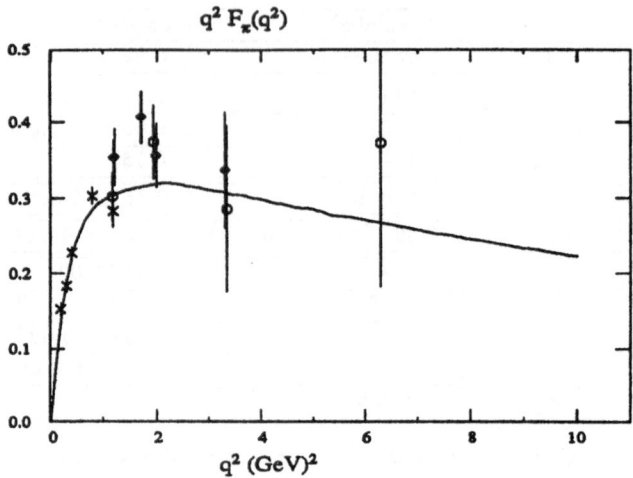

Fig. 4. Calculated form of $F_\pi(q^2)$. The experimental data are from Refs.[18] (crosses), [19] (diamonds) and [20] (circles).

experimental measurements to determine the infrared form of the gluon propagator.

In the DSE approach the pion has an intrinsic size; i.e., it is not pointlike, and its dominant determining characteristic is its dressed quark core, which is described by the proper pion-quark vertex function in Eq. (2.6). The calculations reported herein are *tree-level* calculations, which here actually means that they are obtained with the minimal number of dressed-quark loops. The agreement between theory and experiment is at the 10% level, which is consistent with Refs. [21,22] where *nonpointlike* π-loop contributions are shown to provide < 15% corrections.

Low Energy Constant	*tree-level* DSE	1-π^{point}-loop ChPT[23]
$\bar{l}_1$	-6.1	-2.3 ± 3.7
$\bar{l}_2$	6.6	6.0 ± 1.3
$\bar{l}_6$	11.7	16.5 ± 1.1

Table 2. Low-energy constants of ChPT inferred from *tree-level* DSE calculations of experimental observables.

Clearly, experimental observables can be calculated directly in the DSE approach. From these calculations, however, one might infer values of the low-energy constants used in Chiral Perturbation Theory (ChPT) to parametrise the solution of the chiral Ward identities in QCD. This will provide implicit, nonlinear relations between these constants and the few, underlying parameters of QCD. The results inferred for $\bar{l}_1$, $\bar{l}_2$ and $\bar{l}_6$ from DSE-*tree-level* calculations are presented in Table. 2 and compared with the 1-point-pion-loop corrected values used in ChPT. (The 1-*nonpointlike*-pion-loop correction to r_π estimated in Ref. [22] yields $\bar{l}_6 \simeq 13.0 - 13.5$.) This comparison indicates that *tree-level* DSE calculations incorporate those effects of point-pion-loops in ChPT that serve merely to mock-up the pion's internal structure. Calculations from which the value of other of these constants can be inferred are underway.[24]

Acknowledgments. I would like to thank and congratulate the organisers, Aron Bernstein and Barry Holstein, for bringing about this timely and stimulating meeting, and the workshop secretary, Joanne Gregory, for ensuring that it ran smoothly. This work was supported by the Department of Energy, Nuclear Physics Division under contract number W-31-109-ENG-38.

References

1. C. D. Roberts and A. G. Williams: *Dyson-Schwinger Equations and their Application to Hadronic Physics*, in *Progress in Particle and Nuclear Physics*, Vol. 33, pp. 477-575, ed. by A. Fäßler (Pergamon Press, Oxford, 1994).
2. C. J. Burden: *How can QED_3 help us understand QCD_4*, in Proceedings of "The Workshop on QCD Vacuum Structure", American University of Paris, 1-5 June 1992, Editors H. M. Fried and B. Müller (World Scientific, New York, 1993).
3. C. D. Roberts: *Electromagnetic pion form factor and neutral pion decay*, Argonne National Laboratory preprint # PHY-7842-TH-94 (1994).
4. C. D. Roberts, A. G. Williams and G. Krein: Int. J. Mod. Phys. A **7** 5607 (1992)
5. S. Mandelstam: *Phys. Rev.* **D20** 3223 (1979)
6. U. Bar-Gadda: *Nucl. Phys.* **B163** 312 (1980)
7. M. Baker, J. S. Ball and F. Zachariasen: *Nucl. Phys.* **B186** 531,560 (1981); **B226** 455 (1983); N. Brown and M. R. Pennington: *Phys. Lett.* **B202** 257 (1988); *Phys. Rev.* **D38** 2266 (1988)
8. U. Häbel, *et al*: *Z. Phys.* **A336** 423, 435 (1990)
9. J. E. Mandula and M. Ogilvie: *Nucl. Phys.* **B1A** (Proc. Suppl.) 117 (1987); *Phys. Lett.* **B185** 127 (1987)
10. C. Bernard, C. Parrinello and A. Soni: Phys. Rev. D **49** 1585 (1994)
11. V. N. Gribov: Nucl. Phys. B **139** 1 (1979); D. Zwanziger, Nucl. Phys. **B364** 127 (1991)
12. J. S. Ball and T.-W. Chiu: Phys. Rev. **D22** 2542 (1980)
13. R. Delbourgo and M. D. Scadron: J. Phys. G **5** 1631 (1979)
14. R. T. Cahill and C. D. Roberts: Phys. Rev. D **32** 2419 (1985)
15. C. J. Burden, C. D. Roberts and A. G. Williams: Phys. Lett. **B285** 347 (1992)
16. C. D. Roberts, R. T. Cahill, M. E. Sevior and N. Iannella: Phys. Rev. D **49** 125 (1994)
17. S. Weinberg: Phys. Rev. Lett. **17** 616 (1966).
18. C. N. Brown: *et al*, Phys. Rev. D **8** 92 (1973)
19. C. J. Bebek: *et al*, Phys. Rev. D **13** 25 (1976)
20. C. J. Bebek: *et al*, Phys. Rev. D **17** 1693 (1978)
21. L. C. L. Hollenberg, C. D. Roberts and B. H. J. McKellar: Phys. Rev. C **46**, 2057 (1992)
22. R. Alkofer, A. Bender and C. D. Roberts: *Pion Loop Contribution to the Electromagnetic Pion Charge Radius*, ANL preprint # PHY-7663-TH-93 (1993)
23. J. Gasser and H. Leutwyler: Ann. Phys. (NY) **158** 142 (1984)
24. I. Chappell: *A Calculation of the Pion Scalar Radius*, Argonne National Laboratory DEP report (1994); A. J. Davies, C. D. Roberts and M. J. Thomson: in progress.

Light Hadron Physics from the Lattice[*]

J. W. Negele

Center for Theoretical Physics,
Laboratory for Nuclear Science, and Department of Physics
Massachusetts Institute of Technology, Cambridge, Massachusetts 02139

Abstract

Two aspects of chiral dynamics are explored using lattice QCD. The first is calculation of low-energy hadronic parameters and a brief survey of recent lattice results is presented, including meson decay constants, the quark condensate, pion and nucleon scattering lengths and the nucleon σ-term. The second is a demonstration of the dominant role of instantons in determining light hadron structure and quark propagation in the QCD vacuum. The instanton content of gluon fields sampling the Wilson action is extracted using cooling as a filter to selectively remove essentially all fluctuations of the gluon field except for instantons. Close agreement is demonstrated between quenched lattice QCD results including all gluonic contributions and including only instantons for vacuum correlation functions of hadronic currents and for density-density correlation functions in hadronic bound states.

Introduction

There are two distinct roles that lattice QCD is playing in our understanding of chiral dynamics. The first role is in providing a quantitative tool for the calculation of the parameters of the effective hadronic theory from QCD. In principle, instead of having to fix the parameters of effective field theory and chiral perturbation theory phenomenologically, these parameters should be calculable directly from QCD. At present, lattice field theory is the only quantitative tool at our disposal to calculate these parameters from first principles, and I will show in the first part of this talk that significant progress is being made.

The second role is in providing insight into the structure of the QCD vacuum and light hadrons. To date, the known analytic techniques of theoretical physics have been inadequate to provide even a qualitative understanding of the mechanism responsible for the gross structure of hadrons — whether it is the

---

[*] This work is supported in part by funds provided by the U. S. Department of Energy (D.O.E.) under cooperative research agreement DE-FC02-94ER40818.

Coulomb-like interaction between quarks arising from short-wavelength fluctuations of the gluon field, the behavior at large distances associated with confinement, or physics associated with topological structures in the QCD vacuum corresponding in the semiclassical limit to instantons. Hence, we have turned to numerical solution of nonperturbative QCD for insight into the underlying physics of hadron structure.

In this talk, I will describe results obtained in collaboration with Ming Chu, Jeff Grandy, and Suzhou Huang [7, 8] which provide strong evidence for the dominant role of instantons in determining the structure of light hadrons and the propagation of light quarks in the vacuum. Our strategy is to consider correlation functions which characterize the gross structure of hadrons and quark propagation in the QCD vacuum and which are well described by quenched lattice QCD calculations which sample the full Wilson action. These lattice calculations include all the fluctuations and topological excitations of the gluon field and thus include the full perturbative and non-perturbative effects of the short-range Coulomb and hyperfine interactions, confinement, and instantons. We then use cooling to remove essentially all fluctuations of the gluon field except for the instantons which, because of their topology, cannot be removed by local minimization of the action. Thus, both the Coulomb interaction and confinement are almost completely removed while retaining most of the instanton content. Finally, we use the instantons alone to recalculate the correlation functions characterizing light hadron structure. To the extent to which the gross features of light hadron structure and quark propagation in the QCD vacuum are unaffected by removing all the gluonic modes except instantons, we have strong evidence for the dominant role of instantons.

This work is strongly motivated by the physical arguments and instanton models of Shuryak and others [28, 27, 29, 13] in which the zero modes associated with instantons produce localized quark states and quark propagation takes place primarily by hopping between these states. The additional information which lattice calculations bring to bear are a quantitative determination of the instanton content of the QCD vacuum and a direct comparison of the effects of all gluon contributions versus those of instantons alone.

Calculation of Low Energy Parameters on the Lattice

In principle, all the parameters arising in chiral perturbation theory, as well as other low energy parameters such as scattering lengths and the pion-nucleon sigma term can be calculated from first principles on the lattice. In practice, there are significant practical and computational limitations, and in the space available here I will briefly discuss the salient issues and survey some representative results.

As in all lattice calculations, there are two major categories of approximation. At the cost of omitting dynamical excitation of quark-antiquark pairs from the Fermi sea, making the quenched approximation by omitting the Fermion determinant simplifies the numerical calculation to the point that it is now practical to control extrapolation errors by using realistically small quark masses, small lattice spacing, and large lattice volume. Conversely, present unquenched calculations which include the Fermion determinant make significant errors associated with large quark masses and lattice spacing and small spatial volume.

Let us begin with the lowest order term in chiral perturbation theory, $\mathcal{L}_2$, which includes f_π and $\langle \bar{\psi}\psi \rangle$. In a comprehensive set of quenched calculations on lattices up to $32^2 \times 30 \times 40$ and β ranging from 5.7 to 6.1 Weingarten has calculated the pion decay constant [32]. Using

$$f_n = Z_A \frac{\langle 0|\bar{\psi}\gamma_4\gamma_5\psi|\pi(p)\rangle}{E_\pi(p)}$$

where Z_A is calculated perturbatively with mean field improvement, quenched QCD yields $f_\pi/m_\rho = 0.106(14)$ which is one standard deviation below the experimental value 0.121 (as is F_K). This is to be expected, since in the quenched approximation g^2 falls faster with momentum than in full QCD which implies that $V(r)$ is weaker at the origin so that the wave function is smaller at the origin yielding a smaller decay constant. Consistent with this argument, Fukugita et al. obtain $f_\pi = 94-105$ MeV (where expt = 93 MeV) is an unquenched calculation using two flavors of staggered fermions on lattice up to 20^4 at $\beta = 5.7$ [12].

Calculation of the chiral condensate, $\langle \bar{\psi}\psi \rangle$, is complicated by the fact it is scale dependent and hence one must pick a specific scheme to define it, conventionally the $\overline{\text{MS}}$ scheme at $\mu = 1$ GeV. As discussed by Daniel et al. [12] there are several alternative ratios of correlation functions which yield $Z_P^{-1}Z_A^{-1}\langle \bar{\psi}\psi \rangle$, where Z_P and Z_A are the pseudoscalar and axial renormalization constants. Four different methods of calculation for quenched Wilson fermions at $\beta = 6.0$ on a $16^3 \times 4$ lattice yield consistent results, which in physical units correspond to [12].

$$\langle \bar{\psi}\psi \rangle_{\text{lattice}} = -Z_P Z_A (0.034 \pm 0.011) \text{ GeV}^3$$

to be compared with the phenomenological value [16]

$$\langle \bar{\psi}\psi \rangle_{\text{phenom.}} = -0.0114 \text{ GeV}^3$$

At sufficiently large β, Z_P and Z_A could be calculated perturbatively in the $\overline{\text{MS}}$ scheme and thereby determine $\langle \bar{\psi}\psi \rangle$. Unfortunately, perturbation theory is unreliable at $\frac{6}{g^2} \equiv \beta = 6.0$, so at best, we must regard the combination of the nonperturbative measurement [22] $Z_A = 0.86$ and the one-loop perturbative result $Z_P = 1 - 0.221g^2$ as estimates. The resulting renormalized value

$$\langle \bar{\psi}\psi \rangle_{\text{lattice}} \sim -(0.017 \pm 0.006) \text{ GeV}^3$$

is reasonably consistent with the phenomenological value -0.0114 GeV^3.

An unquenched calculation has been reported by Fukugita et al. [14] using two flavors of staggered fermion on a $20^3 \times 40$ lattice at $\beta = 5.4$. Converting from their normalization and correcting for the four continuum flavors yields

$$\langle \bar{\psi}\psi \rangle = -\frac{3}{4}0.0066(3)a^{-3} = -(0.055 \pm .003)$$

which is significantly above the empirical value.

Thus far, the parameters L_1, through L_{10} of the next order term in the effective Lagrangian $\mathcal{L}_4$ have not been calculated numerically on the lattice. However, recently, a significant step in that direction has been taken by Myint and Rebbi [23] who have formulated the problem on the lattice and calculated the coefficients in the strong coupling expansion. As might be expected, the strong coupling results are typically an order of magnitude too large, so effort must now be directed to numerical evaluation.

We now consider the calculation of $\pi - \pi$, $\pi - N$ and $N - N$ scattering lengths. An excellent review of the basic approach is given by Sharpe [26]. In a one-dimensional potential, calculation of the energy of a particle in a box of length L clearly specifies the phase shift as a function of k. Generalizing to three spatial dimensions, Lüscher has shown for the case of two particles in a cubic box of length L that [21]

$$E - 2m = \frac{-4\pi a_0}{mL^3}\left(1 - 2.837\frac{a_0}{L} + 6.375\left(\frac{a_0}{L}\right)^2\right) + \mathcal{O}(L^{-6})$$

$$\sim \frac{-T}{L^3} + \mathcal{O}(L^{-4})$$

Hence, the $\pi - \pi$ scattering length may be extracted from the ratio

$$R = \frac{\langle\pi^+(t)\pi^0(t)\pi^-(0)\pi^0(0)\rangle}{\langle\pi^+(t)\pi^-(0)\rangle\langle\pi^0(t)\pi^0(0)\rangle} \sim e^{-(E-2m_\pi t)}$$

$$\sim 1 + \frac{T_2}{L^3}t$$

The last expression is particularly transparent physically, since in lowest order time dependent perturbation theory the π^+ and π^0 would interact once at some intermediate t' yielding the t-matrix T_2, integration over t' yields the factor t and the factor $\frac{1}{L^3}$ reflects the fact that the pions must overlap to interact. Data summarized in [26, 15] show excellent agreement of quenched calculations using both staggered and Wilson fermions with the results of chiral perturbation theory,

$$\frac{32\pi f_\pi^2}{m_\pi}a_0^{I=0} = 7 \qquad \frac{32\pi f_\pi^2}{m_\pi}a_0^{I=2} = -2$$

Particularly striking, and not presently understood, is the fact that the $I = 2$ scattering length is essentially independent of quark mass over the range $0.2 < \frac{m_\pi}{m_\rho} < 0.8$. Similar success with slightly greater error bars is obtained using the same methods for the pion-nucleon scattering length

$$\frac{8\pi f_\pi^2}{\mu_{\pi N}}a_0^{I=3/2} = -0.95 \pm .13 \qquad \text{(current algebra = -1)}$$

$$\frac{8\pi f^2}{\mu_{\pi N}}a_0^{I=1/2} = 1.6 \pm 0.7 \qquad \text{(current algebra = +2)}$$

In contrast, the nucleon-nucleon scattering length is far too sensitive to be calculated with current lattice technology. Note that the singlet scattering length is extremely large, $a_0^{^1S_0} = +20.1(4)$ fm because the two nucleon system is nearly bound, and the negative triplet scattering length $a_0^{^3S_1} = -5.432(5)$ fm reflects the loosely bound deuteron whose wave function is non-negligible at 10 fm. Hence, a lattice calculation with heavy quarks such that $\frac{m_\pi}{m_\rho} = 0.86$ surely underestimates the range of the nucleon-nucleon interaction and thus plausibly loses sufficient attraction to render the deuteron unbound and move the singlet amplitude substantially away from the $^1S_0$ pole. Further, a 2.8 fm box is inadequate both for the 10 fm deuteron wave function and 20 fm $^1S_0$ scattering length, so the qualitatively unrealistic result of $a_0^{^1S_0} = 1.0(3)$ fm and $a_0^{^3S_1} = 1.2(2)$ fm of Ref. [15] are understandable and quite difficult to improve.

Finally, we consider calculations of the famous pion-nucleon σ-term

$$\sigma_{\pi N} = \frac{1}{2}(m_\mu + m_d)\langle N|\bar{u}u + \bar{d}d|N\rangle$$

Here the basic issue is that there are two contributions: a valence contribution which is easy to calculate in which the operator $\bar{q}q$ acts on one of the three valence propagators connecting the nucleon source and sink, and a sea contribution in which the operator connects to a closed $\bar{q}q$ loop which must be integrated over the entire lattice and is prohibitive to calculate directly and noisy to calculate by indirect means. To circumvent this problem, Gupta et $al.$ [17] calculate the ratio of the full σ-term, using $\frac{\partial M_N}{\partial m_{quark}} \propto \langle N|\bar{u}u + \bar{d}d|N\rangle$ and an unquenched calculation of the nucleon mass, to the valence diagram, obtaining

$$\frac{\sigma_{\pi N}}{\sigma_{\pi N}^{valence}} \sim 2-3$$

for Wilson fermions with $\beta = 5.4$ through 5.6 on a $16^3 \times 32$ lattice. A comparable calculation by Altmeyer et al [2] with staggered fermions on a $16^3 \times 24$ lattice at $\beta = 5.35$ yields a ratio of approximately 1.5. These ratios are consistent, albeit with large errors, with the phenomenological value $\sigma_{\pi N} \sim 45 - 56$ MeV and the well established valence result $\sigma_{\pi N}^{val} \sim 25$ MeV.

In summary, one sees great promise that lattice QCD will ultimately provide accurate ab $initio$ results for the low energy parameters of QCD. One already observes reasonable precision for the necessary extrapolations $a \to 0$, $m_q \to 0$, and $L \to \infty$ for quenched calculations. Full unquenched QCD results are admittedly primitive by quenched standards. However, I am confident that the great progress which is being made in developing "improved" or "perfect" actions and operator and the prospect of Teraflop-scale computers will provide the tools to calculate accurate quantitative results in the next few years.

Correlation Functions

As in the case of other strongly interacting many-body systems, to understand the structure of the vacuum and light hadrons in nonperturbative QCD it is instructive to study appropriately selected ground state correlation functions, to calculate their properties quantitatively, and to understand their behavior physically.

The vacuum correlation functions we consider are the point-to-point equal time correlation functions of hadronic currents

$$R(x) = \langle\, \Omega\,|\, T\, J(x)\, \bar{J}(0)\,|\, \Omega \rangle$$

discussed in detail by Shuryak [28] and recently calculated in quenched lattice QCD [7]. The motivation for supplementing knowledge of hadron bound state properties by these correlation functions should be obvious if one considers the analogous case of the deuteron. If your goal were to understand the nucleon-nucleon interaction in each spin, isospin and angular momentum channel as a function of spatial separation, the limited information provided by ground state deuteron properties would be frustratingly incomplete, and you inevitably would be led to study nucleon-nucleon scattering phase shifts. Although, regrettably, our experimental colleagues have been most inept in providing us with quark-antiquark phase shifts, the same physical information is contained in the vacuum hadron current correlation functions $R(x)$. As shown by Shuryak [28], in many channels these correlators may be determined or significantly constrained from experimental data using dispersion relations. Since numerical calculations on the lattice agree with empirical results where available, we regard them as valid solutions of QCD in all channels and thus use them to obtain information comparable to scattering phase shifts.

The correlation functions we calculate in the pseudoscalar, vector, nucleon and Delta channels are

$$R(x) = \langle\Omega|T\, J^P(x)\, \bar{J}^P(0)|\Omega\rangle \ ,$$
$$R(x) = \langle\Omega|T\, J_\mu(x)\, \bar{J}_\mu(0)|\Omega\rangle \ ,$$
$$R(x) = \tfrac{1}{4}\,\mathrm{Tr}\ \left(\langle\Omega|T\, J^N(x)\, \bar{J}^N(0)|\Omega\rangle\, x_\nu\, \gamma_\nu\right) \ ,$$

and

$$R(x) = \tfrac{1}{4}\,\mathrm{Tr}\ \left(\langle\Omega|T\, J^\Delta_\mu(x)\, \bar{J}^\Delta_\mu(0)|\Omega\rangle\, x_\nu\, \gamma_\nu\right) \ ,$$

where

$$J^P = \bar{u}\,\gamma_5\, d \ ,$$
$$J_\mu = \bar{u}\,\gamma_\mu\,\gamma_5\, d \ ,$$
$$J^N = \epsilon_{abc}[u^a\, C\,\gamma_\mu\, u^b]\,\gamma_\mu\,\gamma_5\, d^c \ ,$$
$$\text{and} \quad J^\Delta_\mu = \epsilon_{abc}[u^a\, C\,\gamma_\mu\, u^b]\, u^c \ .$$

As in Refs. [7] and [28], we consider the ratio of the correlation function in QCD to the correlation function for non-interacting massless quarks, $\frac{R(x)}{R_0(x)}$, which approaches one as $x \to 0$ and displays a broad range of non-perturbative effects for x of the order of 1 fm.

As discussed in Refs. [7, 28], these correlators already show strong indications of instanton dominated physics. As shown by 't Hooft [18], the instanton induced interaction couples quarks and antiquarks of opposite chirality leading to strong attractive and repulsive forces in the pseudoscalar and scalar channels respectively, and just this behavior is observed in the vacuum correlation functions.

To characterize the gross properties of hadrons, in addition to the mass, we consider quark density-density correlation functions [3, 11, 20] $\rho(x) = \langle h|\hat{\rho}(x)\,\hat{\rho}(0)|h\rangle$. In contrast to wave functions, which have large contributions from the gluon wave functional associated with the gauge choice or definition of a gauge invariant amplitude [30], the density-density correlation function is a gauge-invariant physical observable which directly specifies the spatial distribution of quarks.

Instantons

The Feynman path integral for a quantum mechanical problem with degenerate minima is dominated by paths which fluctuate around stationary solutions to the classical Euclidean action connecting these minima [24]. In the case of the double well potential, a typical Feynman path is composed of segments fluctuating around the left and right minima joined by segments crossing the barrier. If one had such a trajectory as an initial condition, one could find the nearest stationary solution to the classical action numerically by using an iterative local relaxation algorithm. In this method, which has come to be known as cooling, one sequentially minimizes the action locally as a function of the coordinate on each time slice and iteratively approaches a stationary solution. In the case of the double well, the trajectory approaches straight lines in the two minima joined by kinks and anti-kinks crossing the barrier and the structure of the trajectory can be characterized by the number and positions of the kinks and anti-kinks.

In QCD, the corresponding classical stationary solutions to the Euclidean action for the gauge field connecting degenerate minima of the vacuum are instantons [4], and we apply below the analogous cooling technique to identify the instantons corresponding to each gauge field configuration.

Two features of instantons are of particular interest to light hadron physics. The first is the fact that although the fermion spectrum is identical at each minimum of the vacuum, quarks of opposite chirality are raised or lowered one level between adjacent minima, so that instantons give rise to vertices connecting $\bar{q}_L$ and q_R. The second fact is that each instanton gives rise to a localized zero mode of the Dirac operator $D_\mu \gamma_\mu \phi_0(x) = 0$, so that the propagator for light quarks is dominated by these zero modes. This gives rise to a physical picture in which $\bar{q}q$ pairs propagate by "hopping" between localized zero modes associated with instantons.

Lattice Calculations

Cooling [5, 31] is used as a filter to extract the instanton content of 19 gluon configurations obtained by sampling the standard Wilson action on a $16^3 \times 24$ lattice at $\frac{6}{g^2} = 5.7$. We used the Cabibbo-Marinari [6] algorithm with three $SU(2)$ subgroups and $\beta = \infty$ to minimize the action sequentially on each link of the lattice in each cooling step.

To monitor the filtering of different degrees of freedom as a function of cooling steps, we measure several gluonic observables. Short wavelength fluctuations giving rise to the Coulomb and hyperfine interactions are reflected in the total action S. Confinement is monitored by measuring the string tension extracted from a 4×7 Wilson loop, and we refer to results from an earlier calculation at the same value of β with the same cooling algorithm [10]. Finally, to monitor the instanton content, we measure the topological charge $\langle Q \rangle$ and topological susceptibility $\langle Q^2 \rangle$, using the simplest expression for the topological charge density $Q(x_n) = -\frac{1}{32\pi^2}\epsilon_{\mu\nu\rho\sigma} \operatorname{ReTr} [U_{\mu\nu}(x_n) U_{\rho\sigma}(x_n)]$. Note that for a random ensemble of Poisson distributed instantons and anti-instantons, $\langle Q \rangle = 0$ and $\langle Q^2 \rangle = I + A$, the number of instantons plus anti-instantons.

The ratios of vacuum point to point correlation functions, $\frac{R(x)}{R_0(x)}$, are calculated as described in [7]. The effects of lattice anisotropy are removed by calculating $R_0(x)$ on the same lattice as $R(x)$ and measuring the ratio for a cone of lattice sites concentrated around the diagonal. Finite lattice volume effects are corrected by subtracting the contributions of first images, and the correlation functions are fit by a spectral function parameterized by a resonance mass, the coupling to the resonance, and the continuum threshold. Hadron density correlation functions, $\langle h|\rho_u(x)\rho_d(0)|h \rangle$, are calculated as in Refs. [11, 20, 9] for the pion, rho and nucleon, where $\rho_u = \bar{u}\gamma_0 u$ and similarly for ρ_d. Image corrections for finite volume effects are applied as in Ref. [9].

A significant conceptual issue in comparing observables calculated using cooled configurations with uncooled results is how to change the renormalization of the bare mass and coupling constant as the gluon configurations are cooled. We use the physical pion and nucleon masses to determine κ and a for the cooled configurations, with the result that a changes by $\sim 16\%$ after 25 cooling steps. It is significant that the rho mass remains unchanged within errors with this value of a.

Instanton Content of the Gluon Vacuum

To provide a clear picture of how cooling extracts the instanton content of a thermalized gluonic configuration, we display in Fig. 1 the action density $S(1,1,z,t)$ and topological charge density $Q(1,1,z,t)$ for a typical slice of a gluon configuration before cooling and after 25 cooling steps. As one can see, there is no recognizable structure before cooling. Large, short wavelength fluctuations of the order of the lattice spacing dominate both the action and topological charge density. After 25 cooling steps, three instantons and two anti-instantons can be identified clearly. The action density peaks are completely correlated in position and shape with the topological charge density peaks for instantons and with the topological charge density valleys for anti-instantons. Note that both the action and topological charge densities are reduced by more than two orders of magnitude, so that the fluctuations removed by cooling are several orders of magnitude larger than the topological excitations that are retained. Further cooling to 50 steps results in the annihilation of the nearby instanton–anti-instanton pair but retains the well separated instantons and anti-instanton.

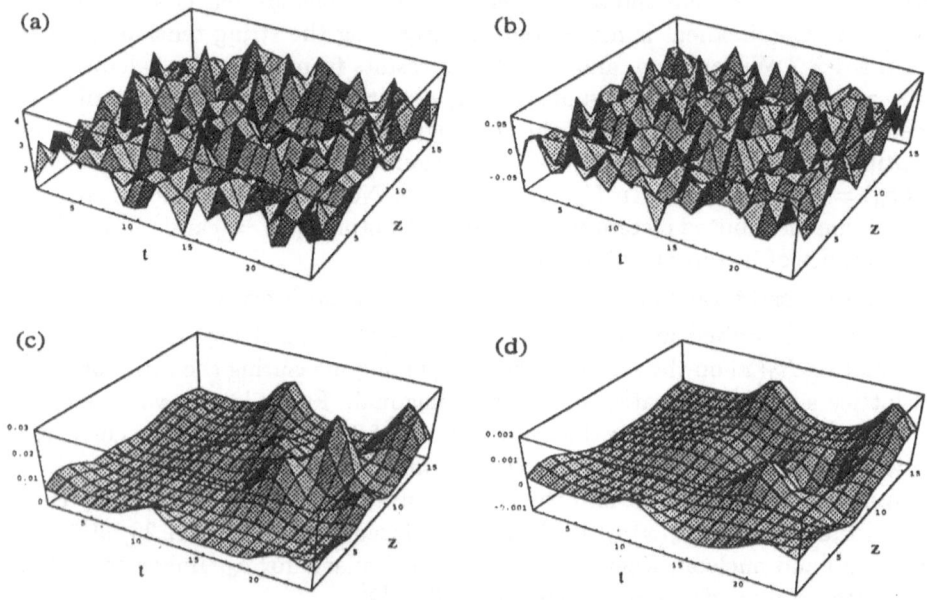

Fig. 1. Cooling history for a typical slice of a gluon configuration at fixed x and y as a function of z and t. The left column shows the action density $S(1,1,z,t)$ before cooling (a) and after cooling for 25 steps (c). The right column shows the topological charge density $Q(1,1,z,t)$ before cooling (b) and after cooling for 25 steps.

As expected, the action is dominated by the short range modes and is therefore very strongly affected by cooling. Denoting the action of a single instanton by $S_0 = 8\pi^2/g^2$, $\langle S \rangle / S_0$ decreases from 20,211 before cooling to 64 and 31 at 25 and 50 sweeps respectively. The topological charge is less sensitive to short range modes, and Q^2 is essentially constant at $\sim 25 \pm 10$ throughout the cooling. The string tension in lattice units, σa^2, is 0.18, 0.05, and 0.03 at 0, 25, and 50 cooling steps. At cooling step 25, the action and string tension have dropped to 0.3% and 27% of the uncooled values, indicating a dramatic reduction in perturbative and confining effects. The difference between $\langle S \rangle / S_0 \sim 65$ and $\langle Q^2 \rangle \sim 25$ indicates that there are sufficient nearby instanton–anti-instanton pairs in each configuration that we have not yet reached the dilute regime where $\langle Q^2 \rangle \sim A + I$, which only sets in beyond 50 steps. We regard the configurations cooled with 25 steps as providing a more complete description of the instanton content of the original configurations, and will therefore emphasize them in our subsequent calculation of hadronic properties.

To estimate the instanton size we measure the topological charge density correlation function $f(x) = \sum_y Q(y)Q(x+y)$ where $Q(y)$ is the topological charge density at point y and the sum is over the whole lattice. The ensemble average of $f(x)$ at cooling step 25 is displayed in Fig. 2. The strong peak at small x is the correlation of a single instanton or anti-instanton with itself. The vanishing of $\langle f(x) \rangle$ at large x implies that the topological charge is uncorrelated at this larger distance and thus averages to zero.

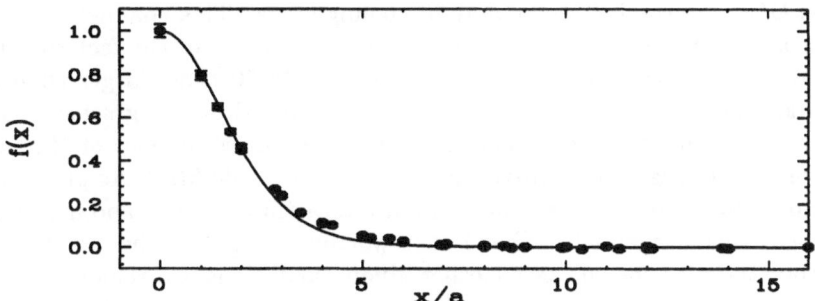

Fig. 2. Topological charge density–density correlation function after 25 cooling steps. Lattice measurements are denoted by solid points with error bars. The curve shows the best fit obtained using a convolution of the topological charge density for a single instanton of size $\rho = 2.5a$.

If we assume that all instantons are well separated, we would expect that each individual peak can be approximated by the analytic instanton topological charge density $Q_\rho(x) = \frac{6}{\pi^2 \rho^4} \left(\frac{\rho^2}{x^2 + \rho^2} \right)^4$ where ρ is the size parameter. Although in principle one should fit with a distribution of values of ρ, a first approximation is obtained by using a single value of ρ which we will interpret as an average value. A convolution of $Q_\rho(x)$ with itself defines a function which can be used to fit the lattice data with ρ as the fitting parameter. The continuous curve in Fig. 2 is the fitted result with $\rho = 2.5a$ for 25 cooling steps and we similarly obtain $\rho = 2.8a$ for 50 cooling steps. The fit fails to reproduce the detailed shape for $x/a \sim 5$, both because of the range in instanton sizes and the nonlinear overlap of instantons as observed in Fig. 1. By analyzing the sizes of lumps in the topological charge density, we also obtain rough histograms of the distribution of instantons with ρ which are consistent with the average values quoted above.

To obtain results in physical units, we use the scale a determined from the nucleon mass, which decreases from 0.168 fm before cooling to 0.142 fm and 0.124 fm at 25 and 50 cooling steps respectively. The 16% decrease in a after 25 cooling steps is quite modest, so that our qualitative conclusions are relatively insensitive to the scale change. The values at 25 (50) cooling steps for the instanton size of 0.36 fm (0.35 fm), for the instanton density of 1.64 fm^{-4} (1.33 fm^{-4}), and for topological charge susceptibility, χ, of [177 MeV]^4 ([200 MeV]^4) compare well with the values 0.33 fm, 1.0 fm^{-4}, and [180 MeV]^4 used in instanton models by Shuryak and collaborators [3].

A similar analysis of cooled configurations has previously been carried out for SU(2) with smaller lattices and slightly different techniques [25], where the positions and magnitudes of peaks in $S(x, y, z, t)$ were used to determine the distribution of sizes of instantons.

Hadronic observables in the cooled vacuum

In the top panels of Fig. 3, we show the ratio of vacuum correlation functions for interacting to non-interacting quarks $\frac{R(x)}{R_0(x)}$ in the pseudoscalar channel for uncooled QCD, 25 cooling steps, and 50 cooling steps. This channel is by far the most attractive of all the meson channels, as reflected in the fact that the correlation function for interacting quarks is roughly 50 times larger than for free quarks, and is thus the only channel to be plotted on a log scale. Since the pion mass is used to determine the bare quark mass, masses of the pion resonance term in Fig. 3 are constrained to be fixed at 140 MeV. As previously noted and discussed in Ref. [28], the 't Hooft interaction is strong and attractive in the pseudoscalar channel. What is noteworthy in Fig. 3 is the fact that to within the statistics shown, as all other gluon excitations are removed by 25 to 50 cooling sweeps, the instantons alone quantitatively produce the observed attraction. To assure that this behavior is not a statistical artifact, in this and every other channel we analyzed two independent sets of 9 and 10 configurations separately and verified that the same behavior occurred in both cases.

Analogous results for $\frac{R(x)}{R_0(x)}$ in the nucleon channel are shown in the bottom panels of Fig. 3, where again the nucleon mass is constrained to be constant because it is used to determine the lattice spacing. The behavior is similar to that in the pseudoscalar channel. After 25 sweeps, the correlation function is qualitatively similar to the uncooled result. In detail, the peak also appears lower after cooling, although it agrees within errors. After an additional 25 sweeps the peak height increases again, agreeing even more closely with the uncooled result.

The ratios of correlation functions $\frac{R(x)}{R_0(x)}$ for the vector channel are shown in the upper panels of Fig. 4. The ρ mass governing the resonance peak is unconstrained, but does not change significantly with cooling. Furthermore, there is virtually no change in the correlation function ratio with cooling. Similarly, in the Δ channel the peak position does not shift significantly with cooling and while its height appears to increase somewhat, the errors are consistent with its staying constant.

Density-density correlation functions in the ground state of the pion, rho, and nucleon are shown in Fig. 5. The errors for the uncooled results have been suppressed for clarity since they are comparable to those for the cooled results.

The striking result for both the rho and the nucleon is the fact that the spatial distribution of quarks is essentially unaffected by cooling — instantons alone govern the gross structure of these hadrons, as indeed they also governed vacuum correlation functions of hadron currents in these same channels. The only case in which a noticeable change is brought about by cooling is in the short distance behavior of the ground state of the pion. This is understandable since in the physical pion, in addition to instanton induced interactions, there is also a strong attractive hyperfine interaction arising from perturbative QCD which, combined with the $1/r$ interaction, gives rise to the central peak in the density. In the rho, the hyperfine interaction has much less effect, both because it is repulsive and is 3 times weaker.

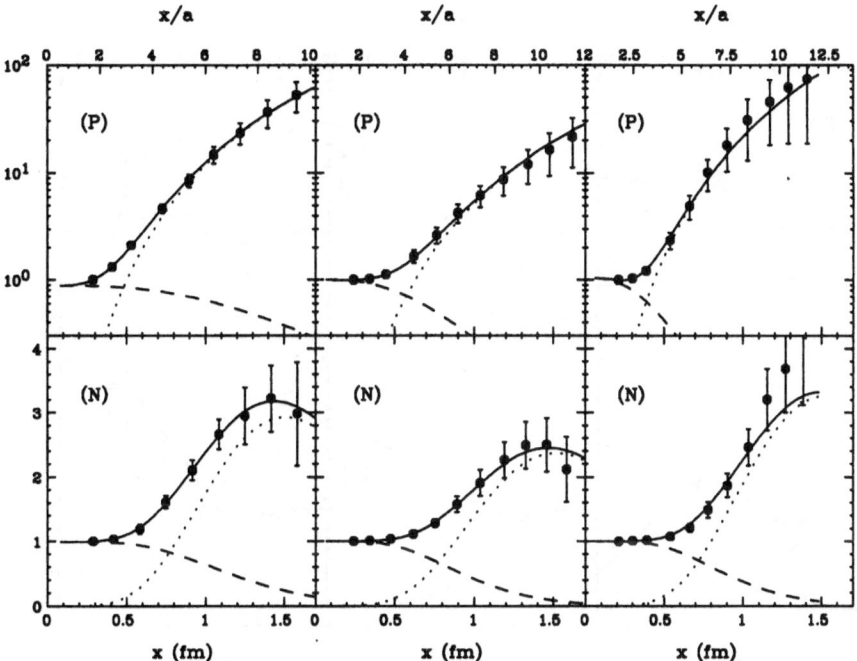

Fig. 3. Comparison of uncooled and cooled vacuum correlation function ratios, $\frac{R(x)}{R_0(x)}$, for pseudoscalar currents (P) and nucleon currents (N). The left, center, and right panels show results for uncooled QCD, 25 cooling steps, and 50 cooling steps respectively. The solid points with error bars denote lattice correlation functions extrapolated to $m_\pi = 140$ MeV. The solid lines denote fits to the correlation functions using a three-parameter spectral function, and the dashed and dotted curves show the contributions of the continuum and resonance components of the spectral functions respectively. The upper scale shows the spatial separation in lattice units and the lower scale shows the separation in physical units.

It is also noteworthy that the cooled density-density correlation functions shown in Fig. 5 for the π, ρ, and nucleon are comparable (within error bars), strongly suggesting that instantons set the overall spatial scale of these hadrons.

Conclusion and Discussion

In conclusion, the close agreement between hadronic observables with cooled and uncooled configurations provides strong evidence for the dominant role of instantons in determining hadron structure and quark propagation in the QCD vacuum.

I should emphasize that the physical picture I have presented of light quarks hopping between zero modes of size of the order of $\sim \frac{1}{3}$ fm is very different from the physical picture of heavy quarks interacting via an adiabatic potential. Because zero modes are so important for light quarks and the characteristic size

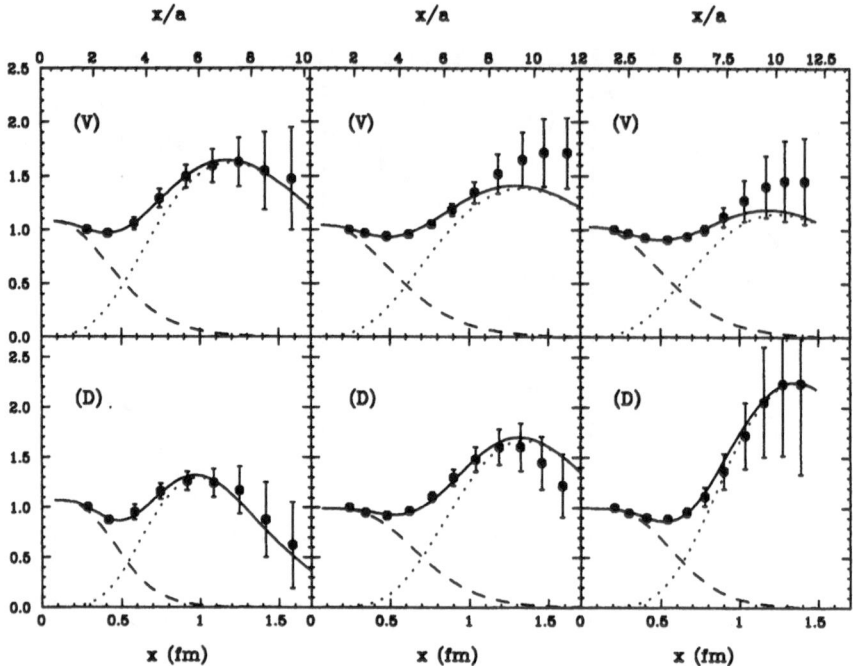

Fig. 4. Comparison of uncooled and cooled vacuum correlation function ratios, $\frac{R(x)}{R_0(x)}$, for vector currents (V) and Delta currents (D). The notation is the same as in Fig. 3.

of instantons is $\sim \frac{1}{3}$ fm, it is essential for light quark physics to work with a lattice spacing a small compared with $\frac{1}{3}$ fm to include the instantons and zero modes explicitly. In contrast, there is no corresponding argument for the relevance of instantons for the heavy quark interaction potential. Hence, although $\frac{1}{3}$ fm instantons cannot be described on the lattice with a ~ 0.5 fm used by Alford, Dimm and Lepage [1], there is no reason why with the inclusion of higher order operators the gluonic excitations corresponding to the $\frac{1}{r}$ potential, the hyperfine interaction, and confinement cannot be adequately described. A strong implication of our results, however, is that an analogous attempt to describe light quark physics on a 0.5 fm mesh cannot describe the essential physics. One simply cannot expect a set of perturbative corrections to describe the nonperturbative instanton physics which is excluded from such a coarse mesh. This distinction is far from academic, since it implies that very large scale calculations on meshes of the order $32^3 \times 64$ will be essential to understand the physics of light hadrons.

Finally, since the present calculations were performed in the quenched approximation, which omits the contributions of quark-antiquark excitations, I should point out two significant limitations of this work. Clearly, when nearly-zero modes are playing an important role in quark propagation, it is also important to include the small weights arising from the small eigenvalues in the fermion

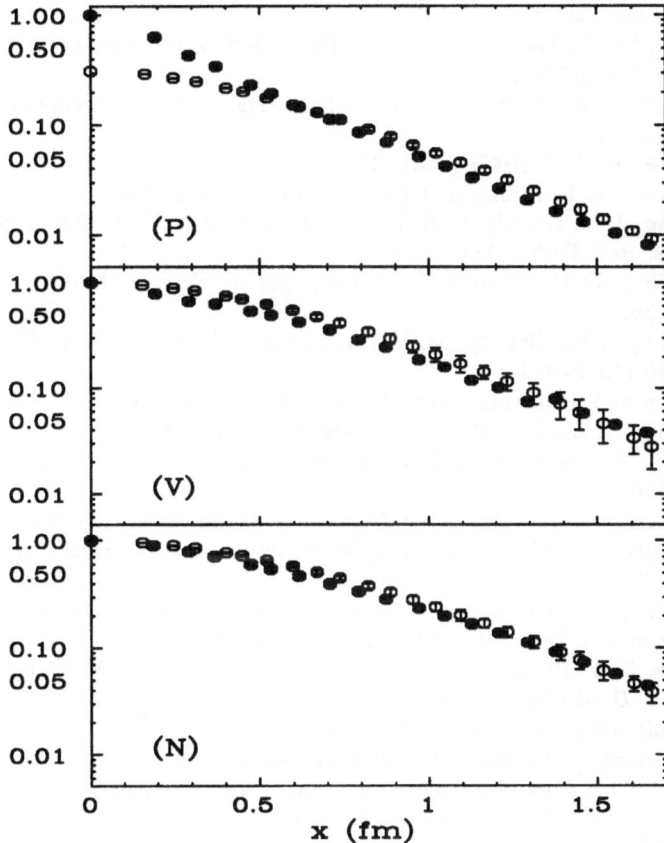

Fig. 5. Comparison of uncooled and cooled density-density correlation functions for the pion, rho, and nucleon. The solid circles denote the correlation functions calculated with uncooled QCD, and the open circles with error bars show the results for 25 cooling steps. The rho and pion results are compared for $m_\pi^2 = 0.16$ GeV^2 and the nucleon results are compared for $m_\pi^2 = .36$ GeV^2. As in Figs. 3 and 4, the separation is shown in physical units. All correlation functions are normalized to 1 at the origin except for the cooled pion correlator, which is normalized to have the same volume integral as the uncooled correlator.

determinant. In addition, in studying the instanton content of the vacuum, it is important to include fermion feedback so that, for example, the tendency of $q\bar{q}$ pairs to bind instanton — anti-instanton pairs is included. Hence, the next step in studying the role of instantons in hadron structure is obviously to include the effect of dynamical fermions in full unquenched lattice calculations.

References

1. M. Alford, W. Dimm and G. P. Lepage, Proceedings of Lattice '94, *Nucl. Physics* **B** (Proc. Suppl.) in press (1994).

2. Altmeyer *et al.* hep-lat/9311012.

3. K. Barad, M. Ogilvie and C. Rebbi, *Phys. Lett.* **B143** (1984) 222; *Ann. Phys.* (NY) **168** (1986) 284 .

4. A. A. Belavin, A. M. Polyakov, A. P. Schwartz, and Y. S. Tyupkin, *Phys. Lett.* **59B** (1975) 85.

5. B. Berg, *Phys. Lett.* **104B** (1981) 475.

6. N. Cabibbo and E. Marinari, *Phys. Lett.* **119B** (1982) 387.

7. M.-C. Chu, J. M. Grandy, S. Huang and J. W. Negele, *Phys. Rev. Lett.* **70** (1993) 225; *Phys. Rev.* **D48** (1993) 3340.

8. M. C. Chu, J. M. Grandy, S. Huang, and J. W. Negele *Phys. Rev.* **D 49** (1994) 6039.

9. M.-C. Chu, J. M. Grandy, M. Lissia, and J. W. Negele, *Nucl. Phys.* **B** (Proc. Suppl.) **26** (1992) 412.

10. M.-C. Chu and S. Huang, *Phys. Rev.* **D45** (1992) 2446.

11. M.-C. Chu, M. Lissia, and J. W. Negele, *Nucl. Phys.* **B360** (1991) 31.

12. D. Daniel, R. Gupta, G. Kilcup, A. Patel, and S. Sharpe, *Phys. Rev.* **D 46** (1992) 3130.

13. D. I. Dyakanov and V. Yu Petrov, *Nucl. Phys.* **B245** (1984) 259; **B272** (1986) 457.

14. M. Fukugita, N. Ishizuka, H. Mino, M. Okawa, and A. Ukawa, *Phys. Rev.* **D 47** (1993) 4739.

15. M. Fukugita, Y. Kuramashi, H. Mino, M. Okawa, and A. Ukawa, hep-lat/9407012.

16. J. Gasser and H. Leutwyler, *Phys. Rep.* **87** (1982) 77.

17. R. Gupta, C. F. Baillie, R. G. Brickner, G. W. Kilcup, A. Patel, and S. Sharpe, *Phys. Rev.* **D 44** (1991) 3272.

18. G. 't Hooft, *Phys. Rev.* **14D** (1976) 3432.

19. Y. Kuramashi, M. Fukugita, H. Mino, M. Okawa, and A. Ukawa, *Phys. Rev. Lett.* **71** (1993) 2387; *Nucl. Phys.* **B** (Proc. Suppl.) **34** (1994) 117.

20. M. Lissia, M.-C. Chu, J. W. Negele and J. M. Grandy, *Nucl. Phys.* **A555** (1993) 272.

21. M. Lücher, *Comm. Math Phys.* **105** (1986) 153.

22. L. Maiani and G. Martinelli, *Phys. Lett.* **B178** (1986) 265.

23. S. Myint and C. Rebbi, hep-lat/9401009, *Nucl. Phys.* **B** (in press).

24. J. W. Negele and H. Orland, *Quantum Many-Particle Systems*, Addison–Wesley (1987).

25. M. I. Polikarpov and A. I. Veselov, *Nucl. Phys.* **B297** (1988) 34.

26. S. Sharpe, *Nucl. Phys.* **B** (Proc. Suppl.) **20** (1991) 406.

27. E. V. Shuryak, *Nucl. Phys.* **B203** (1982) 93; **B203** (1982) 116; **B203** (1982) 140; **B302** (1988) 559; **B302** (1988) 599; **B319** (1989) 521; **B319** (1989) 541; **B328** (1989) 85; **B328** (1989) 102.

28. E. V. Shuryak, *Rev. Mod. Phys.* **65** (1993) 1; and *Nucl. Phys.* **B** (Proc. Suppl.) **34** (1994) 107.

29. E. V. Shuryak and J. J. M. Verbaarschot, Stony Brook Preprint SUNY-NTG-92/40; T. Schäfer, E. V. Shuryak, and J. J. M. Verbaarschot, Stony Brook Preprint SUNY-NTG-92/45.

30. K. Teo and J. W. Negele, *Nucl. Phys.* **B** (Proc. Suppl.) **34** (1994) 390.

31. M. Teper, *Nucl. Phys.* **B** (Proc. Suppl.) **20** (1991) 159.

32. D. Weingarten, *Nucl. Phys.* **B** (Proc. Suppl.) **34** (1994) 29.

Part III: $\pi - \pi$ Interaction and the Chiral Anomaly

Summary of π–π Scattering Experiments

Dinko Počanić

Department of Physics, University of Virginia
Charlottesville, VA 22901-2458, USA

Abstract: The $\pi\pi$ scattering amplitude at threshold is fully determined by the chiral symmetry breaking part of the strong interaction, and, thus, directly constrains the form of the low energy effective lagrangians. Current status of the study of the low energy $\pi\pi$ interaction is discussed, particularly the recent results on reactions $\pi N \to \pi\pi N$ near threshold. Present levels of experimental uncertainties and limitations inherent to the available analysis methods leave ample room for improvements in the determination of the s-wave $\pi\pi$ scattering lengths. Experimental improvements are expected from new measurements of K_{e4} decays and from attempts to study $\pi^+\pi^-$ atoms, while further theoretical work is required in order to make full use of the extensive near-threshold $\pi N \to \pi\pi N$ data that has recently become available.

1 Introduction

Effects of chiral symmetry breaking in low energy interactions of pions, and light hadrons in general, have been studied for over thirty years. Strong interactions break chiral symmetry both "spontaneously" and explicitly. Spontaneous breaking of chiral symmetry is well understood and has led to new concepts and methods that have transcended the domain of intermediate energy physics. On the other hand, the precise mechanism of explicit chiral symmetry breaking (ChSB) remained largely unresolved until the establishment of QCD as the correct theory of strong interactions. The study of explicit ChSB, however, remains relevant even today, owing to the failure of the full QCD to describe strong interaction phenomena at energies below a few GeV. At these energies, QCD becomes nonperturbative and intractable in practice by available calculational methods. In order to overcome this problem, a broad theoretical effort is under way to develop phenomenological lagrangian models based solely on the symmetry properties of the full QCD, as suggested by Weinberg [1]. Chiral symmetry plays a particularly important role at low energies, since it is violated only slightly in the SU(2) realization, and is essential for the understanding of the lightest hadrons. For this reason, chiral symmetry provides either the basic framework or the essential constraints for all modern effective lagrangian models at low energies, such as the chiral perturbation theory (ChPT) [2], and the various realizations of the Nambu–Jona-Lasinio model [3].

It has been long established [4] that low energy $\pi\pi$ scattering provides a particularly sensitive tool in studying the explicit breaking of chiral symmetry, since $a(\pi\pi)$, the $\pi\pi$ scattering lengths, vanish exactly in the chiral limit. To the extent that they differ from zero, $a(\pi\pi)$ provide a direct measure of the symmetry breaking term in the pion sector. Stated in more contemporary language, detailed knowledge of the $\pi\pi$ interaction scattering lengths and phase shifts provides much needed input in fixing the parameters of ChPT and other effective models.

2 Experimental Determination of $\pi\pi$ Scattering Lengths

Experimental evaluation of $\pi\pi$ scattering observables is difficult, primarily because free pion targets are not available. Scattering lengths are especially hard to determine since they require measurements close to the $\pi\pi$ threshold, where the available phase space strongly limits measurement rates. Over the years several reactions have been studied or proposed as a means to obtain near-threshold $\pi\pi$ phase shifts, such as $\pi N \rightarrow \pi\pi N$, K_{e4} decays, $\pi^+\pi^-$ atoms, $e^+e^- \rightarrow \pi\pi$, etc. In practice, only the first two reactions have so far proven useful in studying threshold $\pi\pi$ scattering. The three main methods are discussed below in the order of decreasing reliability.

2.1 Analysis of $K \rightarrow \pi^+\pi^-e^+\nu$ decays

The $K^+ \rightarrow \pi^+\pi^-e^+\nu$ decay (called K_{e4}) is in several respects the most suitable process for the study of near-threshold $\pi\pi$ interactions. The interaction takes place between two real pions on the mass shell, the only hadrons in the final state. The dipion invariant mass distribution in K_{e4} decay peaks close to the $\pi\pi$ threshold, and $l = I = 0$ and $l = I = 1$ are the only dipion quantum states contributing to the process. These factors, and the well understood $V - A$ nature of the weak decay, favor the K_{e4} process among all others in terms of theoretical uncertainties. On the other hand, measurements are impeded by the low branching ratio of the decay, $\sim 3.9 \times 10^{-5}$.

The most recent K_{e4} measurement was made by a Geneva–Saclay collaboration in the mid-1970's [5]. Using a detector system consisting of pion Čerenkov counters, wire chambers, a bending magnet and plastic scintillator hodoscopes around a 4 m decay zone, Rosselet and coworkers detected some 30,000 K_{e4} decays. Analysis of this low-background, high-statistics data illustrates well the difficulties encountered in extracting experimental $\pi\pi$ scattering lengths. Figure 1. summarizes the $\pi\pi$ phase shift information below 400 MeV derived from all K_{e4} measurements to date. The curves in Fig. 1 correspond to three different values of $a_0^0(\pi\pi)$, and illustrate the relative insensitivity of the data at the present level of experimental accuracy.

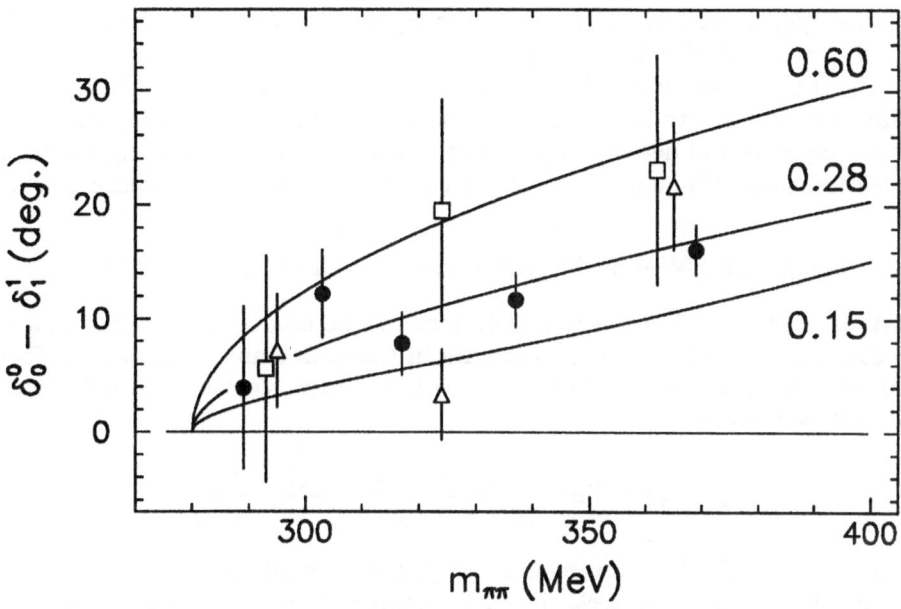

Fig. 1. $\pi\pi$ phase shift information extracted from studies of K_{e4} decays, after Rosselet *et al.* [5]. Phase-shift difference $\delta_0^0 - \delta_1^1$ is plotted against $m_{\pi\pi}$, the dipion invariant mass. Full circles represent results of Rosselet *et al.* [5], while open squares and triangles represent the results of Zylberstejn [41] and Beier *et al.* [42], respectively. The three curves correspond to phase shift solutions assuming three different values of a_0^0, as noted.

By itself, the Geneva–Saclay experiment determines a_0^0 with a $\sim$35% uncertainty, and constrains b:

$$a_0^0 = 0.31 \pm 0.11 \ \mu^{-1} \quad , \qquad b = b_0^0 - a_1^1 = 0.11 \pm 0.16 \ \mu^{-1} \quad ,$$

where b_0^0 is the s-wave $I = 0$ slope parameter defined in the usual way by

$$\frac{\mathrm{Re} A_l^I}{q^{2l}} = a_l^I + b_l^I q^2 + O(q^4) \quad , \qquad \text{with} \qquad q = \frac{1}{2}\sqrt{s - 4\mu^2} \quad , \tag{1}$$

where A_l^I is the $\pi\pi$ partial wave amplitude, s is the center of mass energy of the two pions, and $\mu \equiv m_\pi$ is the pion mass.

$\pi\pi$ scattering phase shifts are, however, further constrained by unitarity, analyticity, crossing and Bose symmetry, extensively studied by Roy [6], and Basdevant *et al.* [7,8]. These constraints are expressed in a set of dispersion relations known as the "Roy equations", which have been evaluated on the basis of existing peripheral $\pi N \to \pi\pi N$ data (see Sect. 2.2). By applying the Roy equation constraints of Basdevant, *et al.* [8], and thus combining the K_{e4} and peripheral pion production results, more accurate values for a_0^0 and b_0^0 were obtained [5], as well as values of scattering lengths and slope parameters for $(l = 0, I = 2)$ and $(l = 1, I = 1)$ [9].

The present experimental accuracy of the K_{e4} measurement of the $\pi\pi$ phase shifts clearly needs to be improved. It is also evident that an independent accurate experimental determination of a_0^2, the $I = 2$ s-wave scattering length, is called for, since this quantity is not directly constrained by K_{e4} measurements.

2.2 Peripheral $\pi N \to \pi\pi N$ Reactions: the Chew–Low Method

Particle production in peripheral collisions can be used to extract information on the scattering of two of the particles in the final state, as shown by Chew and Low in 1959 [10]. Applied to the $\pi N \to \pi\pi N$ reaction, the well-known Chew–Low formula,

$$\sigma_{\pi\pi}(m_{\pi\pi}) = \lim_{t \to \mu^2} \left[\frac{\partial^2 \sigma_{\pi\pi N}}{\partial t \partial m_{\pi\pi}} \cdot \frac{\pi}{\alpha f_\pi^2} \cdot \frac{p^2(t - \mu^2)^2}{t m_{\pi\pi} k} \right] \quad , \tag{2}$$

relates $\sigma_{\pi\pi}(m_{\pi\pi})$, the cross section for pion-pion scattering, to double differential $\pi N \to \pi\pi N$ cross section and kinematical factors: p, momentum of the incident pion, $m_{\pi\pi}$, the dipion invariant mass, t, the Mandelstam square of the 4-momentum transfer to the nucleon, $k = (m_{\pi\pi}^2/4 - \mu^2)^{1/2}$, momentum of the secondary pion in the rest frame of the dipion, $f_\pi \approx 93$ MeV, the pion decay constant, and $\alpha = 1$ or 2, a statistical factor involving the pion and nucleon charge states. The method relies on an accurate extrapolation of the double differential cross section to the pion pole in order to isolate the one pion exchange (OPE) pole term contribution. Since the exchanged pion is off-shell in the physical region $(t < 0)$, this method requires measurements under conditions which maximize the OPE contribution and minimize all background contributions. Thus, suitable measurements require peripheral pion production at values of t as close to zero as possible, which become available at incident momenta typically above ~ 3 GeV/c.

The Chew–Low method has been refined considerably over time, particularly by Baton and coworkers [11], whose approach enables extraction of $\pi\pi$ phase shifts through appropriate treatment of the angular dependence of the $\pi N \to \pi\pi N$ exclusive cross sections. Crossing, Bose and isospin symmetries, analyticity and unitarity, provide dispersion relation constraints on the $\pi\pi$ phase shifts, the "Roy equations" [6-8]. These constraints are particularly useful in evaluating $\pi\pi$ scattering lengths because available phase space restricts severely the statistics of peripheral $\pi N \to \pi\pi N$ measurements below $m_{\pi\pi} \approx 500$ MeV, while accurate data are available at higher $\pi\pi$ energies.

The data base for these analyses has essentially not changed since the early 1970's, and is dominated by two experiments, performed by the Berkeley [12] and CERN-Munich [13] groups. The latter of the two measurements has much higher statistics (300 k events compared to 32 k in the Berkeley experiment). These data were subsequently analyzed by numerous authors, too many to review here; ultimately, the peripheral pion production results were combined with the

Geneva–Saclay K_{e4} data in a comprehensive dispersion-relation analysis [9], as discussed in Sect. 2.1.

There have been independent Chew–Low type analyses since the 1970's; the last published one, performed by the Kurchatov Institute group in 1982, was based on a set of some 35,000 events recorded in bubble chambers [14]. The same group has recently updated their analysis [15].

2.3 $\pi N \rightarrow \pi\pi N$ Reactions near Threshold

Early on, Weinberg pointed out that the OPE graph dominates the $\pi N \rightarrow \pi\pi N$ reaction at threshold [4]. Subsequently, Olsson and Turner constructed a soft-pion lagrangian containing only the OPE and contact terms at threshold [16], and derived a straightforward relation between the $\pi\pi$ and $\pi N \rightarrow \pi\pi N$ threshold amplitudes with only one parameter, ξ, the chiral symmetry breaking parameter. Thus, in Olsson–Turner's model, it is sufficient to measure total $\pi N \rightarrow \pi\pi N$ cross sections, from which quasi-amplitudes can be calculated and extrapolated to threshold. In spite of recent strong criticism for incompatibility with QCD and oversimplified dynamical assumptions, Olsson and Turner's work to date provides the sole direct relation between $\pi N \rightarrow \pi\pi N$ observables and the $\pi\pi$ lagrangian. In this way, soft-pion theory has provided the main inspiration for the near-threshold $\pi N \rightarrow \pi\pi N$ measurements and, in spite of its shortcomings, is still being used by experimentalists to relate the results from different reaction channels in a systematic way.

Unlike the methods described in Sects. 2.1 and 2.2, the last ten years have witnessed a great deal of experimental activity on exclusive and inclusive near-threshold $\pi N \rightarrow \pi\pi N$ measurements. As in Chew–Low peripheral pion production, there are 5 charge channels accessible to measurement,

$$\pi^- p \rightarrow \pi^- \pi^+ n \quad , \qquad \pi^- p \rightarrow \pi^0 \pi^0 n \quad , \qquad \pi^- p \rightarrow \pi^- \pi^0 p \quad ,$$

$$\pi^+ p \rightarrow \pi^+ \pi^0 p \quad , \qquad \text{and} \qquad \pi^+ p \rightarrow \pi^+ \pi^+ n \quad .$$

For brevity, we label them with their final state charges as $(-+n)$, ..., $(++n)$, respectively. Recent experiments reporting total $\pi N \rightarrow \pi\pi N$ cross sections are summarized below, while data available before 1984 is reviewed in Ref. [17].

The OMICRON group at CERN has measured cross sections in the $(+-n)$, $(-0p)$, $(++n)$ and $(+0p)$ channels [18]. They detected coincident pairs of charged particles in a two-sided magnetic spectrometer, restricted to in-plane kinematics. (Limitations inherent in the extraction of total cross sections from data restricted to in-plane kinematics were recently discussed by the Erlangen group [19].) The thin gas target, limited magnetic spectrometer acceptance, and background subtraction, result in large error bars for some of the low-energy OMICRON data points, particularly in the $(++n)$ channel.

At TRIUMF, Sevior and coworkers measured inclusive cross sections for the reaction $(++n)$ using a novel technique involving an active plastic scintillator target combined with neutron detection [20]. Total cross sections were evaluated assuming s-wave dominance due to the proximity of the threshold (5 MeV below their lowest energy measurement). It is significant that the TRIUMF results disagree with the (++n) OMICRON data.

J. Lowe and coworkers measured the (00n) channel at Brookhaven using the Crystal Box detector [21]. Due to the large solid-angle coverage of the detector, this was a kinematically complete measurement of 4 photons following the decays of the two π^0's in the final state. As all particles in the final state are neutral, the lowest point was also about 5 MeV above the threshold.

Finally, a Virginia-Stanford-LAMPF team studied the (+0p) channel using the LAMPF π^0 spectrometer and an array of plastic scintillation telescopes for π^+ and p detection [22,23]. Three classes of exclusive events were recorded simultaneously: $\pi^+\pi^0$ and $\pi^0 p$ double coincidences, and $\pi^+\pi^0 p$ triple coincidences. Since the acceptance of the apparatus and the backgrounds were significantly different for the three classes of events, this experiment had a strong built-in consistency check.

Published total cross sections for all five channels are summarized in Fig. 2., shown in the form of quasi-amplitudes, $|a(\pi\pi N)|$, extracted from the total cross sections following the prescription of Olsson and Turner [16]

$$\sigma(\pi N \rightarrow \pi\pi N) = |a(\pi\pi N)|^2 \cdot \alpha \cdot p_\pi^2 \times \text{phase space} \quad , \qquad (3)$$

where p_π is the c.m. incident pion momentum, and the statistical factor $\alpha = 1/2$ for the (++n) and (00n) channels, and $\alpha = 1$ for the other three channels.

Apart from the pronounced disagreement in the (++n) channel between the TRIUMF [20] and OMICRON [18] data, represented in Fig. 2. by full circles and full triangles, respectively, the entire body of data appears globally consistent within the quoted uncertainties. The straight lines drawn through the data in Fig. 2 are the result of a constrained soft-pion analysis following Olsson and Turner performed by the Virginia group, with the result of $\xi = -0.25 \pm 0.10$ [22,23]. A similar global fit was performed earlier by Burkhardt and Lowe using a slightly different fitting procedure, yielding $\xi = -0.60 \pm 0.10$ [24]. In both analyses the quoted uncertainty of ξ is determined only by the experimental errors and the statistical quality of the global fit. We can interpret the spread between the two values as due to the systematic uncertainties of the method, and take the mean as a representative soft-pion analysis result.

The body of near-threshold $\pi N \rightarrow \pi\pi N$ data keeps growing. Several experiments presently under way at TRIUMF and LAMPF are expected to yield new results shortly, on both exclusive and inclusive cross sections in the (+−n), (++n), (+0p) and (−0p) channels. We also note the high-statistics angular correlation measurements in the (+−n) channel performed by the Erlangen group, who did not report total cross sections [25].

Closer analysis of the exclusive cross sections is only now beginning, as high-statistics data have not been available until recently, and the methods of analysis are still being refined. In a series of recent papers, the Erlangen group has focused on the main graphs contributing to the continuum $\pi N \rightarrow \pi\pi N$ amplitude [26]. On the other hand, the St. Petersburg group [27] has derived the most general

GLOBAL FIT OF $\pi\pi$N THRESHOLD AMPLITUDES

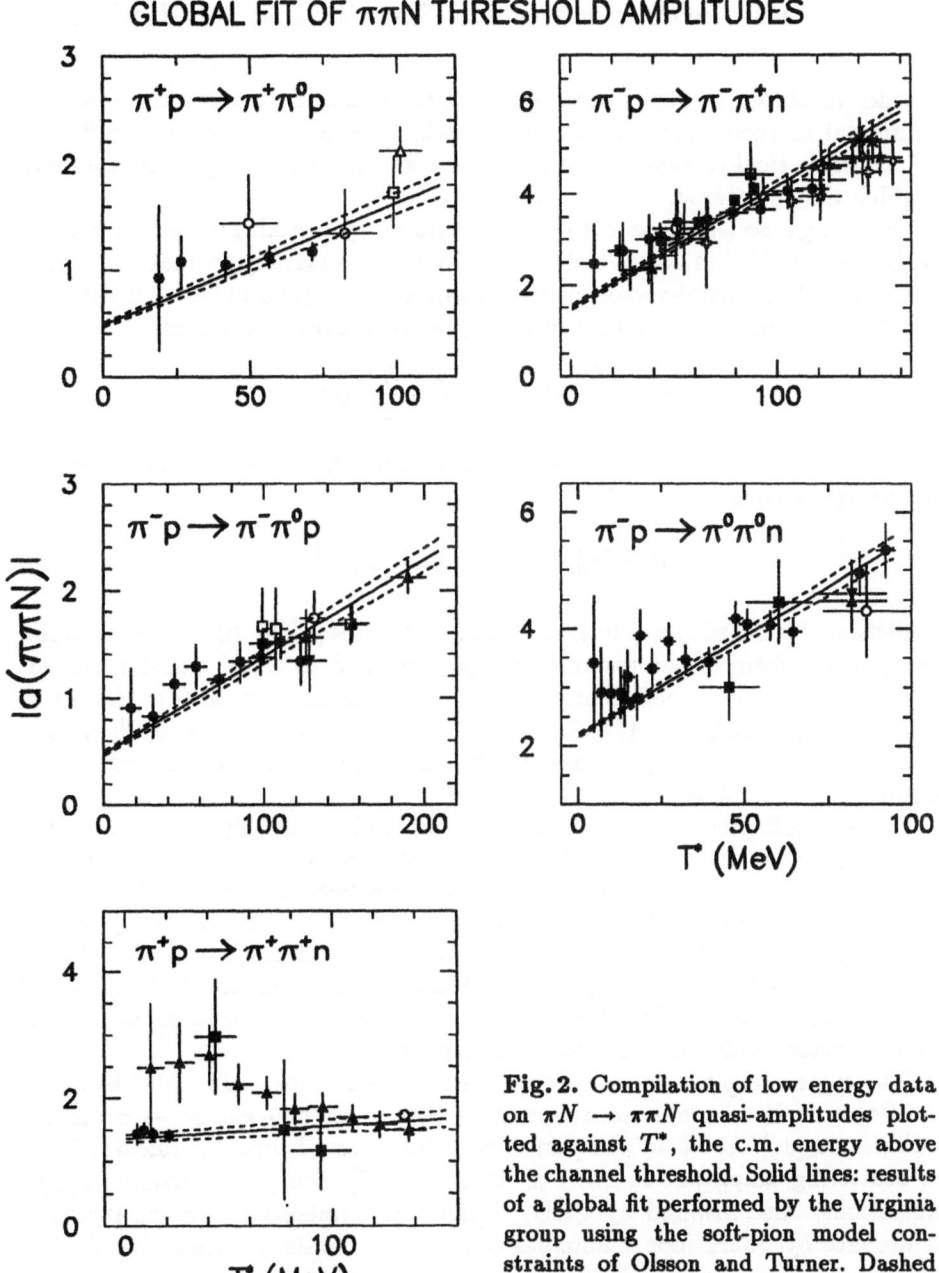

Fig. 2. Compilation of low energy data on $\pi N \rightarrow \pi\pi N$ quasi-amplitudes plotted against T^*, the c.m. energy above the channel threshold. Solid lines: results of a global fit performed by the Virginia group using the soft-pion model constraints of Olsson and Turner. Dashed lines correspond to fits with $\chi^2_{\min} + 1$.

constraints on the kinematical dependence of the background and the OPE amplitudes in the physical region, as dictated by basic symmetries: crossing, Bose, isospin, analyticity and unitarity.

3 Comparison of Experimental Results and Predictions

In order to obtain a proper perspective on the existing data and the three experimental methods discussed in the preceding sections, we review briefly the various theoretical calculations of the s-wave $\pi\pi$ scattering lengths, in the order in which they appeared.

Weinberg's soft-pion model of chiral symmetry breaking relied on current algebra and PCAC [4,28]. Weinberg required of his lagrangian that $\partial^\mu A_\mu$, the divergence of the axial-vector current, form a chiral quadruplet with the pion field. Consequently, the $\pi\pi$ part of the lagrangian assumes the form:

$$\mathcal{L}_{\pi\pi} = -\frac{1}{4f_\pi^2} \cdot [\phi^2(\partial_\mu\phi)^2 - \frac{1}{2}\mu^2(\phi^2)^2] \quad , \tag{4}$$

where ϕ is the pion field. Weinberg also noted that the s-wave scattering lengths a_0^0 and a_0^2 are constrained linearly:

$$2a_0^0 - 5a_0^2 = 6 \frac{\mu}{8\pi f_\pi^2} \approx 0.56 \; \mu^{-1} \quad . \tag{5}$$

Predictions by Schwinger [29] and Chang and Gürsey [30] differed from Weinberg's in the form of the response of the pion field to chiral transformations, resulting in different coefficients of the ϕ^4 pion mass term in (4). Instead of the Weinberg's coefficient 1/2, Schwinger's model suggests 1/4, and Chang and Gürsey's 1/3, respectively. Consequently, calculated values for the $\pi\pi$ s-wave scattering lengths differ.

Since in soft-pion theory (5) constrains a_0^0 and a_0^2 linearly, different models need only to fix the ratio a_0^2/a_0^0, i.e., only one degree of freedom remains. In this respect, Olsson and Turner's parameter ξ (see Sect. 2.3 and Ref. [16]) determines the magnitude of the ϕ^4 pion mass term in (4), and, consequently, the $\pi\pi$ scattering lengths. Thus, Weinberg's prediction is equivalent to setting $\xi = 0$, Schwinger's and Chang–Gürsey's to $\xi = 1$ and 2/3, respectively. Although QCD has confirmed Weinberg's choice as the correct leading-order term, we include the two latter results for historical completeness.

The current-algebra calculation of Weinberg was improved in 1982 by Jacob and Scadron who introduced a correction due to the non-soft $S^* \to \pi\pi$ isobar background [31]. At about the same time, Gasser and Leutwyler calculated the $\pi\pi$ scattering amplitude to order p^4 in ChPT [32], and gave scattering length predictions. Also inspired by QCD, Ivanov and Troitskaya used the model of dominance by quark loop anomalies (QLAD) to obtain $\pi\pi$ scattering lengths [33]. On the other hand, the Jülich group constructed a dynamical model of pseudoscalar-pseudoscalar meson scattering based on meson exchange, and applied it to calculate a number of $\pi\pi$ and $K\pi$ scattering observables at low and intermediate energies [34]. A somewhat complementary approach to ChPT is the Nambu–Jona-Lasinio model [3]. Calculations of $\pi\pi$ scattering lengths within the SU(2)×SU(2) and SU(3)×SU(3) realizations of the NJL model are found in Refs. [35] and [36], respectively.

Table 1. The s-wave $\pi\pi$ scattering lengths: compilation of theoretical model predictions and of experimental analysis results. The fourth column lists the chiral symmetry breaking "offset" $2a_0^0 - 5a_0^2$ defined in Eq. (5). Uncertainties were not entered for theoretical predictions, as they are normally not quoted by authors. For the few calculations that did estimate uncertainties, they ranged from ~ 5 to $\sim 10\%$. All values are listed in units of inverse pion mass, μ^{-1}.

Model/Method	a_0^0 (μ^{-1})	a_0^2 (μ^{-1})	$2a_0^0 - 5a_0^2$	Ref.
(a) Theoretical model predictions				
Weinberg	0.16	-0.046	0.56	[4]
Schwinger	0.10	-0.069	0.56	[29]
Chang-Gürsey	0.12	-0.062	0.56	[30]
Jacob-Scadron	0.20	-0.029	0.56	[31]
ChPT	0.20	-0.042	0.61	[32]
QLAD	0.21	-0.060	0.72	[33]
Meson exch'ge	0.31	-0.027	0.76	[34]
NJL – SU(2)	0.22	-0.074	0.81	[35]
NJL – SU(3)	0.26	-0.062	0.83	[36]
Lattice QCD	0.22	-0.042	0.65	[37]
GCM (fit 1)	0.16	-0.042	0.53	[38]
GCM (calc.)	0.17	-0.048	0.58	[38]
(b) Results of experimental analyses				
K_{e4} + Roy eq.	0.26 ± 0.05	-0.028 ± 0.012	0.66 ± 0.12	[5,9]
Chew–Low PSA	0.24 ± 0.03	-0.04 ± 0.04	0.68 ± 0.21	[14]
Soft-pion/O-T	0.188 ± 0.016	-0.037 ± 0.006	0.56 ± 0.04	[24,22]

Most recently, Kuramashi and coworkers successfully applied quenched lattice QCD on a $12^3 \times 20$ lattice, and obtained $I = 0$ and 2 $\pi\pi$ scattering length values in the neighborhood of the older current-algebra calculations [37]. Finally, Roberts *et al.* have recently developed a model field theory, referred to as the global color-symmetry model (GCM), in which the interaction between quarks is mediated by dressed vector boson exchange [38]. The model incorporates dynamical chiral symmetry breaking, asymptotic freedom, and quark confinement, and was applied to calculate a number of low-energy observables in the pion sector of QCD.

Table 1. summarizes quantitatively the theoretical model predictions and the experimental determinations of the s-wave $\pi\pi$ scattering lengths. The same quantities are also displayed in Fig. 3.

Considerable scatter of predicted values of a_0^0 and a_0^2 is evident in Fig. 3. Even disregarding the 1960's calculations of Schwinger [29] and Chang and Gürsey [30] which were superseded by QCD, a significant range of predicted values remains. However, in view of the present experimental uncertainties (see Fig. 3.), most authors claim that their results are supported by the data. This is not a

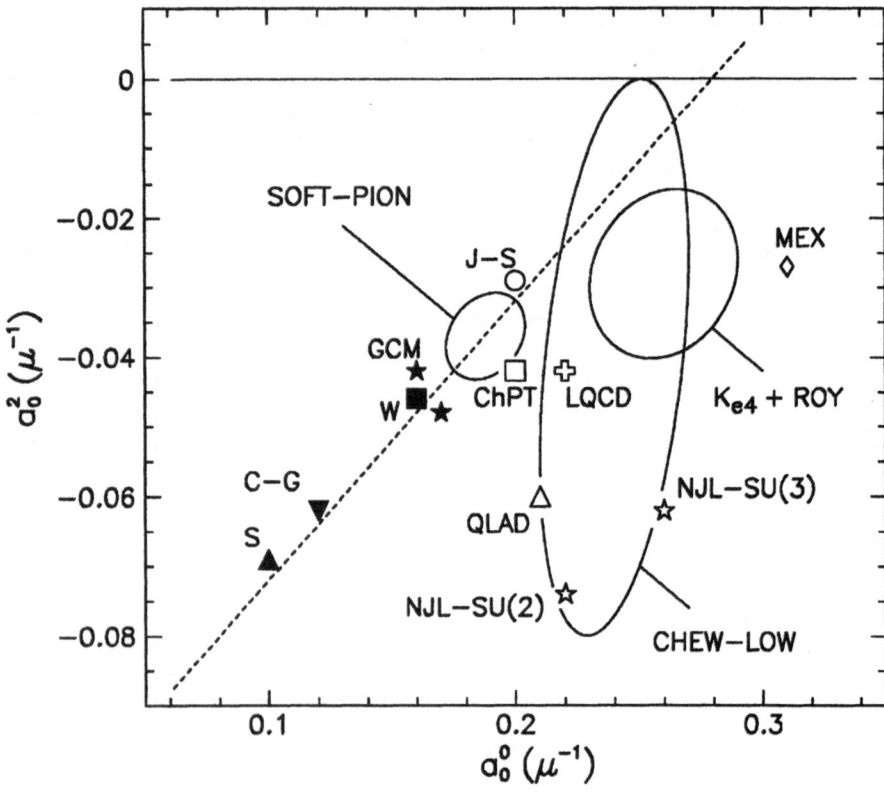

Fig. 3. Summary of the $\pi\pi$ scattering length predictions (symbols) and experimental results (contour limits). Dashed line: Weinberg's constraint given in Eq. (5). Numerical values of $\pi\pi$ scattering lengths, experimental limits, and corresponding references are listed in Table 1. Model calculations: Weinberg (full square), Schwinger (filled triangle), Chang and Gürsey (filled inverted triangle), Jacob and Scadron (open circle), Gasser and Leutwyler – ChPT (open square), Ivanov and Troitskaya – QLAD (open triangle), Lohse *et al.* – Meson Exchange (open rhomb), Ruivo *et al.* and Bernard *et al.* – NJL (open stars), Kuramashi *et al.* – quenched lattice gauge QCD (open cross), Roberts *et al.* – GCM (filled stars).

satisfactory state of affairs. Progress is required on two fronts in order to improve the present uncertainties in the experimentally derived $\pi\pi$ scattering lengths.

First, more accurate data on K_{e4} decays are needed, in order to reduce the current error limits in the analysis. However, K_{e4} data alone will not suffice because of their insensitivity to the $I = 2$ $\pi\pi$ channel. Measurements of the $\pi^+\pi^-$ atom proposed at several laboratories, if feasible with reasonable statistics, could provide the much needed additional theoretically unambiguous information. The quantity to be measured is the decay rate of the $\pi^+\pi^-$ ground state into the $\pi^0\pi^0$ channel, which is proportional to $|a_0^0 - a_0^2|^2$ [39].

Second, the mounting volume and accuracy of the exclusive and inclusive near-threshold $\pi N \to \pi\pi N$ data contain, in principle, valuable information regarding threshold $\pi\pi$ scattering. Theoretical interpretation of this data requires much improvement in order to make full use of this information. The work of Bernard, Kaiser and Meißner [40] appears to be a promising step in that direction.

This work has been supported by a grant from the United States National Science Foundation.

References

1. S. Weinberg, Physica **96A** 327 (1979)
2. J. Gasser and H. Leutwyler, Ann. Phys. (N.Y.) **158** 142 (1984); Nucl. Phys. **B250** 465 (1985)
3. Y. Nambu and G. Jona-Lasinio, Phys. Rev. **122** 345 (1961); *ibid.* **124** 246 (1961); For a review of recent work based on the NJL model see, e.g., S. P. Klevansky, Rev. Mod. Phys. **64** 649 (1992)
4. S. Weinberg, Phys. Rev. Lett. **17** 616 (1966), *ibid.* **18** 188 (1967)
5. L. Rosselet, *et al.*, Phys. Rev. D **15** 574 (1977)
6. S. M. Roy, Phys. Lett. **35B** 353 (1971)
7. J. L. Basdevant, J. C. Le Guillou and H. Navelet, Nuovo Cim. **7A** 363 (1972)
8. J. L. Basdevant, C. G. Froggatt and J. L. Peterson, Nucl. Phys. **B72** 413 (1974)
9. M. M. Nagels, *et al.*, Nucl. Phys. **B147** 189 (1979)
10. G. F. Chew and F. E. Low, Phys. Rev. **113** 1640 (1959)
11. J. B. Baton, G. Laurens and J. Reignier, Phys. Lett. **33B** 525 (1970)
12. S. D. Protopopescu *et al.*, Phys. Rev. D **7** 1279 (1973)
13. G. Grayer *et al.*, Nucl. Phys. **B75** 189 (1974)
14. E. A. Alekseeva *et al.*, Zh. Eksp. Teor. Fiz. **82** 1007 (1982) [Sov. Phys. JETP **55** 591 (1982)]
15. O. O. Patarakin and V. N. Tikhonov, Kurchatov Institute of Atomic Energy preprint IAE-5629/2 (1993)
16. M. G. Olsson and L. Turner, Phys. Rev. Lett. **20** 1127 (1968); Phys. Rev. **181** 2141 (1969), L. Turner, Ph. D. Thesis, Univ. of Wisconsin, 1969 (unpublished)
17. D. M. Manley, Phys. Rev. D **30** 536 (1984)
18. G. Kernel, *et al.*, Phys. Lett. **B216** 244 (1989); *ibid.* **B225** 198 (1989); Z. Phys. C **48** 201 (1990); *ibid.* **51** 377 (1991); in *Particle Production Near Threshold* Nashville, 1990, edited by H. Nann and E. J. Stephenson, (AIP, New York, 1991)
19. H.-W. Ortner *et al.* Phys . Rev. C **47** R447 (1993)
20. M. E. Sevior, *et al.* Phys. Rev. Lett. **66** 2569 (1991)
21. J. Lowe, *et al.* Phys. Rev. C **44** 956 (1991)
22. D. Počanić, *et al.* Phys. Rev. Lett. **72** 1156 (1994)
23. E. Frlež, Ph. D. Thesis, Univ. of Virginia, 1993 (Los Alamos Report LA-12663-T, 1993)
24. H. Burkhardt and J. Lowe, Phys. Rev. Lett. **67** 2622 (1991)
25. H.-W. Ortner *et al.*, Phys. Rev. Lett. **64** 2759 (1990); R. Müller *et al.*, Phys. Rev. C **48** 981 (1993)

26. O. Jäkel, H.-W. Ortner, M. Dillig and C. A. Z. Vasconcellos, Nucl. Phys. **A511** 733 (1990); O. Jäkel, M. Dillig and C. A. Z. Vasconcellos, *ibid.* **A541** 675 (1992); O. Jäkel and M. Dillig, *ibid.* **A561** 557 (1993)

27. A. A. Bolokhov, V. V. Vereshchagin and S. G. Sherman, Nucl. Phys. **A530** 660 (1991)

28. S. Weinberg, Phys. Rev. **166** 1568 (1968)

29. J. Schwinger, Phys. Lett. **24B** 473 (1967)

30. P. Chang and F. Gürsey, Phys. Rev. **164** 1752 (1967)

31. R. Jacob and M. D. Scadron, Phys. Rev. D **25** 3073 (1982)

32. J. Gasser and H. Leutwyler, Phys. Lett. **125B** 325 (1983)

33. A. N. Ivanov and N. I. Troitskaya, Yad. Fiz. **43** 405 (1986) [Sov. J. Nucl. Phys. **43** 260 (1986)]

34. D. Lohse, J. W. Durso, K. Holinde and J. Speth, Nucl. Phys. **A516** 513 (1990)

35 M. C. Ruivo, C. A. de Sousa, B. Hiller and A. H. Blin, Nucl. Phys. **A575** 460 (1994)

36 V. Bernard, U.-G. Meißner, A. H. Blin and B. Hiller, Phys. Lett. **B253** 443 (1991)

37. Y. Kuramashi, M. Fukugita, H. Mino, M. Okawa and A. Ukawa, Phys. Rev. Lett. **71** 2387 (1993)

38. C. D. Roberts, R. T. Cahill, M. E. Sevior and N. Iannella, Phys. Rev. D **49** 125 (1994)

39. J. R. Uretsky and T. R. Palfrey, Phys. Rev. **121** 1798 (1961)

40. V. Bernard, N. Kaiser and Ulf-G. Meißner, preprint hep-ph/9404236 (1994)

41. A. Zylberstejn, Ph.D. thesis, University of Paris, Orsay, 1972 (unpublished)

42. E. W. Beier *et al.*, Phys. Rev. Lett. **29** 511 (1972); *ibid.* **30** 399 (1973)

Working group on $\pi\pi$ scattering

J. Gasser [1] and M.E. Sevior [2]
Co-chairs

[1] Institut für Theoretische Physik, Universität Bern, Siderstrasse 5, 3012 Bern, Switzerland.
[2] School of Physics, University of Melbourne, Parkville, 3052, Australia.

There were four days of stimulating talks and informal discussions on theoretical and experimental aspects of low-energy $\pi\pi$ scattering. Predictions for the $\pi\pi$ phase shifts, reviews of experimental data and methods to evaluate the phase shifts from the data were all presented. In our summary talks on the working group, given at the end of the conference, we underlined in particular the following points:

- Low-energy theorems for the $\pi\pi$ scattering amplitude provide precision tests of QCD. Presently available data agree with the predictions within the errors–which are, in some cases, substantial.

- Therefore, to fully test these predictions, new experiments are needed. These must allow one to extract in a theoretically clean manner the $\pi\pi$ amplitude at the 5% level.

- It appears that the data on K_{e4} decays, to be taken by KLOE at DAΦNE, meet this requirement.

- The planned measurement of the lifetime of $\pi^+\pi^-$ atoms has furthermore the opportunity to provide us with a theoretically clean and experimentally independent determination of the scattering length $|a_0 - a_2|$.

- There are already precise low-energy data for the $\pi N \to \pi\pi N$ reactions that almost meet the required *experimental* precision. On the other hand, a *theoretically* sound method which would allow one to extract the $\pi\pi$ phase shifts with the required precision is still lacking. The interested reader is strongly encouraged to work on this problem.

- Full kinematics, $\pi N \to \pi\pi N$ data at energies of 300 MeV or greater, when analyzed with the Chew-Low technique, may also help to construct precise low-energy phases.

A dense accumulation of very useful information was provided in the talks, and we felt uneasy writing an extended overview of what had been presented. We therefore asked the speakers to write on their own a short summary of their contribution. It seemed to us that this procedure is the best guarantee that no important information is lost.

The interested reader may find these summaries below, in the order they were presented in the working group. Together with the references provided, they will allow her or him to further study this fascinating subject.

1 $\pi\pi$ scattering in QCD

Ulf-G. Meißner, Physique Théorique, Centre de Recherches Nucléaires et Université Louis Pasteur de Strasbourg, B.P. 28, F-67037 Strasbourg Cedex 2, France

The purest process to test the predictions of CHPT is the elastic scattering of pions in the threshold region, $\pi^a + \pi^b \to \pi^c + \pi^d$ at $\sqrt{s} \geq 2M_\pi$. The $\pi\pi$ amplitude takes the chiral expansion $A(s,t,u) = A^{(2)}(s,t,u) + A^{(4)}(s,t,u) + \mathcal{O}(p^6)$. The first term in this series was given by Weinberg in 1966 [1], $A^{(2)}(s,t,u) = (s - M_\pi^2)/F_\pi^2$, and to one–loop accuracy in 1983 by Gasser and Leutwyler [2]. The most pertinent result of their study is the accurate prediction for the S–wave isospin zero scattering length, $a_0^0 = 0.20 \pm 0.01$. The comparison of the corresponding phase shifts $\delta_l^I(s)$ with the data can be found in ref.[3], which shows that the one loop corrections reach 50% of the tree result between $\sqrt{s} = 450\ldots 500$ MeV depending on the channel. In that reference, the phase of ϵ' is accurately predicted, $\phi(\epsilon') = 45 \pm 6\,°$. The phase difference $\delta_0^0 - \delta_1^1$ can also be inferred from K_{l4} decay form factors for 250 MeV $\leq \sqrt{s} \leq 350$ MeV. A thorough study of the available data of ref.[4] including the calculation of higher loop effects via a dispersive representation complemented by CHPT constraints is given in [5]. The beautiful consistency between the K_{l4} and the available $\pi\pi$ data gives further credit to the 1983 results of ref.[2] together with an improved description of the D–waves. The improved data on these decays from DAΦNE are eagerly awaited for. Another experimental method to get at the phase difference $|a_0^0 - a_0^2|$ is the measurement of the lifetime of pionic molecules, $\tau^{-1} = C|a_0^0 - a_0^2|^2$ with C known within better than 1%. Such an experiment is proposed at CERN [6]. Determining τ within 10% would give $|a_0^0 - a_0^2|$ with 5% accuracy. Furthermore, studies of the reaction $\pi N \to \pi\pi N$ beyond tree level accuracy (which is needed to analyze the many novel accurate data) have only started [7]. At present, it is not clear how good a tool this reaction is to determine the S–wave scattering lengths. Finally, in the framework of one–loop CHPT the πK scattering lengths, range parameters and phase shifts together with a discussion of resonance effects have been given in [8]. More references can be found in [9].

References

1. S. Weinberg: Phys. Rev. Lett. **17** (1966) 616
2. J. Gasser and H. Leutwyler: Phys. Lett. B **125** (1983) 325
3. J. Gasser and Ulf-G. Meißner: Phys. Lett. B **258** (1991) 219
4. L. Rosselet et al.: Phys. Rev. D **15** (1977) 574
5. H. Bijnens, G. Colangelo and J. Gasser: preprint BUTP-94/4, Nucl. Phys. B, in print
6. G. Czapek et al.: Letter of intent CERN-SPSLC 92–44, 1992
7. V. Bernard, N. Kaiser and Ulf-G. Meißner: Phys. Lett. B **332** (1994) 415
8. V. Bernard et al.: Nucl. Phys. B **357** (1991) 129; Nucl. Phys. B **364** (1991) 283
9. Ulf-G. Meißner: Rep. Prog. Phys. **56** (1993) 903

2 Axiomatic field theory versus chiral perturbation theory

G. Wanders, B. Ananthanarayan and D. Toublan, Institut de physique théorique, Université de Lausanne, CH 1015, Lausanne, Switzerland

A large effort has been devoted in the sixties and seventies to the derivation of properties of scattering amplitudes that follow rigorously from the principles of quantum field theory. This undertaking has been pioneered by André Martin [1] and was particularly successful for pion-pion scattering. Axiomatic field theory implies analyticity, crossing symmetry and unitarity, and as a consequence, the Froissart bound and the existence of twice subtracted dispersion relations. Combining these properties one gets detailed constraints on the shape of the pion-pion amplitudes in the triangle $0 < s < 4m_\pi^2$, $0 < t < 4m_\pi^2$, $0 < u < 4m_\pi^2$ of the Mandelstam plane [2]. We started a confrontation of the chiral approximation of the pion-pion amplitudes [3] with these axiomatic constraints. Conditions on the S- and P-waves in the interval $0 < s < 4m_\pi^2$ turn out to be relevant in the sense that they constrain the coupling constants $\bar{l}_i$ of the second order chiral lagrangian. The combination $\bar{l} = 2\bar{l}_1 + 4\bar{l}_2$, whose actual value is approximately 21, has a lower bound of the order of 5. A remarkable feature is that the $\pi^0 - \pi^0$ S-wave nearly fixes the D-wave for $0 < s < 4m_\pi^2$. The constraints become tighter if the chiral partial waves are unitarized: one finds indications that $\bar{l}$ has to be as large as $O(20)$ and one gets information on the size of the third order corrections.

References

1. A. Martin: *Scattering Theory: Unitarity, Analyticity and Crossing* (Springer Verlag, Berlin, Heidelberg, New York, 1969)
2. For a review giving the constraints we use, see S.M. Roy: Helv. Physica Acta **63** (1990) 627
3. J. Gasser and H. Leutwyler: Ann. Phys. (N.Y.) **158** (1984) 142

3 How experimental $\pi\pi$ scattering reflects chiral dynamics

M.R. Pennington, Centre for Particle Theory, University of Durham, Durham DH1 3LE, U.K.

In the theorists' world low energy pion dynamics is chiral dynamics, but what is it about the real world, accessible to experiment, that actually reflects these properties ? Is it something more than the lightness of the pion ? Low energy theorems, exact in a world of massless pions, require the $\pi\pi$ scattering amplitude to vanish at the centre of the Mandelstam triangle. Because these amplitudes are analytic functions of several complex variables, this Adler zero is not an isolated zero, but one point on a whole surface of zeros on which the $\pi^+\pi^- \rightarrow \pi^0\pi^0$ amplitude must vanish. PCAC works because the world of massless pions is not very different from the physical world for which $s + t + u = 4m_\pi^2$. Thus the physical $\pi^+\pi^- \to \pi^0\pi^0$ amplitude must also vanish inside the Mandelstam triangle. Such a subthreshold zero remarkably requires that the $\pi\pi$ amplitudes grow asymptotically, which indeed we believe they do. Moreover, the surface on which the Adler zero lies near the Mandelstam triangle, produces a dip in the $\pi^-\pi^0 \to \pi^-\pi^0$ differential cross-section, which becomes the Legendre zero of the ρ-resonance, enforcing the appearance of a spin-one resonance. $\pi\pi$ amplitudes with no chiral zero have no such resonance. Thus in the real world the existence of the ρ produces a near threshold zero. Indeed, where this zero crosses the axis of symmetry in the $\pi^+\pi^- \to \pi^0\pi^0$ amplitude (at $s = s_A$) determines the ratio of the $I = 0$ to $I = 2$ S-wave scattering lengths. In Weinberg's model of $\pi\pi$ scattering, this is at $s_A = m_\pi^2$.

In a world in which isospin two $\pi\pi$ interactions are weaker than those with $I = 0$ or 1, the strength of the low energy $\pi\pi$ interaction is naturally controlled by the strength of the $\rho\pi\pi$ coupling through the ratio Γ_ρ/m_ρ^3. In chiral dynamics, this parameter is $1/96\pi f_\pi^2$; numerically very much the same. The ρ-dominance of the low energy $I = 1$ $\pi\pi$ amplitude and the weakness of exotic channels are features of the real world. These properties lead to a relation between the $I = 0$ and $I = 2$ $\pi\pi$ scattering lengths, a_0^0, a_0^2 — a relation known as the Morgan-Shaw universal curve, on which all known phase shift solutions lie. How these solutions are obtained from high energy data on $\pi N \to \pi\pi B$ ($B=$ baryon) at small momentum transfers, where the pion pole dominates the production processes, is described. The Roy equations, a system of rigorous partial wave dispersion relations that embody 3-channel crossing symmetry, allow these phase shift results above 500 MeV to be extrapolated to threshold to determine the S-wave scattering lengths a_0^0, a_0^2. Present data allow a whole range corresponding to the Adler zero in the $\pi^+\pi^- \to \pi^0\pi^0$ amplitude passing near to the Mandelstam triangle from $s_A = -0.5m_\pi^2$ to $3.5m_\pi^2$, whereas Chiral Perturbation Theory makes the definite prediction of $s_A = 0.99m_\pi^2$. It is stressed that the many recent experiments on $\pi N \to \pi\pi N$ near threshold teach us just about low energy $\pi N \to \pi\pi N$ scattering and are only related to $\pi\pi$ scattering in a model-dependent way. Precise measurements of near threshold $\pi\pi$ phases are needed from K_{e4} decays to fix the chiral symmetry breaking parameter, encoded in the position of the on-shell appearance of the Adler zero, better.

Experimental $\pi\pi$ scattering and chiral dynamics are inextricably tied by the continuation of the Adler zero into the scattering regions.

4 The $\pi\pi$ amplitude in generalized chiral perturbation theory

M. Knecht, Division de Physique Théorique[1], Institut de Physique Nucléaire, F-91406 Orsay Cedex, France

[1]Unité de Recherche des Universités Paris XI et Paris VI associée au CNRS.

The fundamental order parameter $F_0 = F_\pi|_{m_u=m_d=m_s=0} \simeq 90$ MeV defines a characteristic scale of spontaneous χSB in QCD. The fact that F_0 is about ten times smaller than the typical mass scale $\Lambda_H \sim 1$ GeV of hadronic bound states remains unexplained so far and raises a question: What is the natural size of other order parameters, such as the quark condensate $B_0 = -F_0^{-2} < \bar{q}q >_0$, $q = u, d, s$, which describe χSB ? Generalized χPT [1] is motivated by the fact that our present theoretical or experimental knowledge is compatible with a value of B_0 as large as Λ_H or as small as F_0. In the latter case, the standard expansion of the low energy effective lagrangian has to be reorganized differently. Although standard χPT and GχPT define the same effective lagrangian, they differ at a given finite order: explicit symmetry breaking terms, and the corresponding low energy constants, which standard χPT would consider only at higher orders, appear already at lowest orders in GχPT. At leading order, there is essentially one additional parameter, the quark mass ratio $r = m_s/\hat{m}$, which can take any value between $r_1 = 6.3$ (for which B_0 vanishes) and $r_2 = 25.9$ (the standard value). The deviations from the Goldberger-Treiman relation yield a lowest order measurement of r. Present values of F_π, $g_{\pi NN}$,... lead to the bound $r \leq 10.6 \pm 4.2$. Differences between the predictions of the two alternatives will be most important in processes which are sensitive to quark mass effects already at leading order. For instance, the low energy behaviour of the spectral function associated with the divergence of the axial current is enhanced by a factor of 10 for $r \lesssim 10$, which would considerably affect the existing estimates of $\hat{m}$ using the QCD sum rules [2]. In GχPT, the one loop predictions for S-wave $\pi\pi$ scattering depend on r and differ substancially from the standard case for $r \lesssim 10$ [3]. Precise determination of low energy $\pi\pi$ phase shifts from future high statistics K_{l4} data would allow to disentangle the two alternatives.

References
1. M. Knecht and J. Stern: preprint IPNO-TH 94-53, to appear in the second edition of the DaΦne Handbook (L. Maiani, G. Pancheri and N. Paver, eds.)
2. J. Stern, N. H. Fuchs and M. Knecht: preprint IPNO-TH 93-38 (hep-ph/9310299), and in: "Proceedings of the Third Workshop on the Tau-Charm Factory, Marbella, Spain, 1-5 June 1993", Eds. J. Kirkby and R.Kirkby (Editions Frontieres, 1994)
3. M. Knecht, B. Moussallam and J. Stern: preprint IPNO-TH 94-54

5 Testing chiral perturabtion theories of $\pi\pi$ scattering

M.G. Olsson, Department of Physics, University of Wisconsin–Madison, Madison, WI 53706
e-mail:Olsson@phenom.physics.wisc.edu

An expansion of the $\pi\pi$ scattering amplitude about the subthreshold point

$\nu = t = 0$ is considered. We first address the evaluation of the eighteen expansion coefficients, including all quadratic terms in ν^2 and t, from the available experimental data. We use fixed-t dispersion relations along with scattering lengths from threshold pion production, Ke$_4$ data, and the phase shifts from pion pole extrapolation in high energy pion production.

Theoretical expressions for each of the subthreshold expansion coefficients are computed explicitly from ChPT predictions. Eight of the subthreshold coefficients receive contributions exclusively from the loop correction to the leading amplitude. These coefficients are particularly sensitive to the structure of the chiral theory.

A comparison of prediction with experiment shows that although the standard ChPT is qualitatively correct, many of the loop checking coefficients are off by factors of two (and many standard deviations). We also consider an "improved" ChPT scheme (Stern, Sazdijan, Fuchs) which seems to work much better.

6 Pion observables and the qq interaction in the infrared

Craig Roberts, Physics Division 203, Argonne National Laboratory, Argonne IL 60439-4843
e-mail: cdroberts@anl.gov

The nonperturbative Dyson-Schwinger equation (DSE) approach has been widely used in studies of QCD [1] and to further develop the Global Colour-symmetry Model (GCM) [2], a QCD-based field theory, defined by the action:

$$S[\bar{q}, q] =$$
$$\int d^4x\, d^4y\, \left\{ \bar{q}(x)\left[\gamma \cdot \partial + m\right]\delta^4(x - y)q(y) + \tfrac{1}{2} j_\mu^a(x)g^2 D_{\mu\nu}(x - y)j_\nu^a(y) \right\} ,$$

where $j_\mu^a(x) = \bar{q}(x)\frac{\lambda^a}{2}\gamma_\mu q(x)$ and $D_{\mu\nu}(x - y)$ is the nonperturbatively-dressed gluon propagator obtained from DSE studies. This model incorporates asymptotic freedom, dynamical chiral symmetric breaking and confinement.

In this presentation, the complimentary use of the QCD DSEs and the GCM, to calculate the quantities in the table [3] and $F_\pi(q^2)$, is described. As illustrated, the agreement between experiment and theory is excellent in all cases. [$\langle\bar{q}q\rangle_{\text{Exp.}}$ is the typical QCD sum-rules value.]

In this approach the pion is a fully dressed q-$\bar{q}$ bound-state, with a momentum-dependent Bethe-Salpeter amplitude, and therefore an intrinsic size, which entails that pion-loops are strongly suppressed away from resonances, providing contributions of $< 15\%$ to observables.

It is argued that precision measurements of π-π scattering and $F_\pi(q^2)$, for spacelike-$q^2 > 2$ GeV^2, can be used to place tight constraints on $D_{\mu\nu}(k \simeq 0)$; i.e., the effective quark-quark interaction in the infrared, the form of which is presently unknown.

Acknowledgment: This work was supported by the US Department of Energy, Nuclear Physics Division, under contract number W-31-109-ENG-38.

References
1. C. D. Roberts and A. G. Williams: *Dyson-Schwinger Equations and their Application to Hadronic Physics*, in: Progress in Particle and Nuclear Physics, Vol. 33, pp. 477-575, ed. by A. Fäßler (Pergamon Press, Oxford, 1994)
2. R. T. Cahill and C. D. Roberts: Phys. Rev. D **32**, 2419 (1985)
3. C. D. Roberts, R. T. Cahill, M. E. Sevior and N. Iannella: Phys. Rev. D **49**, 125 (1994)

	Calculated	Experiment
f_π	0.0839 GeV	0.0931 ± 0.001
$-\langle \bar{q}q \rangle^{\frac{1}{3}}_{1\,\text{GeV}^2}$	0.211	0.220 ± 0.01
$m^{\text{ave}}_{1\,\text{GeV}^2}$	0.0061	0.0075 ± 0.004
m_π	0.127	0.138
r_π	0.596 fm	0.663± 0.006
$g_{\pi^0\gamma\gamma}$	0.497	0.504 ± 0.019
a_0^0	0.174	0.21 ± 0.02
a_0^2	-0.0496	-0.040 ± 0.003
a_1^1	0.0307	0.038 ± 0.002
a_2^0	0.00161	0.0017 ± 0.0003
a_2^2	-0.000251	

7 Pion production on polarized targets: Evidence for $\sigma(750)$ state and consequences for $\pi\pi$ phase shifts

M. Svec, Physics Department, McGill University, Montreal, QC, Canada H3A 2T8.

We present a new and improved model independent amplitude analysis of reactions $\pi^+ n_\uparrow \to \pi^+\pi^- p$ at 5.98 and 11.85 GeV/c and $\pi^- p_\uparrow \to \pi^-\pi^+ n$ at 17.2 GeV/c measured with transversely polarized targets at the CERN Proton Synchrotron. For dipion masses below 1000 MeV the pion production process is described by two S-wave and six P-wave production amplitudes. Previous analyses suffered from the presence of unphysical solutions for moduli of amplitudes or cosines of their relative phases, causing uncertainties regarding the signal for scalar state $I = 0$ $0^{++}(750)$. To remove the unphysical solutions we use a Monte Carlo approach to amplitude analysis. In each (m,t) bin we randomly varied the input spin density matrix elements 30 000 times within their experimental errors and performed amplitude analysis for each selection. Unphysical solutions were rejected and the physical solutions produced a continuous range of values for moduli, cosines of relative phases and for partial wave intensities. A clear signal for $\sigma(750)$ resonance emerges in all four solutions for S-wave intensity I_S at all three energies.

Its $\pi^+\pi^-$ decay width is estimated to be in the range 200–300 MeV. All determinations of $\pi\pi$ phase shifts from unpolarized data on $\pi^-p \to \pi^-\pi^+n$ rely on the absence of A_1-exchange amplitudes.

This essential assumption leads to predictions for polarized spin density matrix elements which are clearly ruled out by data on polarized targets. We conclude that the usual determinations of $\pi\pi$ phase shifts cannot be correct.

References

1. M. Svec, A. de Lesquen and L. van Rossum: Phys. Rev. D **45** 55 (1992)
2. M. Svec, A. de Lesquen and L. van Rossum: Phys. Rev. D **46** 949 (1992)
3. M. Svec: Phys. Rev. D **47** 2132 (1993)
4. M. Svec: McGill University preprint 1993

8 Preliminary results of a new measurement of the reaction $\pi^+p \to \pi^+\pi^+n$ near threshold

M.E. Sevior, School of Physics, University of Melbourne, Parkville, 3052, Australia

The pioneering work of Olsson and Turner [1] showed that the amplitude for $\pi p \to \pi\pi N$ reactions at threshold is largely determined by the One Pion Exchange (OPE) process, which in turn is simply related to the amplitude for $\pi - \pi$ scattering. Although the exact relationship given by Olsson and Turner for the connection between the threshold pion production amplitude and $\pi - \pi$ scattering has recently been questioned, there is still general agreement that the OPE process is a major contribution to the threshold pion production amplitude. Modern calculations of this process are encouraged.

TRIUMF experiment 561, measurements of the total cross sections for $\pi^+ + p \to \pi^+\pi^+n$ and $\pi^- + p \to \pi^+\pi^-n$, aimed to provide precise cross section values for these reactions very close to threshold (within 5 MeV). The first phase of the experiment concentrated on the $\pi^+ + p \to \pi^+\pi^+n$ reaction. These data have been fully analyzed and published [2] [3]. At nearly the same time the results of the OMICRON group for $\pi^+ + p \to \pi^+\pi^+n$ became available in the literature [4]. The OMICRON data are in serious disagreement with our work and in fact their lowest data point is almost a factor of 3 higher than our measurements. The second phase of 561 was to concentrate on the $\pi^- + p \to \pi^+\pi^-n$ reaction, however we were also able to repeat a couple of our energies for the $\pi^+ + p \to \pi^+\pi^+n$ reaction to check our earlier results.

The details of the original experiment are given in detail in [3]. We made a number of improvements for the second phase of the measurements. These were, an improved active target design that transmits more of the light to the photomultiplier tubes, operating the phototubes at lower gain to prevent signal degradation over the course of the run, a second level trigger to eliminate our reliance on a hardware detection of $\pi^+ \to \mu^+$ decays in the first level trigger, and more channels of 500 Mega-Sample per second transient digitizers through the use of the TRIUMF developed GaAs 500 Mhz CCD's.

Our preliminary numbers for the cross sections at pion kinetic energies of 200 MeV and 184 MeV are $1.42 \pm 0.20 \pm 0.18$ μb and $0.30 \pm 0.05 \pm 0.04$ μb respectively. These results are in excellent agreement with our previous measurement and remain almost a factor of 3 below the OMICRON results.

References
1. M. G. Olsson and L. Turner: Phys. Rev. Lett. **20** 1127 (1968); M.G. Olsson, E.T. Osypowski and L. Turner: Phys. Rev. Lett. **38** 296 (1977)
2. M.E. Sevior et al.: Phys. Rev. Lett. **66** 2569 (1991)
3. M.E. Sevior et al.: Phys. Rev D **48** 3968 (1993)
4. G. Kernal et al.: Z. Phys. C **48**, 201 (1990)

9 First results of exclusive measurements of $\pi N \to \pi\pi N$ with the CHAOS detector

M.A. Kermani, University of British Columbia, Vancouver British Columbia, V6T 1Z1 Canada

The Canadian High Acceptance Orbit Spectrometer (CHAOS) was used to make measurements of four fold differential cross sections for $\pi N \to \pi\pi N$ reactions with both positive and negative polarity incident pions at $200 \leq T_\pi \leq 300$ MeV. On the order of 10^4 events were recorded for both the $\pi^\pm p \to \pi^+\pi^\pm n$ and $\pi^\pm p \to \pi^\pm\pi^0 p$ channels at each incident pion energy. The last stage of the experiment was completed in July 1994, and the analysis is currently underway. A brief introduction to the detector was provided and some preliminary results were discussed. Once the analysis is complete, the measured cross sections will be compared to the predictions of the model first developed by Oset et al. [1] and later improved by Johnson et al. [2]. In addition, Chew-Low analysis [3] will be performed on the higher energy data in order to obtain $\pi\pi$ scattering cross sections.

References
1. E. Oset et al.: Nucl. Phys. A **466** (1985) 584
2. R.R. Johnson et al.: πN Newsletter, Oct 1993
3. G.F. Chew et al.: Phys. Rev. **113** (1959) 1640

10 Interpretation of $\pi N - \pi\pi N$ measurements

A.A. Bolokhov, Sankt-Petersburg State University, Sankt-Petersburg, 198904, Russia, and
M.E. Sevior, University of Melbourne, Parkville, Vic., 3105 Australia

The predictions of Chiral perturbation theory [1] explain the very sharp interest in the values of the $\pi\pi$–scattering lengths. There are several approaches for extracting the latter from $\pi N \to \pi\pi N$ ($\pi 2\pi$) data.

1. The Chew–Low extrapolation procedure [2] is an apparently model-independent approach which can provide complete information on the $\pi\pi$ cross section provided the OPE dominates and the interval of the nucleon momentum transfer τ allows a unique extrapolation. The last condition and the need for sufficient statistics shifts the region of application of the Chew–Low procedure to rather large values of energy ($\approx 1 - 4$ GeV). When comparing the phase space of momentum transfer $-20\mu^2 < \tau < -0.2\mu^2$ at $P_{Lab} = 500$ MeV/c with the distance of extrapolation $\approx \mu^2$ it becomes obvious that provided there are enough statistics, the kinematics of the $\pi 2\pi$ reaction itself does not prevent the use of the Chew–Low procedure at much smaller energies. It is the presence of contributions like that of Δ and N^* isobars which makes a straightforward extrapolation difficult at moderate energies due to the perturbation of the simple τ–dependence of the OPE graph.

2. Since at intermediate energies there are a number of contributions to the $\pi 2\pi$ amplitudes, the approach might be changed to determining the OPE parameters directly in the physical region of the reaction — this was implemented by the model of Oset and Vicente–Vacas [3]. In the energy region below $P_{Lab} = 500$ MeV/c, the variation of τ is sufficient to detect the OPE contribution while the contributions of the concurrent processes (being nonnegligent) are smooth enough. The model [4] takes these features into account and this naturally completes the Oset–Vicente approach and its multiple derivatives.

3. The investigations by Olsson and Turner [5] of Chiral Dynamics at the threshold of $\pi 2\pi$ reactions, where the application of Chew–Low procedure is impossible, resulted in the formulae expressing the $\pi\pi$–scattering lengths in terms of the threshold characteristics of these reactions. These formulae have gained a broad scale of application, especially in the recent years when new data on the $\pi N \to \pi\pi N$ reactions in the close–to–threshold became available [6,7,8]. The evidence of the importance of next–to–leading order terms of the Chiral Lagrangian for the $\pi 2\pi$ amplitude, in particular, the D–wave parameters of the OPE graph, makes it necessary to modify the Olsson–Turner method.

References

1. J. Gasser and H. Leutwyler: Ann. Phys. (NY) **158** (1984) 142; Nucl. Phys. B **250** (1985) 465

2. G.F. Chew and F.E. Low: Phys. Rev. **113**, 1640(1959)

3. E. Oset and M.J. Vicente–Vacas: Nucl. Phys. A **446** 584(1985)

4. A.A. Bolokhov, V.V. Vereshagin and S.G. Sherman: Nucl. Phys. A **530** (1991) 660

5. M.G. Olsson and L. Turner: Phys. Rev. Lett. **20** (1968) 1127; Phys. Rev. D **6** (1972) 3522

6. G. Kernel et al.: Phys. Lett. B **216** (1989) 244; Z. Phys. C **51** (1991) 377

7. M.E. Sevior et al.: Phys. Rev. Lett. **66** (1991) 2569

8. J. Lowe et al.: Phys. Rev. C **44** (1991) 956

11   Measuring K_{e4} parameters at DAϕNE

Paula J. Franzini, Paul Scherrer Institute, CH-5232 Villigen PSI, Switzerland
e-mail: Franzini@psiclu.psi.ch

In this talk I examine the most frequent K_{l4} decay mode, $K^+ \to \pi^+\pi^-e^+\nu_e$ (K_{e4+}), as it will be studied at the Φ factory DAϕNE due to become operational in Frascati at the end of 1996 with an anticipated yearly luminosity of $5-10\times10^{32}$ cm^2/s. With the lower value, 3×10^5 K_{e4+} events are expected, a factor of ten more than at the previous experiment studying this mode (Rosselet in 1977). In addition to the statistical improvement, systematic improvements are also expected due to the very clean experimental environment. I briefly compare the previous and future experiments, as regards systematic errors. I discuss the use of the maximum likelihood method (MLM) and asymmetries. While the $\pi\pi$ phase shift δ can be written as a ratio of asymmetries and thus determined in a theoretically clean and experimentally quickly calculated manner, the MLM gives an error on δ 1.6 times smaller, and can be used to determine all parameters (with the smallest possible errors), not just δ. Since this is a very abbreviated writeup, I refer the reader to our paper in Ref. 1 for details and references.

For an event sample of 300,000 events, we anticipate the following accuracies:

$$\Delta f_{s0} = 0.014, \quad \Delta g_0 = 0.038, \quad \Delta h_0 = 0.14,$$

$$\Delta f_{p0} = 0.014, \quad \Delta\lambda_\pi = 0.004, \quad \Delta\lambda_l = 0.011,$$

to be compared with $\Delta f_{s0} = 0.14$, $\Delta g_0 = 0.27$, $\Delta h_0 = 0.68$, $\Delta\lambda_\pi = 0.02$ for the previous experiment. Here the subscript 0 denotes value for dipion and dilepton invariant masses (s_π and s_l) of zero. f_s, f_p, g and h are the lowest order coefficients in a partial wave expansion of the K_{l4} form factors F, G, and H. λ_π and λ_l are the slopes of these coefficients (taken to be the same for all of them) with respect to $q^2 = (s_\pi - 4m_\pi^2)/4m_\pi^2$ and $s_l/4m_\pi^2$. These accuracies were obtained with the central values (based on the previous experiment, and on theory)

$$f_{s0} = 5.6 \quad g_0 = 4.8 \quad h_0 = -2.7 \quad f_{p0} = 3.3 \quad \lambda_\pi = 0.08 \quad \lambda_l = 0.08.$$

The *relative* accuracies are essentially independent of central values taken, for f, g, and h; the *absolute* accuracies remain constant for λ (and δ, below). For other numbers of events N, the errors should just be scaled by $1/\sqrt{N}$.

For the phase shift $\delta = \delta_s - \delta_p$ we calculate the accuracies, in five equal event bins (in s_π) of 60,000 events each, of 0.02, 0.013, 0.011, 0.009, and 0.008. These are to be compared with 0.13, 0.07, 0.05, 0.04 and 0.04 for the previous experiment. I consider different parameterizations of δ in terms of the scattering lengths and slopes a_0^0, a_1^1, b_0^0, and b_1^1. Using for example the parameterization and theoretical information used by the previous experiment, we find an accuracy $\Delta a_0^0 = 0.01$, to be compared with the previous value of 0.05, which should thus allow us to determine if the present discrepancy between theory and experiment is statistically significant.

References
1. M. Baillargeon and P. J. Franzini: preprint PSI-PR-94-25, hep-ph/9407277, contribution to the DAΦNE physics handbook, 2nd edition

12 On the Pais–Treiman method to extract $\pi\pi$ phases in K_{e4} decays

<u>M. Knecht</u>, Division de Physique Théorique[1], Institut de Physique Nucléaire F-91406 Orsay Cedex, France

The measurement of the $\pi\pi$ phase shifts near threshold is a key issue of low energy hadronic physics. It is known since a long time that the main source of model independent experimental information on low energy $\pi\pi$ phases comes from K_{e4} decays. In these decays, a complete $\pi\pi$ phase-shift analysis could be performed in principle, assuming nothing more than unitarity or the Watson final state interaction theorem. However, such a complete procedure would require a detailed amplitude analysis of K_{e4} to be performed with respect to all five kinematical variables. To avoid this task, which is rather problematic in practice, Pais and Treiman have proposed a method to measure $\pi\pi$ phases in a much simpler way, which is based on the observation that $\tan(\delta_0^0 - \delta_1^1)$ can be simply expressed as the ratio of two angular moments ($\bar{I}_4$ and $\bar{I}_7$) of the K_{e4} differential decay rate, provided all partial waves beyond the S and P waves are neglected. While the corrections Δ to the Pais–Treiman formula,

$$\bar{I}_7/2\bar{I}_4 = \tan(\delta_0^0 - \delta_1^1)\,\{1 + \Delta\}$$

coming from higher partial waves might indeed be small, it is however important to estimate their size. This has been done [1] using the explicit representation of K_{e4} form factors at order one loop in χPT, which gives $\Delta = \Delta^{(2)} + O(E^4)$. Numerically, $\Delta^{(2)}$ is small over the whole phase space (both in standard and in generalized χPT), it barely reaches 0.5%. Higher orders will certainly modify the above number, but not its order of magnitude, which is essentially given by the kinematics of the K_{e4} decay. Our conclusion is then that the Pais–Treiman formula

$$\tan(\delta_0^0 - \delta_1^1) = \bar{I}_7/2\bar{I}_4$$

is free of corrections up to the percent level, over the whole accessible range of energy.

References
1. G. Colangelo, M. Knecht and J. Stern: preprint IPNO-TH 94-36, BUTP-94/11, ROM2F-94/19 and Phys. Lett. B, in press

[1]Unité de Recherche des Universités Paris XI et Paris VI associée au CNRS.

13 $K_S \to \pi^+\pi^-\pi^0$ and 3π rescattering in interferometry machines

Giancarlo D'Ambrosio INFN - Sezione di Napoli, 80125 NAPOLI, ITALY
e-mail:dambrosio@axpnal.na.infn.it

The CP conserving $K \to 3\pi$ decay amplitudes are well described by Chiral Perturbation Theory (ChPT). They have been calculated, including the next to leading order corrections, in Ref. [1] and turn out to be in good agreement with the experimental data. $K_S \to \pi^+\pi^-\pi^0$ decay receives both CP conserving and CP violating contributions; the CP conserving decay amplitude is odd under $\pi^+ - \pi^-$ momenta exchange and thus, neglecting high angular momenta states, is induced by a $\Delta I = 3/2$ transition. The ChPT calculation of Ref. [1] leads the prediction:

$$\text{Br}(K_S \to \pi^+\pi^-\pi^0)_{CP+} = (2.4 \pm 0.7) \cdot 10^{-7}, \tag{1}$$

However, this width is difficult to observe due to the small branching ratio. Final state interactions in $K \to 3\pi$ are important since they determine the size of CP violation in this channel and they also furnish an independent test of ChPT. Indeed the pions are near-threshold and rescattering in $K \to 3\pi$ is related, at leading order, to $\pi - \pi$ scattering. Due to the smallness of phase space they are expected to be small (~ 0.1). In typical width measurements the phases (δ) appear through the $\cos\delta \simeq 1 - \delta^2/2$ and thus difficult to detect. This is different if you measure the time interference among K_S and K_L [2] . For instance at LEAR tagged K^0 and $\overline{K^0}$ are produced [3], and the simplest means to observe interference is represented by the asymmetry

$$A_f^{+-0}(t) = \frac{\int d\phi^a \left[|A(K^0 \to \pi^+\pi^-\pi^0)|^2 - |A(\overline{K^0} \to \pi^+\pi^-\pi^0)|^2 \right]}{\int d\phi \left[|A(K^0 \to \pi^+\pi^-\pi^0)|^2 + |A(\overline{K^0} \to \pi^+\pi^-\pi^0)|^2 \right]} \tag{2}$$

where $d\phi^a$ indicates an antisymmetric integration in the charged pion energies. Up to first order in ε, the decay amplitude squared as a function of time is given by

$$|A\left(K^0(\overline{K^0}) \to \pi^+\pi^-\pi^0\right)|^2 \simeq \frac{1}{2}\left(1 \mp 2\Re e\, \varepsilon\right)\left\{ \exp\left(-\Gamma_S t\right)|A_S|^2 + \exp\left(-\Gamma_L t\right)|A_L|^2 \right.$$
$$\left. \pm 2\exp\left(-\Gamma t\right)\left[\Re e\,\left(A_L A_S^*\right)\cos\left(\Delta m t\right) + \Im m\,\left(A_L A_S^*\right)\sin\left(\Delta m t\right)\right] \right\}, \tag{3}$$

where $\Delta m = m_L - m_S$, $\Gamma = (\Gamma_L + \Gamma_S)/2$ and $A_{S,L} \equiv A_{+-0}^{S,L}$. Then Eq. (2) can be rewritten as

$$A_f^{+-0}(t) = \frac{2e^{-\Gamma t}\int d\phi^a\, \Re e A_L\, \Re e A_S}{\int d\phi\,[e^{-\Gamma_S t}|A_S|^2 + e^{-\Gamma_L t}|A_L|^2]}\left[\cos\left(\Delta m t\right) + \tilde{\delta}\sin\left(\Delta m t\right) + O(\tilde{\delta})\right], \tag{4}$$

The phase $\tilde{\delta}$ of eq.(4) can be written as:

$$\tilde{\delta} \simeq \delta_{1S} - \delta_2, \tag{5}$$

where δ_2 and δ_{1S} are the 3π strong phases of the $I = 2$ and of the symmetric $I = 1$ final states, respectively. With a different integration over the Dalitz Plot also the phase of the non-symmetric $I = 1$ final state could be selected [2]. We stress that by accurate time dependent study both interference terms in (3) can be measured. Also ϕ-factories are useful in this respect. Indeed ϕ decays in a very well defined quantum mechanical state: $K_S K_L - K_L K_S$, which respectively decay into f_1, f_2 at the times t_1, t_2; choosing $f_1 = l\pi\nu$ and $f_2 = \pi^+\pi^-\pi^0$, integrating the flux $I(l\pi\nu, \pi^+\pi^-\pi^0, t_1, t_2)$ over $t_1 + t_2$ one can then study

$$
\begin{aligned}
R^\pm(t) &= \frac{\int I(\pi^+\pi^-\pi^0, l^\pm\pi^\mp\nu, t) d\phi_{3\pi}^a \, d\phi_{l\pi\nu}}{\int [I(\pi^+\pi^-\pi^0, l^\pm\pi^\mp\nu, t)] \, d\phi_{3\pi}^a \, d\phi_{l\pi\nu}}, \\
&= \pm 2 \frac{\int \Re \left(A_L^{+-0} A_S^{+-0*} \right) d\phi_{3\pi}^a}{\Gamma_L^{+-0} e^{+\frac{\Delta\Gamma}{2}t} + \Gamma_S^{+-0} e^{-\frac{\Delta\Gamma}{2}t}} \left[\cos(\Delta mt) + \tilde{\delta}\sin(\Delta mt) \right], \quad (6)
\end{aligned}
$$

where $t = t_1 - t_2$. As discussed in ref. [2] the study of $R^\pm(t)$ will certainly lead to a determination of the $(A_S^{+-0})_{CP+}$ amplitude, performing an interesting test of ChPT in the $\Delta I = 3/2$ transitions, and perhaps could also lead to a direct measurement of the $\pi^+\pi^-\pi^0$ rescattering functions. Preliminary results at CPLEAR [3], also report already evidence of non-vanishing $A(K_S \to \pi^+\pi^-\pi^0)$. Furthermore fixed target experiments with regenerators could attack the problem.

References

1. J. Kambor, J. Missimer and D. Wyler Phys. Lett. B261 (1991) 496.
2. 2. G. D'Ambrosio and N. Paver, Phys. Rev. D46 (1992) 252; G. D'Ambrosio and N. Paver, Phys. Rev. D49 (1994) 4560; G. D'Ambrosio, G. Isidori, A. Pugliese and N. Paver, Roma preprint 998 (1994), to be published in Phys. Rev. D.
3. CPLEAR Collaboration, T. Ruf, talk given at the 27th Int. Conf. on High Energy Physics, Glasgow, July 1994, to appear in the proceedings. KLOE Collaboration A. Aloisio et al.

14 Non perturbative methods in χPT

A. Dobado[1] and J. R. Peláez, Departamento de Física Teórica, Universidad Complutense, 28040 Madrid, Spain, and
J. Morales, Departamento de Física Teórica, Universidad Autónoma, 28049 Madrid, Spain (On leave of absence from Centro Internacional de Física, Colombia)

Recently we have been working on non-perturbative methods applied to χPT [1] in order to improve its unitarity behavior and enlarge its energy applicability region. In particular, using dispersion relations for the inverse elastic scattering

---

[1] Remark by co-chairs: We invited A. Dobado to give a talk at the working group. As he could not attend the conference, A. Dobado has kindly provided the following abstract of his planned presentation.

amplitudes we have found that, under general assumptions, they lead to the previously considered formal Padé approximants [2], thus giving them a more sound basis. Then with this result we have made a three parameter fit of seven $\pi\pi$ and πK scattering channels, which improves and extends the previous one-loop computations without enlarging the number of parameters nor explicitly introducing new fields. Moreover, we obtain, in the appropriate channels, poles in the second Riemann sheet that can be understood as the ρ and the K^* resonances which are predictions of our approach if one fits the three parameters using other channels. The details and main results of this work can be found in [3].

As a second and independent approach we have also defined and obtained the pion scattering amplitudes in the χPT large N expansion [4] (N being the number of Goldstone bosons). In particular we have computed the effective action at the leading order including the effects due to the pion mass up to order m^2/f_π^2. We use this action to reproduce the Linear Sigma Model and to fit the $\pi\pi$ scattering with just one parameter (see [5] for detais). Within this framework, the fit is not as good as in the previous case but still it is competitive with the standard one-loop fit, both failing in the description of the ρ resonance. However, it can improve when next to leading ($1/N^2$) corrections are included. Note that due to the structure of the large N amplitudes it is possible to find poles in the second Riemann sheet that could reproduce resonances as it happens in the large N description of the LSM. We are presently also working in that direction.

In conclusion, the great phenomenological success of χPT can be enlarged using well defined non-perturbative techniques. The improvement affects unitarity, the energy applicability range, and remarkably in order to make predictions, without increasing the number of free parameters.

References

1. S. Weinberg: Physica **96A** (1979) 327; J. Gasser and H. Leutwyler: Ann. of Phys. **158** (1984) 142; Nucl. Phys. B **250** (1985) 465 and 517
2. Tran N. Truong: Phys. Rev. D **61** (1988) 2526; A. Dobado, M.J. Herrero and J.N. Truong: Phys. Lett. B **235** (1990) 129
3. A. Dobado and J.R. Peláez: Phys. Rev. D **47**(1992) 4883
4. A. Dobado and J.R. Peláez: Phys. Lett. B **286** (1992) 136; M. J. Dugan and M. Golden: Phys. Rev. D **48** (1993) 4375; A. Dobado, A. López and J. Morales: preprint FT/UCM/15/94
5. A. Dobado and J. Morales: preprint FT/UCM/18/94, hep-ph 9407321

Summary of the Two Pion Threshold and Chiral Anomaly Working Group

R. A. Miskimen

Department of Physics and Astronomy, University of Massachusetts, Amherst, MA
01003-4525

1 Introduction

The Two Pion Threshold and Chiral Anomaly Working Group focused on the chiral
anomaly in the low energy interactions of pseudo-scalar mesons, threshold two pion
photoproduction, and the $N \rightarrow \Delta$ axial form factor. The broad scope of physics
in the working group is reflected in the range of beam energies for the proposed
measurements, from approximately 300 MeV to 600 GeV. The reports in this summary
represent relatively new research directions in the field of chiral dynamics, motivated
largely by the development of new experimental facilities at Mainz, Frascati, CEBAF,
and Fermi Lab. The next three years should see considerable progress on the chiral
anomaly and threshold two pion production.

Theoretical and experimental investigations of the chiral anomaly are important
because they test fundamental principles of chiral quantum field theory and the Standard Model (see sections by P. Venugopal and G. Ecker in this working group summary). The chiral anomaly has been used to calculate amplitudes for radiative nonleptonic kaon decays such as

$$K_L \rightarrow \pi^+\pi^-\gamma$$
$$K^+ \rightarrow \pi^+\pi^0\gamma$$

which share the feature that the bremsstrahlung amplitude is suppressed, making
verification of the anomalous amplitude substantially easier (see sections by G. Ecker
and G. D'Ambrosio). Measurements of these decays are planned for $DA\phi NE$ (see
invited paper by R. Baldini).

The chiral anomaly also predicts the amplitude for

$$\gamma \rightarrow \pi^+\pi^-\pi^0.$$

This amplitude has been measured at Serpukhov, however, comparison of the data
with theory is inconclusive. Fermi Lab experiment E781 will make a new measurement
of the $\gamma \rightarrow 3\pi$ amplitude using 600 GeV/c pions (see section of M. Moinester and
Ref. [12]). A new measurement has also been proposed at CEBAF using the CLAS
detector with 2 GeV unpolarized photons initially, and later with linearly polarized
photons (see sections by R. Miskimen and K. Wang).

While threshold single pion photo- and electroproduction on the nucleon has been
an extremely active research area for chiral perturbation theory (CPT), studies of
threshold two pion photoproduction are just beginning. A calculation based on an
effective Lagrangian model coupling pions to the nucleon and Δ indicates that the

Δ contribution can be substantial at threshold (see section by M. Benmerrouche). Another calculation using heavy baryon chiral perturbation theory indicates that for the first 10 MeV above threshold loop contributions make the $\pi^0\pi^0$ final state the dominant two pion final state (see invited paper by V. Bernard). The predicted threshold cross sections are small but measureable, and the TAPS collaboration at Mainz has observed $\pi^0\pi^0$ production at threshold (see invited paper by T. Walcher).

There has been considerable interest in measurements of $N \to \Delta_{33}$ electromagnetic form factors as a test of baryon structure. Recently a measurement of the $N \to \Delta$ axial-vector form factor has been proposed at CEBAF using the soft pion limit to relate $ep \to e'\pi^-\Delta^{++}$ to the axial matrix element (see section by R. Hicks). Measurements at low Q^2 can test CPT predictions for the axial radius of the $N \to \Delta$ transition.

2 Phenomenology of the Chiral Anomaly

Eswara P. Venugopal

Department of Physics and Astronomy, University of Massachusetts, Amherst, MA
01003-4525

The chiral anomaly [1] is a well established component of effective lagrangians describing the interactions of pseudoscalar mesons at low energies [2]. As first given by Witten [3], the anomalous lagrangian can be written in the form

$$\mathcal{L}_{anom} = -\frac{N_c}{48\pi^2}\epsilon^{\mu\nu\alpha\beta}[eA_\mu\ Tr(QL_\nu L_\alpha L_\beta - QR_\nu R_\alpha R_\beta)-$$

$$ie^2 F_{\mu\nu}A_\alpha\ Tr(Q^2 L_\beta - Q^2 R_\beta + \frac{1}{2}QUQU^\dagger L_\beta - \frac{1}{2}QU^\dagger QU\ R_\beta)]$$

where $F_{\mu\nu} = \partial_\mu A_\nu - \partial_\nu A_\mu$, $L_\mu = \partial_\mu UU^\dagger$, $R_\mu = \partial_\mu U^\dagger U$ with A_μ being the photon field, U being a nonlinear function of the pseudoscalar fields, and Q the quark charge matrix. The lagrangian $\mathcal{L}_{anom}$ has no free parameters and depends only on N_c, the number of colors.

The best known manifestation of the chiral anomaly is the decay $\pi^0 \to \gamma\gamma$, for which the predicted amplitude

$$F_\pi^{th} = \frac{\alpha N_c}{3\pi f_\pi} = 0.025\ Gev^{-1}$$

for $N_c = 3$ is in excellent agreement with the measured value of

$$F_\pi^{exp} = 0.025 \pm 0.001\ Gev^{-1}$$

thus confirming strongly that the number of colors is three.

However, in the case of $\gamma \to 3\pi$, things are not as clear. The measured amplitude [4]

$$F_{3\pi}^{exp} = 12.9 \pm 0.9 \pm 0.5\ Gev^{-3}$$

is not in agreement with the theoretically predicted value

$$F_{3\pi}^{th} = \frac{eN_c}{12\pi^2 f_\pi^3}$$

However, more recent calculations of $F_{3\pi}$ [5] which included $O(p^6)$ corrections to the amplitude have provided a value more consistent with data. The limited accuracy of the current data also point to the need for a more precise experiment.

In the case of η/η' decays, the situation is somewhat analogous. For example, in the decays $\eta/\eta' \to \gamma\gamma$, we have the theoretically predicted amplitudes [6]

$$F_{\eta\gamma\gamma}(0) = -\frac{\alpha N_c}{3\sqrt{3}\pi}\left(\frac{\cos\theta}{f_8} - 2\sqrt{2}\frac{\sin\theta}{f_0}\right)$$

$$F_{\eta'\gamma\gamma}(0) = -\frac{\alpha N_c}{3\sqrt{3}\pi}\left(\frac{\sin\theta}{f_8} + 2\sqrt{2}\frac{\cos\theta}{f_0}\right)$$

Comparing these with the experimental numbers

$$F_{\eta\gamma\gamma}(0) = 0.0249 \pm 0.001 \; Gev^{-1}$$

$$F_{\eta'\gamma\gamma}(0) = 0.0328 \pm 0.0024 \; Gev^{-1}$$

we predict values for the octet and singlet decay constants of $f_8 \approx 1.25 f_\pi$ and $f_0 \approx 1.04 f_\pi$ with the mixing angle between the η and the η' set to $\theta \approx -20°$. The value for f_8 agrees well with that obtained by a full one-loop calculation in χPT [2,7].

However, the analysis of the $\eta/\eta' \to \pi^+\pi^-\gamma$ decays is complicated by the strong ρ^0-meson dominance in the $\pi^+\pi^-$ spectrum measured at various facilities [8]. The anomalous amplitudes in this case have to be supplemented with higher order corrections, particularly from ρ-exchange. Previous attempts, in this regard, to fit the observed ρ^0-spectrum have not been very successful. However, there has been some recent progress in the calculations [9].

Recently, there has been considerable interest in calculating higher order corrections to anomalous processes [10]. At order $O(p^6)$, the contributions are of two types, 1) Loop diagrams involving one anomalous vertex and one vertex from $\mathcal{L}^{(2)}$, and 2) Counterterms at $O(p^6)$. A counterterm lagrangian including all possible terms has been constructed [11]. The low-energy constants that parametrise the $\mathcal{L}^{(6)}$ lagrangian serve to absorb the divergences arising from the loop diagrams. The finite parts of these counterterms are assumed to be saturated by vector meson contributions in a manner similar to the non-anomalous sector [10]. There is a need for more phenomenological input in this regard.

Experimentally, tests of the chiral anomaly are being proposed or conducted at various locations. Experiments at Fermilab [12] and CEBAF [13] will probe the $\gamma \to 3\pi$ vertex with greater accuracy. The coming online of the e^+e^- collider at Daϕne will provide a wealth of information on radiative and nonleptonic kaon decays. There have also been a series of experiments at CERN, IHEP and other facilities on η/η'-decays [14].

These theoretical and experimental attempts to test the chiral anomaly will provide a greater understanding of the role of symmetries in theories of the fundamental interactions.

3 The Chiral Anomaly in Non-Leptonic Kaon Decays

G. Ecker

Institut für Theoretische Physik, Universität Wien, Boltzmanngasse 5, A-1090
Wien, Austria

The standard model is a chiral quantum field theory. The chiral structure is responsible for the existence of the chiral anomaly [1]. Although its theoretical origin and mathematical properties are well understood, experimental tests of this important ingredient of modern particle physics are relatively rare.

The chiral anomaly manifests itself most directly in the low–energy interactions of the pseudoscalar mesons. The appropriate framework to study these effects is chiral perturbation theory [2]. For the strong, electromagnetic and semileptonic weak interactions, all anomalous Green functions can be obtained from the Wess–Zumino–Witten (WZW) functional [3]. The chiral anomaly also appears in the non–leptonic weak interactions starting at $O(p^4)$. There are two different manifestations of the anomaly: the reducible amplitudes [15], which can again be derived directly from the WZW functional, and direct contributions [16,17] that are subject to some theoretical uncertainties.

The reducible amplitudes arise from the contraction of meson lines between a weak $\Delta S = 1$ vertex and the WZW functional. The so–called pole contributions can be given in closed form [15] as a local Lagrangian. There are other reducible contributions that cannot be written in local form. However, in the octet limit all reducible anomalous amplitudes of $O(p^4)$ can be predicted in terms of the single lowest–order coupling G_8.

The direct anomalous contributions arise from the contraction of the W boson field between a strong Green function on one side and the WZW functional on the other side. Using the operator product expansion to integrate out the heavy fields, one gets an effective Hamiltonian in terms of four–quark operators, which must be realized at the bosonic level in the presence of the anomaly. There are both factorizable (leading in $1/N_c$) and non–factorizable (non–leading in $1/N_c$) contributions. The bosonized form of the direct anomalous amplitude can be fully predicted [16]. At $O(p^4)$, the anomaly turns out to contribute to all the possible octet operators proportional to the ε tensor, but it does not produce any additional non–covariant terms. The coefficients of these four operators get also non–factorizable contributions of non–anomalous origin that cannot be computed in a model–independent way.

A complete list can be given [18] of all kinematically allowed non–leptonic K decays that get local contributions from the anomaly at $O(p^4)$. Only radiative K decays are sensitive to the anomaly in the non–leptonic sector. The most frequent "anomalous" decays $K_L \to \pi^+\pi^-\gamma$ and $K^+ \to \pi^+\pi^0\gamma$ share the remarkable feature that the normally dominant bremsstrahlung amplitude is strongly suppressed, making the experimental verification of the anomalous amplitude substantially easier. This suppression has different origins: $K^+ \to \pi^+\pi^0$ proceeds through the small 27–plet part of the non–leptonic weak interactions, whereas $K_L \to \pi^+\pi^-$ is CP–violating. The remaining non–leptonic K decays with direct anomalous contributions are either suppressed by phase space or by the presence of an extra photon in the final state.

For $K_L \to \pi^+\pi^-\gamma$, the direct emission rate is completely dominated by the magnetic amplitude. There is however a strong destructive interference between the $O(p^4)$ contribution and the anomalous reducible amplitude, first appearing at $O(p^6)$. There is also an important VMD contribution at $O(p^6)$, which generates a sizeable dependence of the magnetic amplitude on the photon energy. Although one cannot make absolute predictions for this decay, chiral perturbation theory establishes a correlation between the rate and the photon energy slope in agreement with experiment.

For $K^+ \to \pi^+\pi^0\gamma$, there is a potentially sizeable electric amplitude interfering with bremsstrahlung. This interference must be taken into account in the experimental analysis to extract the contribution of the anomaly to the rate. The VMD contribution to the magnetic amplitude is less important than in $K_L \to \pi^+\pi^-\gamma$. The measured direct–emission rate is in good agreement with the factorization estimate.

Although not as straightforward as for electromagnetic and semileptonic weak processes, non–leptonic K decays offer interesting possibilities for experimental tests of the chiral anomaly.

4 $K \to \pi\pi\gamma$: a Search for Novel Couplings in Kaon Decays

Giancarlo D'Ambrosio

[1] INFN - Sezione di Napoli, 80125 Napoli Italy

Gino Isidori

Dipartimento di Fisica, University of Rome, "La Sapienza", I-00185 Italy

$K \to \pi\pi\gamma$ decays are an interesting channel to test chiral meson structure. Of course to this purpose we have to be able to separate the inner bremsstrahlung (A_{IB}) contribution

$$A_{IB}(K(p_K) \to \pi_1(p_1)\pi_2(p_2)\gamma(q,\varepsilon)) = e\left(\frac{\varepsilon p_b}{q p_b} - \frac{\varepsilon p_a}{q p_a}\right) A(K \to \pi\pi), \qquad (4.1)$$

($(p_a, p_b) \equiv (p_+, p_-)$ for the neutral kaon decays and $(p_a, p_b) \equiv (p_+, p_K)$ or $(p_a, p_b) \equiv (p_K, p_-)$ for the charged ones), predicted by QED, from the direct emission (DE) amplitude, which is the genuine structure dependent contribution.

The $K \to \pi\pi\gamma$ amplitude is usually decomposed also in a different way, separating the electric and the magnetic terms. Defining the dimensionless amplitudes E and M as in [18,19], we can write:

$$A(K \to \pi\pi\gamma) = \varepsilon_\mu(q) \left[E(z_i)(p_1 \cdot q\, p_2^\mu - p_2 \cdot q\, p_1^\mu) + M(z_i)\epsilon^{\mu\nu\rho\sigma}p_{1\nu}p_{2\rho}q_\sigma\right]/m_K^3,$$

where

$$z_i = \frac{p_i \cdot q}{m_K^2}, \qquad \text{and} \qquad z_3 = \frac{p_K \cdot q}{m_K^2} = z_1 + z_2.$$

As it can be seen from eq. 4.1, the inner bremsstrahlung amplitude can contribute only to the $E(z_i)$ term, while the direct emission amplitude can contribute both to the

$E(z_i)$ and the $M(z_i)$ terms. Summing over photon helicities there is no interference among electric and magnetic terms:

$$\frac{\partial^2 \Gamma}{\partial z_1 \partial z_2} = \frac{m_K}{(4\pi)^3}(|(E(z_i)|^2 + |M(z_i)|^2)\left[z_1 z_2 - (1 - 2z_3 - 2r_m^2) - r_m^2(z_1^2 + z_2^2)\right]$$

$(r_m = m_\pi/m_K)$. Thus the two contributions A_{IB} and $A_{DE}^{Electric}$ can interfere in the $E(z_i)$ amplitude, contrary to the case of the amplitude $M(z_i)$ where only a direct emission contribution appears:

$$|A(K \to \pi\pi\gamma)|^2 = |A_{IB}|^2 + 2 \cdot \Re\{A_{IB}^* A_{DE}^{Electric}\} + |A_{DE}^{Electric}|^2 + |A_{DE}^{Magnetic}|^2.$$

The magnetic and the electric direct emission amplitudes can be decomposed in a multipole expansion [19]. In the approximation to consider only the lowest multipole component (the dipole one) the electric amplitude is CP conserving in the K_S decay and CP violating in the K_L one, while the magnetic amplitude is CP conserving in the K_L decay and CP violating in the K_S one. For this reason, since A_{IB} is enhanced by the pole at zero photon energy, the K_L decay is completely dominated by the electric transition, while electric and magnetic contributions are of the same order in the K_L decay. Higher multipoles are suppressed by angular momentum barrier, but they carry informations on the chiral expansion. In ChPT one has only IB contributions at $O(p^2)$. At the next order DE amplitudes appear contributing to E and M. The contributions to M have been discussed in Ref. [18]. Also all the $O(p^4)$ electric contributions have been computed [18,20,21]. Interestingly the $O(p^4)$ counterterm contributions to DE are the same for $K_S \to \pi^+\pi^-\gamma$ and $K^+ \to \pi^+\pi^0\gamma$. Particularly we have been interested to the contributions of two p^6 typical operators to these decays with the goals [21]: i) to test the convergence of ChPT and ii) to test the role of vector mesons. We have searched for kinematically regions where the $O(p^4)$ contributions where less prominent. We just outline the main results of the studied decays.

$K_S \to \pi^+\pi^-\gamma$ Bremsstrahlung $(BR(K_S \to \pi^+\pi^-\gamma)_{IB, E_\gamma^* > 20\ MeV} \simeq 4.80 \cdot 10^{-3})$ is two orders of magnitudes larger than the corresponding interferencial DE contribution [20,21], whose size is largely determined by the $O(p^4)$ counterterm.

$K_L \to \pi^+\pi^-\gamma$ The Bremsstrahlung and magnetic contribution are comparable $(\sim 10^{-5})$. It is shown [21] that $O(p^4)$ loop contribution, which generates a quadrupole component might be smaller than the $O(p^6)$ counterterm contribution.

$K^+ \to \pi^+\pi^0\gamma$ Due to the $\Delta I = 1/2$ rule the bremmsstrahlung is comparable $(\sim 10^{-4})$ to direct emission component and the studied p^6 counterterms are not easy to be distinguished from the p^4 contributions.

$K_L \to \pi^0\pi^0\gamma$ This decay is suppressed, since it requires a quadrupole component, which is generated from the studied p^6 counterterms. Thus it is very sensitive to the size of the coefficients of these operators. However the branching ratio is small $(\sim 10^{-9})$.

5 Chiral Anomaly Tests

Murray A. Moinester

School of Physics and Astronomy, Raymond and Beverly Sackler Faculty of Exact Sciences, Tel Aviv University, 69978 Ramat Aviv, Israel

The Chiral Axial Anomaly can be studied with a 600 GeV pion beam in FNAL experiment E781 [12,22]. For the γ-π interaction, the abnormal intrinsic parity (chiral anomaly) component of the effective chiral lagrangian leads to interesting predictions [23] for the process $\gamma \rightarrow 3\pi$; described by the amplitude $F_{3\pi}$. We consider perturbative expansions of the lagrangian (χPT) to terms quartic in momenta and masses ($O(p^4)$), and higher. The anomaly leads to a prediction for $F_{3\pi}$ in terms of N_c, the number of colors in QCD; and f, the charged pion decay constant. We use $N_c = 3$ and f= 92.4 $\pm$ 0.2 MeV [24] in the equations given previously [4,23] for this amplitude. The $F_{3\pi}$ prediction is: $F_{3\pi} = N_c(4\pi\alpha)^{\frac{1}{2}}/(12\pi^2 f^3) \sim 9.7 \pm 0.2~GeV^{-3}$, $O(p^4)$. We estimate a theoretical uncertainty of 0.2 GeV^{-3} from f and including the accuracy of the $O(p^4)$ prediction. The latter is of order $m_\pi^2/\Lambda^2 \sim 2\%$, where $\Lambda \sim 1~GeV$ sets the scale for the χPT expansion.

The amplitude $F_{3\pi}$ was measured by Antipov et al. [4] at Serpukhov with 40 GeV pions. They studied pion production by a pion in the nuclear Coulomb field via the Primakoff reaction $\pi^- + Z \rightarrow \pi^{-\prime} + \pi^0 + Z'$, where Z is the nuclear charge. In the one-photon exchange domain, this reaction is equivalent to $\pi^- + \gamma \rightarrow \pi^{-\prime} + \pi^0$. The cross section formula for the Primakoff reaction was given in Ref. 5, and depends on $F_{3\pi}^2$, Z^2, and kinematic variables. The Antipov et al. data sample (roughly 200 events) covered the range $s < 10.~m_\pi^2$; where $\sqrt{s}$ is the invariant mass of the $\pi^-\pi^0$ final state. The experiment yielded $F_{3\pi} = 12.9 \pm 0.9(stat) \pm 0.5(sys)~GeV^{-3}$. The cited experimental result differs from the $O(p^4)$ expectation by at least two standard deviations. Therefore, the chiral anomaly prediction at $O(p^4)$ is not confirmed by this data.

Bijnens et al. [23] studied $O(p^6)$ χPT corrections in the anomalous sector. They determine parameters of the lagrangian via vector meson dominance (VMD) calculations. For $F_{3\pi}$, the corrections increase the lowest order value by roughly 10%, which then changes the theoretical prediction by 1. GeV^{-3}. Given the VMD assumption, we make a rough uncertainty estimate of 30% uncertainty for this contribution. The prediction, including the errors given previously is then: $F_{3\pi} \sim 10.7 \pm 0.5~GeV^{-3}$, $O(p^6)$; almost consistent with the data.

The Primakoff cross section at 600 GeV energy is about 100 nb for a C^{12} target for an s interval of 4-10 m_π^2. The expected number of two-pion events for this s-interval is about 2. $\times 10^4$. The large number of events will allow analysis of the data separately in different intervals of s. This is important to control systematic uncertainties.

Another reaction to determine $F_{3\pi}$ is $\pi^- + e \rightarrow \pi^{-\prime} + \pi^0 + e'$, whereby a high energy pion scatters inelasticly from a target electron in an atomic orbit. The number of such events observed by Amendolia et al. [25] was 36 for a Hydrogen target. The experiment did not extract a value for $F_{3\pi}$. For this reaction on a carbon target, as in Fermilab E781, the number of expected events is roughly 2000. The experimental backgrounds for this reaction have been described.

How well does χPT work in the anomalous sector? We described how the Fermilab E781 experiment can give improved answers to this question.

This work was supported by the U.S.-Israel Binational Science Foundation (BSF), Jerusalem, Israel.

6 Study of the Chiral Anomaly in the $\gamma\pi^+ \to \pi^+\pi^0$ Reaction Near Threshold

R. A. Miskimen

Department of Physics and Astronomy, University of Massachusetts, Amherst, MA
01003-4525

A test of the chiral anomaly in the the $\gamma\pi^+ \to \pi^+\pi^0$ reaction near threshold is planned for CEBAF [13]. The $\gamma \to 3\pi$ amplitude, $F^{3\pi}$, first derived from PCAC [26] and later from the Wess-Zumino-Witten Lagrangian [3], is given by

$$F^{3\pi} = \frac{eN_c}{12\pi^2 F_\pi^3} = 9.5 \ GeV^{-3}.$$

It has a form similar to the π^0 decay amplitude in that it depends only on the number of colors N_c and the pion decay constant F_π. While the chiral anomaly is associated with low energy reactions involving pseudo-scalar mesons, it is generally regarded as a short range phenomena. Evidence for this seen in the VMD prediction for this amplitude, which is 50% larger than the chiral perturbation theory (CPT) result [27].

The $\gamma\pi^+ \to \pi^+\pi^0$ reaction will be studied by measuring $\gamma p \to \pi^+\pi^0 n$ cross sections near the pion pole ($t \approx -m_\pi^2$) using tagged photons with energies between 1 and 2 GeV, and the CEBAF Hall B CLAS detector. Define the four-vectors q, k_1, and k_2, as the four-vectors of the photon, and outgoing π^+ and π^0, respectively. Also define $s = (k_1 + k_2)^2$, $t' = (q - k_1)^2$, and $u = (q - k_2)^2$. Then the CPT prediction for $F^{3\pi}$ is strictly valid in the limit where all pairs of invariant masses are zero, $q^2 = s = t' = u = 0$. In the physical region, where $s > 4m_\pi^2$ and $t = (q-k_1-k_2)^2 < 0$, $F^{3\pi}$ acquires momentum dependence on q^2, s, t', and u. It is generally expected that VMD can yield the momentum dependence of $F^{3\pi}$, and Fig. 6.1 shows the s dependence of $F^{3\pi}$ for several different theoretical models [5,28,29].

As discussed by Moinester in this working group summary, there has been a measurement of $F^{3\pi}$ by Antipov *et al.* at Serpukhov using a 40 GeV/c π^- beam [4]. Figure 6.1 shows the value obtained for $F^{3\pi}$ averaged over the range $s = 5 - 11m_\pi^2$ and compared to various models for $F^{3\pi}(s)$. The result obtained by Antipov, $F^{exp} = 12.9 \pm 0.9 \pm 0.5 \ GeV^{-3}$, is higher than the chiral result $F^{theo} = 9.5 \ GeV^{-3}$, and taken literally would imply the number of colors is four. The data point is also in agreement with the VMD model.

While the Serpukhov experiment was statistics limited, with real photons it will be possible to acquire high statistics (several thousand events) in a few weeks of running. The advantage to a photon induced reaction near the pion pole is that both signal and backgrounds will be electromagnetic in origin, which should be optimal for background rejection. Background reactions can be minimized with cuts on s, t, the π^+ center of mass angle, and the πn invariant mass. Linearly polarized photons can also be used to enhance signal-to-background (see section by K. Wang). In addition, the use of real photons will eliminate ω-pole contributions to the amplitude [5,28,29].

The data will be analyzed using the Chew-Low extrapolation technique, where the function $f(t) \equiv (t - m_\pi^2)^2 d^2\sigma/dtds$ is extrapolated to the pion pole ($t = m_\pi^2$) using a polynomial fit to the data. At the pion pole $f(m_\pi^2)$ is related to $\sigma_{\gamma\pi \to \pi\pi}$ through known kinematic factors [30]. In addition, a theoretical calculation of the cross section, including background amplitudes, is planned [31].

Figure 6.1 shows the anticipated errors in $F^{3\pi}$ for the proposed measurement as a function of s. In calculating the errors, a 10% systematic error in the cross section was added in quadrature, and it was assumed that the errors are not background limited. Compared to the data point of Antipov et al., the proposed experiment will measure $F^{3\pi}$ with greater accuracy, and the momentum dependence of $F^{3\pi}$ can be tested.

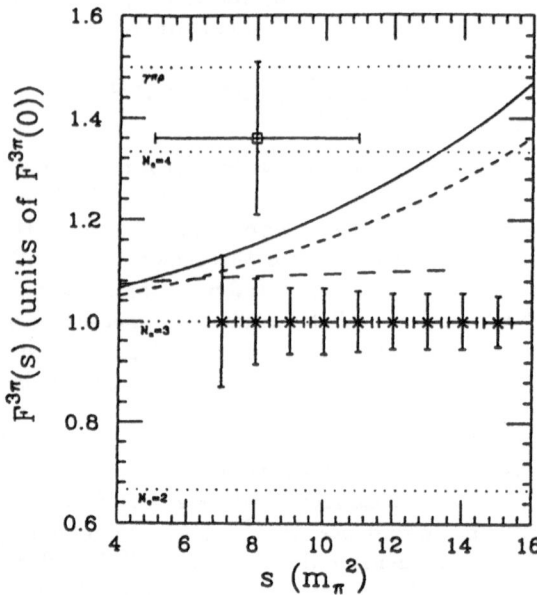

Fig. 6.1 $F^{3\pi}(s, t', u)$ as a function of s for the models of Terent'ev [28] (solid curve), Rudaz [29] (short-dashed curve), and for Bijnens et al. [5] (long-dashed curve). The dotted lines show the amplitudes for $N_c = 4$, $N_c = 3$, and $N_c = 2$. Also shown is the low energy limit of a $\gamma \to \pi\rho \to 3\pi$ calculation [27]. The square data point is from Antipov et al. [4] The other data points show the anticipated statistical and systematic uncertainties for the proposed measurement.

7 Study of the Chiral Anomaly Using Linearly Polarized Photons

K. Wang and B. Norum

Department of Physics, University of Virginia, Charlottesville, VA 22901-2458

As discussed in the previous sections, the chiral anomaly term in the effective Lagrangian of chiral perturbation theory (χPT) predicts the value of $F^{3\pi}$, the coupling constant of the $\gamma - 3\pi$ vertex. Experimental measurement of $F^{3\pi}$ will be a rigorous test of χPT. $F^{3\pi}$ can be measured from the $\gamma p \rightarrow \pi^+\pi^o n$ reaction as has been proposed at CEBAF with the CLAS detector and an unpolarized tagged photon beam[13]. One of the difficulties in this experiment are the backgrounds, which arise mainly from ρ-meson production and Δ-production. This problem can be improved in an effective way by using linearly polarized tagged photons. When the photons are polarized, the angular distribution in the two-pion rest frame can be expressed as [32]

$$W(cos\theta, \phi) = \frac{3}{8\pi} sin^2\theta (1 - P\epsilon \, cos2\phi)$$

where ϵ is the photon polarization, θ and ϕ are the polar and azimuthal angles of the pion relative to the photon incoming direction and the photon polarization plane respectively. $P = -1$ for t-channel pseudoscaler (one-pion exchange) process which is responsible to the $\gamma - 3\pi$ coupling; and $P = 1$ for diffractive process by which the ρ-meson is produced. Therefore the t-channel contribution can be separated from backgrounds by the different dependence on angle ϕ.

Currently there are two proposals for producing polarized tagged photons in Hall B at CEBAF. One uses laser back scattering from electron beams [33], the other bremsstrahlung radiation from a diamond crystal [34]. In the first method, polarized photons from a laser are stored in a resonant optical cavity to reach high intensity. High energy and high flux polarized photons are produced after compton back scattering from energetic electrons. The polarization can be as high as 100%. With a 6 GeV electron beam and frequency doubled laser light, almost 2 GeV polarized photons can be obtained, which is just at the energy range suitable for this experiment. Compared with bremsstrahlung radiation, there is no huge peak at the low energy tail of the photon spectrum, and hence this coincedence measurement can be performed without worrying about the high rate from low energy photons. In the other method, where the polarized photons are produced in bremsstrahlung radiation from a diamond crystal, the polarization is approximately 40% for $E_\gamma = 2$ GeV with 4 GeV electron beam. The drawback to this method is the low tagging efficiency and narrower photon spectrum.

With a polarized tagged photon beam, and with the experimental results and experience which will be obtained using unpolarized photons, a higher quality measurement of $F^{3\pi}$ will be practicable at a later time. In addition, a parallel theoretical calculation of the cross sections with polarized photons should be undertaken.

8 Double-Pion Photoproduction

M. Benmerrouche

Saskatchewan Accelerator Laboratory, University of Saskatchewan, Saskatoon,
Saskatchewan S7N 0W0, Canada

The past few years have seen renewed experimental and theoretical interest in electromagnetic pion production from nucleons and nuclei. The purpose of this contribution is to focus on the recent theoretical work we have begun at the Saskatchewan Accelerator Lab on double pion photoproduction from the nucleon [35]. Such an analysis is required if one is to understand the threshold region in connection with the "LET" and to be able to explore the second resonance region where many resonance parameters are poorly known. Recently [36], there is some suggestion that double pion photoproduction could be used to determine $\gamma \pi^+ \to \pi^+ \pi^0$ from which one hopes to extract the axial anomaly $\gamma \to 3\pi$. The experimental data base on double pion photoproduction is quite primitive. However, there are preliminary results just completed at the tagged photon beam facility MAMI at MAINZ and have greatly improved the quality of the data [37]. We use an effective Lagrangian model based on the coupling of pions to the nucleon and the delta. The electromagnetic interaction is introduced via minimal substitution leading to a set of Feynman diagrams to be evaluated at the tree level [35]. The numerical results of our calculations are valid only in the energy range roughly below $650 MeV$ and agree rather well with the experimental data on $\gamma p \to \pi^+\pi^- p$. At higher photon lab energies, the model is deficient vis-a-vis the inclusion of other resonances such as the $D_{13}(1525)$ and the roper $P_{11}(1440)$ which are expected to play an important role. In Table I, we display our predictions for the threshold values of the quantity X defined by [35]

$$X = \frac{\alpha M_N^2}{(2\pi)^4} \sum_{spins,\epsilon} |\mathcal{M}_{fi}|^2, \tag{8.1}$$

where $\alpha = e^2/(4\pi)$ and M_N the nucleon mass. We have chosen three values of the off-shell parameter Λ necessary in describing an interacting spin-3/2 field. One can see that the delta contribution can be substantial at threshold and the Λ dependence is only minor. In Fig.8.1 we show the total cross section for the reaction $\gamma p \to \pi^+\pi^- p$ for photon lab energies up to $650 MeV$. The data are the results of the bubble chamber experiments from Frascati [38] (indicated by triangles) and the most recent preliminary measurements from Daphne collaboration performed at MAINZ [39] (shown in dots). It is clear that the Δ contribution dominate the reaction and the nonresonant contribution is essentially negligible (indicated by dashed line). Extension of the model to higher energies will be discussed elsewhere.

	Born		Born + $\Delta(1232)$	
		$\Lambda = 0$	$\Lambda = -0.250$	$\Lambda = -0.175$
$p\pi^+\pi^-$	3.234	1.739	1.824	1.816
$n\pi^+\pi^\circ$	1.211	0.644	0.673	0.674
$p\pi^\circ\pi^\circ$	0.059	0.034	0.036	0.035
$n\pi^+\pi^-$	2.226	1.146	1.201	1.203
$p\pi^-\pi^\circ$	1.211	0.644	0.673	0.674
$n\pi^\circ\pi^\circ$	0.004	0.004	0.004	0.004

Contributions to the quantity X as defined by eq. 8.1 in $\mu barn$.

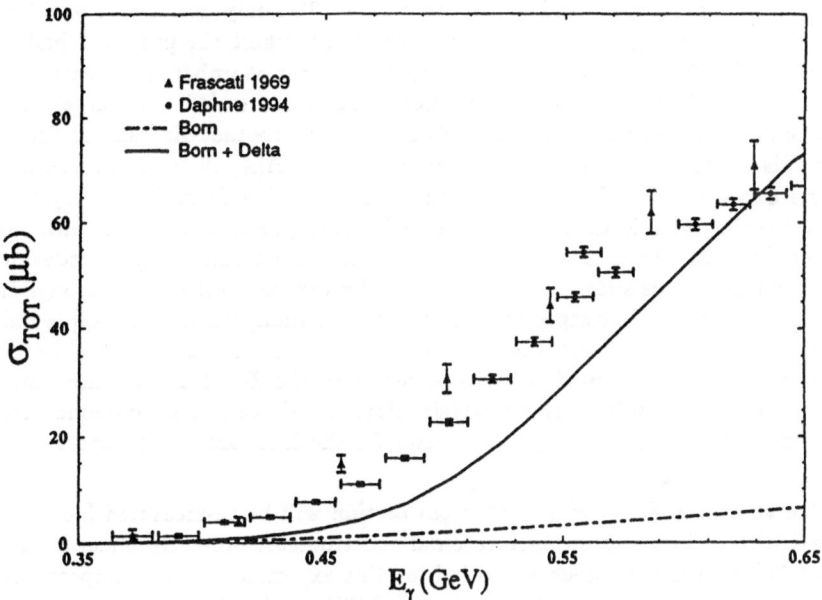

Fig. 8.1 Theoretical predictions for the reaction $\gamma p \rightarrow \pi^+\pi^- p$. The dashed line is the contribution from the nonresonant diagrams alone. The solid line includes the intermediate delta resonance diagrams. The triangles and dots are the experimental data from Frascati and Daphne respectively (see text).

9 Measurement of the $N \rightarrow \Delta$ Axial-Vector Form Factor

R. S. Hicks

Department of Physics and Astronomy, University of Massachusetts, Amherst, MA 01003-4525

A current priority is to determine the electromagnetic form factors of the $N \rightarrow \Delta_{33}$ transition. On the other hand, the corresponding axial-vector form factors remain poorly known. These weak form factors not only provide direct quantitative tests of our understanding of the structure of the light quark baryons, they also constrain the interpretation of other reactions. For example, a good knowledge of the axial vector form factor is essential for a precise measurement of parity-violation in the $N \to \Delta_{33}$ transition.

Although neutrino scattering provides the most direct course of measuring the axial form factors, as indicated by Fig. 9.1, only general trends have been established by this means due to the miniscule neutrino cross sections. An alternative method, proposed in the late sixties [40], relies upon low-energy theory to relate $\pi\Delta$ electroproduction to the axial matrix element $< \Delta|A_\mu|N >$ as well as electromagnetic vertices between the nucleon, Δ, and higher resonances. The only useful data of this type are from a measurement of $ep \to e'\pi^-\Delta^{++}$ at DESY where the produced hadrons were detected with full angular acceptance in a streamer chamber [41]. Features of these data, also confirmed in photoproduction measurements [42], are largely consistent with the dominance near threshold of the $\pi\Delta$ Born contact diagram. In the soft-pion limit this contact term is directly related to the $< \Delta|A_\mu|N >$ matrix element which can be expressed [43] in terms of four axial transition form factors $H_1(t), ...H_4(t)$, where t is the usual Mandelstam variable. However, because measurements cannot be made exactly at threshold, it is necessary to consider terms of higher-order in m_π^2, as well as backgrounds such as $\pi\Delta$ production by various mechanisms through intermediate states. In their interpretation of the DESY data, Bartl et al. [43] found only a slight sensitivity of the cross section to the electromagnetic $< \Delta^{++}|V_\mu|N >$ vertex while, among the intermediate resonances, only the $D_{13}(1520)$ contribution proved significant. As a result of this analysis, Bartl et al. were able to deduce the $t = 0$ intercepts and a common slope parameter for the four axial form factors.

The threshold $ep \to e'\Delta^{++}\pi^-$ cross section will be re-measured [44] at CEBAF using CLAS, a large-acceptance toroidal spectrometer presently under construction. About 10^6 events have been projected for this experiment, a vast improvement over the several thousand obtained in the earlier DESY work. The excellent track resolution achievable in CLAS makes it unnecessary to observe the produced soft π^-, which can be precisely reconstructed. Furthermore, the tracking resolution and large acceptance of CLAS will combine to minimize systematic errors. It has also been shown that background events from the competing non-resonant and $\Delta^0\pi^+$ production can be eliminated, with an uncertainty of a few percent, by Dalitz plot analyses.

Figure 9.1 shows that, in both statistical precision and kinematic range, the projected results for the $ep \to e'\Delta^{++}\pi^-$ cross section far surpass the quality of the existing neutrino data. The high statistical precision of the electroproduction measurements, even very close to threshold, will allow more careful checks of the dominance of the contact term than possible in the previous electroproduction experiment. Hence the contributions of background processes can be established more confidently. Presently under consideration are methods to extend the measurements as close as possible to the soft-pion $Q^2 \to 0$ limit, where Chiral Perturbation theory can make firm predictions regarding the axial vector transition form factors.

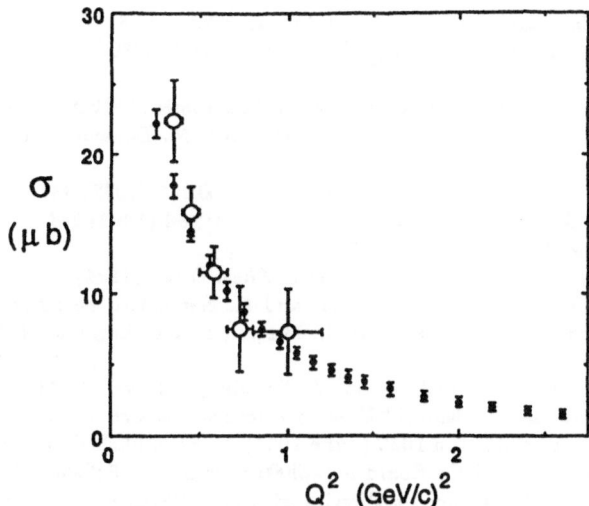

Fig. 9.1 Open circles indicate neutrino results for $\Delta^{++}\pi^-$ production; solid points are projected electroproduction results for a beam energy of 4 GeV and missing mass $W < 1.36$ GeV.

10 Acknowledgements

The working group members thank the organizers of the conference for their encouragement and hospitality.

References

[1] S.L. Adler, *Phys. Rev.* **177** (1969) 2426; J.S. Bell and R. Jackiw, *Nuovo Cim.* **60A** (1969) 47; W.A. Bardeen, *Phys. Rev.* **184** (1969) 1848

[2] S. Weinberg, Physica **96A** (1979) 327; J. Gasser and H. Leutwyler, Ann. Phys. **158** (1984) 142; Nucl. Phys. **B250** (1985) 465

[3] J. Wess and B. Zumino, *Phys. Lett.* **37B** (1971) 95; E. Witten, *Nucl. Phys.* **B223** (1983) 422

[4] Y. M. Antipov, *et al.*, *Phys. Rev.* **D36** (1987) 21

[5] J.Bijnens et al. *Phys.Lett.B* **237** (1990) 488

[6] B.R. Holstein, UMHEP preprint UMHEP-386

[7] J.F. Donoghue et. al., Phys. Rev. Lett. **55** (1985) 2766

[8] M.S. Chanowitz in: Proc. 6th Int. Workshop on Photon-Photon Collisions 1984 (World Scientific, Singapore 1985) and references therein

[9] M. Benayoun et. al. Z. Phys. **C58** (1993) 31 and references therein

[10] J.F. Donoghue and D. Wyler, Nucl. Phys. **B316** (1989) 289; J. Bijnens et. al., Z. Phys. **C46** (1990) 599

[11] D. Issler, SLAC-PUB 4943 (1990)

[12] J. Russ, Carnegie-Mellon U., spokesman, FNAL E781 SELEX proposal; M. A. Moinester, Bulletin Board hep-ph@xxx.lanl.gov/9409307, Preprint 2204-94, these proceedings

[13] R.A.Miskimen, K.Wang, A.Yegneswaran, Spokesmen, CEBAF Proposal PR-94-015

[14] S.I. Bityukov et. al., Z. Phys. C50 (1991) 451; M. Benayoun et. al. Z. Phys. C58 (1993) 31

[15] G. Ecker, H. Neufeld and A. Pich, *Phys. Lett.* **B278** (1992) 337

[16] J. Bijnens, G. Ecker and A. Pich, *Phys. Lett.* **B286** (1992) 341

[17] H.-Y. Cheng, *Phys. Rev.* **D42** (1990) 72

[18] G. Ecker, H. Neufeld and A. Pich, *Nucl. Phys.* **B413** (1994) 321

[19] G. D'Ambrosio, G. Ecker, G. Isidori and H. Neufeld to be published in the "The DAΦNE Physics Handbook", eds. L. Maiani, L. Pancheri and N. Paver, II edition and references therein

[20] G. D'Ambrosio, M. Miragliuolo and F. Sannino, Z. Physik **C 59** (1993) 451

[21] G. D'Ambrosio and G. Isidori "$K \rightarrow \pi\pi\gamma$ decays: a search for novel couplings in kaon decays", Roma Preprint n.1030 (June 1994), to be published in Z. Physik C.

[22] M. A. Moinester, Bulletin Board hep-ph@xxx.lanl.gov - 9409307, Contribution: Proceedings of the Conference on Physics with GeV-Particle Beams, Juelich, Germany, Aug. 1994, World Scientific, Eds. H. Machner and K. Sistemich.

[23] J.Bijnens, *Int. Journal Mod. Phys. A* **8** (1993) 3045

[24] B.R.Holstein, *Phys.Lett. B* 244 (1990) 83 *Phys.Rev. D* **36** (1987) 21

[25] S.R.Amendolia, *et al*, Phys. Lett.B **155** (1985) 457

[26] S. L. Adler, B. W. Lee, S. B. Treiman, and A. Zee, *Phys. Rev.* **D4** (1971) 3497; M. Terent'ev, *JETP Lett.* 14 (1971) 94; R. Aviv and A. Zee, *Phys. Rev.* **D5** (1972) 2372.

[27] T. D. Cohen, *Phys. Lett.* **233** (1989) 467.

[28] M. V. Terent'ev, *Phys. Lett.* **38B** (1972) 419.

[29] S. Rudaz, *Phys. Rev.D* 10 (1974) 3857

[30] T. A. Aibergenov, *et al.*, Proceedings of the Lebedev Physics Institute, Academy of Sciences of the U.S.S.R., **186** (1988) 169.

[31] Private communication, M. Benmerrouche

[32] K. Schilling, and G. Wolf, *Nucl. Phys.* **B61** (1973) 381

[33] Letter of intend to CEBAF PAC6, Spokespersons B. Norum, T.P. Welch, Proposal to CEBAF PAC8, Spokesperson B. Norum, A. Bernstein, and T.P Welch.

[34] Letter of intend to CEBAF PAC8, Spokespersons J.P. Connelly, P.L. Cole.

[35] M. Benmerrouche and E. Tomusiak, *Phys. Rev. Lett.* **73** (1994) 400

[36] R. A. Miskimen, these proceedings and private communication (1994)

[37] T. H. Walcher, these proceedings

[38] G. Gialanella, et. al., Nuovo Cimento **63A**, 892 (1969)

[39] L. Murphy, *Double pion photon production on single nucleons from threshold up to 780MeV*, Ph. D Thesis (1994), Rensselaer, unpublished

[40] T. Ebata, *Phys. Rev.* 154, 1341 (1967); P. Carruthers and K. W. Huang, *Phys. Lett.* **24B**, 464 (1967); *Nuovo Cim.* **107A**, (1968)

[41] P. Joos et al., *Phys. Lett.* 52B, 481 (1974); *Phys. Lett.* **62B**, 230 (1976); K. Wacker et al., *Nucl. Phys.* **B144**, 269 (1978)

[42] Aachen–Berlin–Bonn–Hamburg–Heidelberg–München collaboration, *Phys. Rev.* 175, 1669 (1968)

[43] A. Bartl *et al. Lett. al Nuovo Cim.* 18, 588 (1977); *Il Nuovo Cim.* **45A**, 457 (1978)

[44] CEBAF Proposal PR-94-005, L. Elouadrhiri *et al.* (1994).

Part IV: Hadron Polarizabilities

Low-energy photon-photon collisions to two-loop order

J. Gasser

Institute for Theoretical Physics, University of Bern,
Sidlerstrasse 5, CH-3012 Bern, Switzerland
e-mail: Gasser@itp.unibe.ch

Abstract: I report on a recent calculation of the amplitude for $\gamma\gamma \to \pi^0\pi^0$ to two loops in chiral perturbation theory.

1   Introduction

The cross section for $\gamma\gamma \to \pi^0\pi^0$ and for $\gamma\gamma \to \pi^+\pi^-$ has been calculated some time ago [1, 2] in the framework of chiral perturbation theory (CHPT) [3]-[7] and of dispersion relations. For an early evaluation using effective chiral lagrangians see Ref. [8]. In the case of charged pion-pair production, the chiral calculation [1] at next-to-leading order is in good agreement with the Mark II data [9] in the low-energy region. On the other hand, for $\gamma\gamma \to \pi^0\pi^0$, the one-loop prediction [1, 2] disagrees with the Crystal Ball data [10] and with dispersion theoretic calculations [11]-[17] even near threshold.

In the process $\gamma\gamma \to \pi^+\pi^-$, the leading-order contribution[1] is generated by tree diagrams. One has a control on higher order corrections in this case, in the sense that it is explicitly seen that the one-loop graphs do not modify the tree amplitude very strongly near threshold [1]. Tree diagrams are absent for $\gamma\gamma \to \pi^0\pi^0$ which starts out with one-loop graphs. To establish the region of validity of the chiral representation also in this channel, the amplitude has therefore been evaluated at two-loop order in Ref. [19].

Is a next-to-leading order calculation sufficient in this case? If the corrections are large, the reliability of the result is certainly doubtful. However, a glance at the data shows that the corrections needed to bring CHPT and experiment into agreement are not large–a 25-30% change in amplitude is sufficient. Corrections of this size are rather normal in reactions where pions in an isospin zero S-wave state are present [20]. As an example I mention the isospin zero S-wave $\pi\pi$ scattering length, whose tree-level value [21] receives a 25% correction from one-loop graphs [22]. Corrections of a similar size are present in the scalar form factor of the pion [23].

---

[1]I denote the first nonvanishing contribution to any quantity by "the leading-order contribution", independently of whether it starts out at tree level or at higher order in the chiral expansion.

In the following, I summarize the main result of our calculation, considering in particular the cross section and the pion polarizabilities, referring the reader to Ref. [19] for details. For a discussion of this process in the framework of generalized chiral perturbation theory, see Ref. [24].

2 Kinematics

The matrix element for pion pair production

$$\gamma(q_1)\gamma(q_2) \to \pi^0(p_1)\pi^0(p_2)$$

is given by

$$\langle \pi^0(p_1)\pi^0(p_2)\text{out} \mid \gamma(q_1)\gamma(q_2)\text{in}\rangle = i(2\pi)^4\delta^4(P_f - P_i)T^N \ ,$$

with

$$
\begin{aligned}
T^N &= e^2\epsilon_1^\mu\epsilon_2^\nu V_{\mu\nu} \ , \\
V_{\mu\nu} &= i\int dx e^{-i(q_1x+q_2y)}\langle \pi^0(p_1)\pi^0(p_2)\text{out} \mid Tj_\mu(x)j_\nu(y) \mid 0 \rangle.
\end{aligned}
$$

Here j_μ is the electromagnetic current ($e^2/4\pi = 1/137.036$). The decomposition of the correlator $V_{\mu\nu}$ into Lorentz invariant amplitudes reads with $q_1^2 = q_2^2 = 0$

$$
\begin{aligned}
V_{\mu\nu} &= A(s,t,u)T_{1\mu\nu} + B(s,t,u)T_{2\mu\nu} + \cdots \ , \\
T_{1\mu\nu} &= \frac{s}{2}g_{\mu\nu} - q_{1\nu}q_{2\mu} \ , \\
T_{2\mu\nu} &= 2s\Delta_\mu\Delta_\nu - \nu^2 g_{\mu\nu} - 2\nu(q_{1\nu}\Delta_\mu - q_{2\mu}\Delta_\nu) \ ,
\end{aligned}
\tag{1}
$$

where $\Delta_\mu = (p_1 - p_2)_\mu$, and where

$$
\begin{aligned}
s &= (q_1 + q_2)^2, \ t = (p_1 - q_1)^2, \ u = (p_2 - q_1)^2 \ , \\
\nu &= t - u \ ,
\end{aligned}
$$

are the standard Mandelstam variables. The ellipsis in Eq. (1) denotes terms which do not contribute to the scattering amplitude T^N (gauge invariance). The amplitudes A and B are analytic functions of the variables s, t and u, symmetric under crossing $(t, u) \to (u, t)$. It is useful to introduce in addition the helicity amplitudes

$$
\begin{aligned}
H_{++} &= A + 2(4M_\pi^2 - s)B \ , \\
H_{+-} &= \frac{8(M_\pi^4 - tu)}{s}B \ .
\end{aligned}
$$

The helicity components H_{++} and H_{+-} correspond to photon helicity differences $\lambda = 0, 2$, respectively. The differential cross section for unpolarized photons in the center-of-mass system is

$$\frac{d\sigma^{\gamma\gamma\to\pi^0\pi^0}}{d\Omega} = \frac{e^4s}{1024\pi^2}\beta(s)H(s,t) \ ,$$

$$H(s,t) = |H_{++}|^2 + |H_{+-}|^2 ,$$
$$\beta(s) = (1 - 4M_\pi^2/s)^{1/2}.$$

3 Low-energy expansion

The evaluation of the amplitude $V^{\mu\nu}$ in the framework of CHPT is standard–
I assume that the reader is familiar with the method. For an outline of the
calculation see Ref. [19]. The main points are the following:

1. The underlying effective lagrangian has the structure

$$\mathcal{L} = \mathcal{L}_2 + \mathcal{L}_4 + \mathcal{L}_6 + \cdots ,$$

   where the indices denote the number of derivatives on the pion field, or
   quark mass terms. The leading term $\mathcal{L}_2$ is the nonlinear sigma-model lagrangian. $\mathcal{L}_4$ contains all possible contributions with four derivatives, or
   two derivatives and one quark mass term, or two quark mass insertions.
   These are multiplied with low-energy constants l_i whose divergences absorb the ultraviolet singularities of the one-loop graphs with $\mathcal{L}_2$. Finally,
   $\mathcal{L}_6$ contains terms with six derivatives, four derivatives and one quark
   mass term, etc. [25]. These absorb the ultraviolet divergences of two-
   loop graphs. In the following I consider the isospin symmetric case where
   $m_u = m_d$.

2. The *leading* term in the chiral expansion of $V^{\mu\nu}$ is generated by one-loop
   graphs with $\mathcal{L}_2$. Tree diagrams from $\mathcal{L}_4$ do not contribute to this process–
   the one-loop integrals are therefore finite. The result is [1, 2]

$$H_{++}^{1loop} = \frac{4(s - M_\pi^2)\bar{G}(s)}{sF_\pi^2} , \quad H_{+-}^{1loop} = 0 ,$$

$$\bar{G}(s) = -\frac{1}{16\pi^2}\left\{1 + \frac{2M_\pi^2}{s}\int_0^1 \frac{dx}{x}\ln(1 - \frac{s}{M_\pi^2}x(1-x))\right\} ,$$

(2)

   where $F_\pi \simeq 93.2$ MeV ($M_\pi \simeq 135$ MeV) is the pion decay constant (pion
   mass).

3. The *next-to-leading order* terms are generated by two-loop, one-loop and
   tree-diagrams generated by $\mathcal{L}_2, \mathcal{L}_4$ and $\mathcal{L}_6$, respectively. For an evaluation
   of these and for an explicit expression of the amplitudes A and B see Ref.
   [19]. Here I only note that the amplitudes contain, at this order in the
   low-energy expansion, the renormalized constants $l_i^r, i = 1, \ldots, 6$ from $\mathcal{L}_4$
   and, in addition, three more renormalized parameters $h_\pm^r, h_s^r$ from $\mathcal{L}_6$.

4. The constants l_i^r from $\mathcal{L}_4$ are known [4]. On the other hand, those from
   $\mathcal{L}_6$ have not yet been determined in a systematic manner. The ones which

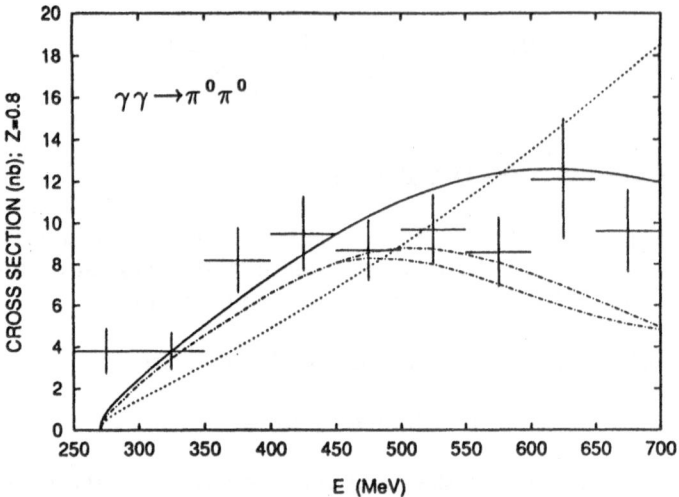

Figure 1: The $\gamma\gamma \to \pi^0\pi^0$ cross section $\sigma(|\cos\theta| \le Z)$ as a function of the center-of-mass energy E at $Z = 0.8$, together with the data from the Crystal Ball experiment [10]. The solid line is the full two-loop result, and the dashed line results from the one-loop calculation [1, 2]. The band denoted by the dash-dotted lines is the result of the dispersive calculation by Pennington (Fig. 23 in Ref. [13]).

contribute to $\gamma\gamma \to \pi^0\pi^0$, namely $h_{\pm}^r$ and h_s^r, have been estimated in [19] in the standard manner [4, 26] using resonance exchange with $J^{PC} = 1^{--}, 1^{+-}, 0^{++}, 2^{++}$. For an estimate of a particular contribution to h_{+}^r by use of sum rules see Ref. [24, 27]. It would be interesting to pin down all three couplings $h_{\pm}^r$ and h_s^r with this technique.

4 The cross section $\gamma\gamma \to \pi^0\pi^0$

Fig. 1 displays the data for the cross section $\sigma(s; |\cos\theta| \le Z = 0.8)$ as determined in the Crystal Ball experiment [10]. They are shown as a function of the center-of-mass energy $E = \sqrt{s}$. The dashed line displays the one-loop result [1, 2], evaluated with the amplitudes (2), whereas the solid line denotes the two-loop result [19]. Finally, the dash-dotted lines show the result of a dispersive analysis (Fig. 23 in Ref. [13]). In that calculation, use was made of the $I = 0, 2$ S-wave $\pi\pi$ phase shifts from Ref. [28] (these phase shifts satisfy [29] constraints imposed by Roy-type equations). The one-loop result is below the data also near the threshold. This fact created some dust in the literature, to the extent that the validity of CHPT in this process was questioned. However, it is seen that the contributions from the two-loop graphs generate the corrections which are needed to bring the calculation into agreement with the present data and with the dispersive calculation. In this connection, I recall that the low-energy constants $h_{\pm}^r$ and h_s^r contribute very little to the cross section below $\sqrt{s} = 450$MeV [30]-[35]. Their exact value is therefore of no concern for the low-energy region in $\gamma\gamma \to \pi^0\pi^0$. In Fig. 2 is shown a comparison of the two-loop result with the calculation of Donoghue and Holstein [14]. These authors use a

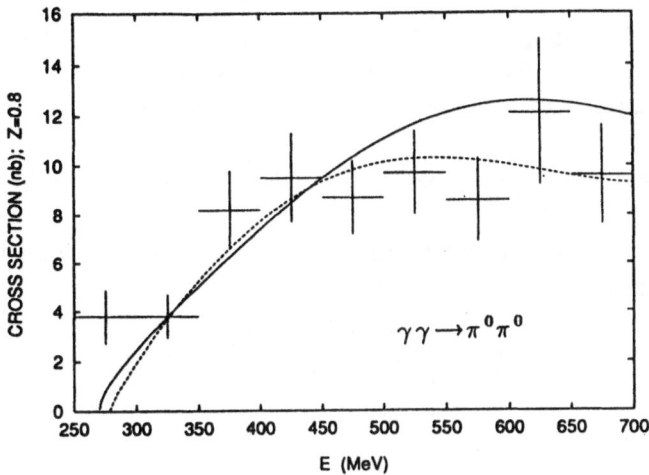

Figure 2: The $\gamma\gamma \rightarrow \pi^0\pi^0$ cross section $\sigma(|\cos\theta| \leq Z)$ as a function of the center-of-mass energy at $Z = 0.8$, with the data from the Crystal Ball [10] experiment. The solid line is the two-loop result, whereas the dashed line is taken from the dispersive analysis of Donoghue and Holstein (Fig. 2 in Ref. [14]).

dispersive representation of the S-wave projected helicity amplitude and fix the subtraction constant from the chiral representation at order E^4. Contributions from resonance exchange which generate the left-hand cut are also added. The final result for the cross section agrees very well with the two-loop calculation below $E = 400$ MeV. The main differences in the two representations are as follows. First, in the dispersive method, higher order terms are partially summed up. I consider the fact that the cross sections agree as an indication that yet higher orders in the chiral expansion indeed do not affect the amplitude in the threshold region very much. (In Ref. [19], the uncertainty due to higher orders is estimated at $15\% - 20\%$ in the cross section below 400 MeV.) Secondly, CHPT reveals that the amplitude contains chiral logarithms, generated by pion loops. All of these effects are not incorporated in the dispersive analysis of Refs. [12, 13, 14]–I refer the reader to Ref. [19] for a more detailed discussion of this point.

5 Pion polarizabilities

Polarizabilities are also treated in the contributions of Moinester [37] and of Baldini and Bellucci to this workshop [38]. I refer the reader to these references for additional material.

5.1 Notation

To set notation, consider Compton scattering for charged pions,

$$\gamma(q_1)\pi^\pm(p_1) \rightarrow \gamma(q_2)\pi^\pm(p_2) \ ,$$

in the laboratory system $p_1^0 = M_\pi$. (The symbol M_π denotes the charged or the neutral pion mass.) The electric ($\bar{\alpha}_\pi$) and magnetic ($\bar{\beta}_\pi$) polarizabilities are obtained by expanding the Compton amplitude at threshold,

$$T^C = 2\left[\epsilon_1 \cdot \epsilon_2^\star \left(\frac{e^2}{4\pi M_\pi} - \bar{\alpha}_\pi \omega_1 \omega_2\right) - \bar{\beta}_\pi \left(q_1 \times \epsilon_1\right) \cdot \left(q_2 \times \epsilon_2^\star\right) + \cdots\right] ,$$

with $q_i^\mu = (\omega_i, q_i)$. For neutral pions, one uses the analogous definition. In terms of the helicity components, one has

$$\bar{\alpha}_{\pi^0} \pm \bar{\beta}_{\pi^0} = \frac{e^2}{4\pi M_\pi} H_{+\mp}(s = 0, t = M_\pi^2) .$$

Below I also use the notation

$$\begin{aligned} (\alpha \pm \beta)^C &= \bar{\alpha}_\pi \pm \bar{\beta}_\pi , \\ (\alpha \pm \beta)^N &= \bar{\alpha}_{\pi^0} \pm \bar{\beta}_{\pi^0} . \end{aligned}$$

5.2 Data on pion polarizabilities

There exist up to now two determinations of charged pion polarizabilities via measurement of the Compton amplitude. At Serpukhov [39], radiative pion-nucleus scattering $\pi^- A \to \pi^- \gamma A$ has been used. Here the incident pion scatters from a virtual photon in the Coulomb field of the nucleus. In the pion production process $\gamma p \to \gamma \pi^+ n$ examined at the Lebedev Institute [40], the incoming photon scatters from a virtual pion. Analyzing the data with the constraint $(\alpha+\beta)^C = 0$ gives[2]

$$(\alpha - \beta)^C = \begin{cases} 13.6 \pm 2.8 & [39] \\ 40 \pm 24 & [40] . \end{cases}$$

The Serpukhov data have been analyzed also relaxing the constraint $(\alpha+\beta)^C = 0$, with the result

$$\begin{aligned} (\alpha + \beta)^C &= 1.4 \pm 3.1(\text{stat.}) \pm 2.5(\text{sys.}) \ [41] , \\ (\alpha - \beta)^C &= 15.6 \pm 6.4(\text{stat.}) \pm 4.4(\text{sys.}) \ [41] . \end{aligned}$$

Here I have converted the value quoted for $\bar{\beta}_\pi$ into a number for $(\alpha - \beta)^C$, adding the errors in quadrature.

Furthermore, also the process $\gamma\gamma \to \pi\pi$ has been used to obtain information on the polarizabilities. In Ref. [17], unitarized S-wave amplitudes have beeen constructed, which contain $(\alpha-\beta)^{C,N}$ as adjustable parameters. A simultaneous fit to Mark II and Crystal Ball data gives

$$\begin{aligned} (\alpha - \beta)^C &= 4.8 \pm 1.0 \ [17] , \\ (\alpha - \beta)^N &= -1.1 \pm 1.7 \ [17] , \end{aligned} \tag{3}$$

[2]I express the polarizabilities in units of 10^{-4}fm^3 throughout.

where I have taken into account that the definition of the polarizabilities in [17] is 4π larger than the one used here (see Ref. [42], Eq. 1). Further, in Ref. [43], the sum $(\alpha + \beta)^{C,N}$ has been determined with similar techniques,

$$(\alpha + \beta)^C = \begin{cases} 0.22 \pm 0.06 & \text{[43]} \quad \text{(Mark II [9])} \\ 0.30 \pm 0.04 & \text{[43]} \quad \text{(CELLO [44])}, \end{cases}$$

$$(\alpha + \beta)^N = 1.00 \pm 0.05 \quad \text{[43]} \quad \text{(Crystal Ball [10])} . \tag{4}$$

Note that the determination of $(\alpha \pm \beta)^{C,N}$ from $\gamma\gamma \to \pi\pi$ suffers from uncertainties which are difficult to estimate in the approach used by Kaloshin et al. [17, 43], because their method does not provide a systematic way to control the inherent uncertainties in the model amplitude used to fit the data. See also Ref. [45] for a critical discussion of this method.

In Ref. [46], the bound $|\bar{\alpha}_{\pi^0}| < 35$ has been obtained from a study of the $e^+e^- \to \pi^0\pi^0\gamma$ reaction.

Finally, information on the charged pion polarizabilities may be obtained from $\gamma\gamma \to \pi^+\pi^-$ data in the following manner [47]. Both the one-loop expression for the transition amplitude and the leading-order expression for $\bar{\alpha}_\pi$ and $\bar{\beta}_\pi$ contain the low-energy constant $2l_6^r - l_5^r$ as the only free parameter. Extracting it from a fit to the cross section determines $\bar{\alpha}_\pi$ and $\bar{\beta}_\pi$ at this order. The same combination of low-energy constants occurs in radiative pion decay [48], and the corresponding value [4] agrees within the errors with the one found from $\gamma\gamma \to \pi^+\pi^-$.

5.3 Chiral expansion

The structure of the quark mass expansion of the polarizabilities is very similar to the one of the threshold parameters in $\pi\pi$ scattering. To illustrate this, I consider, in addition to the polarizabilities, also the chiral expansion of the $I = 0$, S-wave scattering length a_0 [21, 22],

$$
\begin{array}{ccccc}
& tree & 1loop & 2loops & 3loops
\end{array}
$$

$$a_0 = \frac{7M_\pi^2}{32\pi F_\pi^2}\left\{ A + \frac{M_\pi^2 B}{16\pi^2 F_\pi^2} + O(M_\pi^4) + O(M_\pi^6)\right\}$$

$$(\alpha \pm \beta)^{N,C} = \frac{e^2}{64\pi^3 F_\pi^2 M_\pi}\left\{ 0 + A_\pm^{N,C} + \frac{M_\pi^2 B_\pm^{N,C}}{16\pi^2 F_\pi^2} + O(M_\pi^4)\right\}$$

The first line indicates the number of loops required to generate the corresponding term in the quark mass expansion. The similarity of the two expansions is obvious-the only difference being that the expansion for the scattering length starts out with tree graphs [21], whereas the leading order term in the polarizabilities are generated by one-loop diagrams [1, 2, 49],

$$A = 1 \ , \quad A_\pm^N = \begin{pmatrix} 0 \\ -\frac{1}{3} \end{pmatrix} \ , \quad A_\pm^C = \begin{pmatrix} 0 \\ 64\pi^2(2l_5^r - l_6^r) \end{pmatrix} .$$

Table 1: Neutral pion polarizabilities to two loops in units of 10^{-4}fm^3.

	$O(E^{-1})$	$O(E)$			
	1 loop	$h_{\pm}^r$	2 loops	total	uncertainty
$(\alpha + \beta)^N$	0.00	1.00	0.17	$\simeq 1.15$	± 0.30
$(\alpha - \beta)^N$	-1.01	-0.58	-0.31	$\simeq -1.90$	± 0.20
$\bar{\alpha}_{\pi^0}$	-0.50	0.21	-0.07	$\simeq -0.35$	± 0.10
$\bar{\beta}_{\pi^0}$	0.50	0.79	0.24	$\simeq 1.50$	± 0.20

The next-to-leading order terms B and $B_{\pm}^N$ have been determined in Ref. [22] and [19], respectively. $B_{\pm}^N$ contain the low-energy constants $h_{\mp}^r$ from the order E^6 lagrangian. Work to evaluate the corresponding coefficients $B_{\pm}^C$ in the charged channel case is in progress [50].

The numerical value for the *leading-order terms* in the expansion of the polarizabilities is

$$
\begin{aligned}
(\alpha + \beta)^{N,C} &= 0.0 , \\
(\alpha - \beta)^N &= -1.0 , \\
(\alpha - \beta)^C &= 5.3 .
\end{aligned}
$$

The *two-loop* result for the neutral pion case is shown in table 1. The second column contains again the leading order contribution $O(E^{-1})$, whereas the third and fourth ones display the terms of order E. The total values are given in column 5. Finally, the estimates of the uncertainties are shown in the last column. These correspond to the uncertainty with which the low-energy constants were obtained in Ref. [19], and contain neither effects from higher orders in the quark mass expansion nor any correlations.

Turning now to a comparison with the data, it is seen that the two-loop result for $(\alpha \pm \beta)^N$ agrees within the errors with the value found in [17, 43], see Eqs. (3,4)– remember, however, the proviso after Eq. (4). As for the charged channel case, I note that the two-loop contribution B_+^N contains (squares of) chiral logarithms [19],

$$
B_+^N = \frac{2}{9} \left(\ln \frac{M_\pi^2}{\mu^2} - 96\pi^2 l_2^r \right) \ln \frac{M_\pi^2}{\mu^2} + \cdots ,
$$

where μ denotes the scale of renormalization. These are the analogue of the chiral logarithm in the expansion of a_0 [22],

$$
B = -\frac{9}{2} \ln \frac{M_\pi^2}{\mu^2} + \cdots .
$$

The contribution from the chiral logarithms is potentially large. There is no reason why such logarithms should not be present in $B_{\pm}^C$ as well. Therefore, in order to compare the prediction with the data in a meaningful manner, a full two-loop calculation is required also for the charged channel [50].

5.4 On the determination of polarizabilities from data on $\gamma\gamma \to \pi\pi$

A direct measurement of the Compton amplitude $\gamma\pi \to \gamma\pi$ is difficult to achieve, as is illustrated by the scarce data available on this process. Direct experimental information on the pion polarizabilities is therefore difficult to obtain as well. For this reason, one may seek to determine them instead from data on the crossed channel reaction $\gamma\gamma \to \pi\pi$. Let me compare the mathematical situation with pion-nucleon scattering. Here, the value Σ of the amplitude at the Cheng-Dashen point is of considerable theoretical interest, because it is closely related to the sigma term. It has been shown by the Karlsruhe group [51] that Σ can, also in practice, be obtained from the data on $\pi N \to \pi N$ by analytic continuation. In this language, the determination of the polarizabilities from $\gamma\gamma \to \pi\pi$ amounts to the opposite problem: determine from data on $\pi\pi \to N\bar{N}$ the scattering lengths in $\pi N \to \pi N$. It would be interesting to proceed in a manner similar to the case of pion-nucleon scattering in order to find out whether data on $\gamma\gamma \to \pi\pi$ does or does not suffice to pin down in practice the polarizabilities (see also Ref. [45]).

Quite apart from this general setting, one may gain information on the polarizabilities by use of an explicit expression for the amplitude, which serves to interpolate between the Compton threshold and the physical region for $\gamma\gamma \to \pi\pi$. The free parameters in the amplitude are adjusted such that a satisfactory description for $\gamma\gamma \to \pi\pi$ is achieved, which allows one finally to read off the values of the polarizabilities. Examples of this procedure may be found in Refs. [13, 14, 17, 43, 45, 47], for a critical discussion of the method see in paticular Ref. [45]. In the case where the chiral representation [19] is used for the interpolation in the neutral channel, the situation is as follows [14]. As I mentioned above, there are three low-energy constants $h_\pm^r, h_s^r$ which enter the amplitude at order E^6. The two parameters $h_\pm^r$ may be traded for the polarizabilities $(\alpha \mp \beta)^N$, whereas h_s^r may be determined e.g. from resonance exchange. It turns out that the cross section in the low-energy region is not very sensitive to the value of the polarizabilities. To illustrate, Fig. 3 displays the cross section for a fixed value $(\alpha + \beta)^N = 1.15$ and a fixed value of h_s^r, varying $(\alpha - \beta)^N$ between -0.95 and -3.8. See also Fig. 10 in Ref. [14]. The sensitivity of the cross section to $(\alpha + \beta)^N$ is even weaker. The charged channel is discussed in Ref. [14], see in particular Fig. 9 in this reference.

I conclude that, in case the chiral amplitude is used as an interpolation in the neutral channel, low-energy data on the cross section alone will not suffice to pin down the neutral pion polarizabilities at this order in the low-energy expansion [14, 19, 24, 45]. On the other hand, the chiral amplitude contains 3 parameters which may be determined by other means [19, 24, 27]. Once this is achieved, the process $\gamma\gamma \to \pi^0\pi^0$ serves as a consistency check of the calculation.

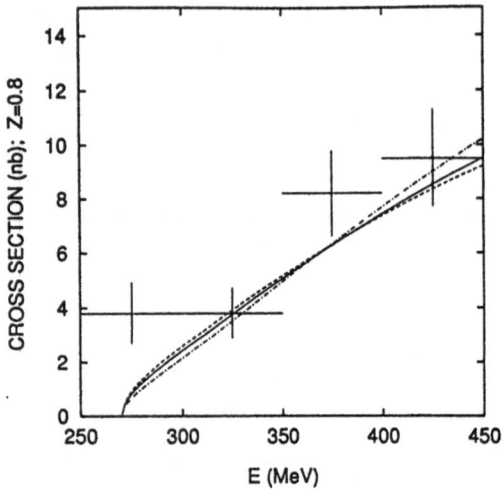

Figure 3: The two-loop result for the $\gamma\gamma \rightarrow \pi^0\pi^0$ cross section, evaluated with $(\alpha+\beta)^N = 1.15$ and $h_s^r = 7$, varying $(\alpha-\beta)^N$ between $(\alpha-\beta)^N = -0.95$ (dashed line) and $(\alpha - \beta)^N = -3.8$ (dash-dotted line). The solid line corresponds to $(\alpha - \beta)^N = -1.9$ [19]. See also Fig. 2 for explaining text.

Using presently available data, we have seen above that the chiral amplitude has successfully passed this check [19].

6   Summary and conclusion

1. At leading order in the chiral expansion, the amplitude for $\gamma\gamma \rightarrow \pi^0\pi^0$ is generated by one-loop graphs [1, 2]. In the case of $SU(2)_L \times SU(2)_R \times U(1)$ considered here, it involves the pion decay constant and the pion mass as the only parameters. The corresponding cross section deviates somewhat from the data and from dispersive calculations already near threshold.

2. The neglected terms in this calculation are related to $\pi\pi$ final-state interactions and to three new low-energy constants $h_{\pm}^r$ and h_s^r which occur at order E^6 in the effective action.

3. To investigate these corrections, we have evaluated the next-to-leading order terms in the chiral expansion (two-loop diagrams), have determined the size of the new couplings by resonance saturation and have estimated the uncertainties in the improved amplitude [19].

4. The improved cross section agrees rather well with the data and with dispersion theoretic calculations at and also substantially above the threshold region, see Fig. 1 and Fig. 2. The enhancement in the cross section is mainly due to $\pi\pi$ rescattering and to the renormalization of the pion decay constant.

5. The two-loop corrections are not unduly large–their size is similar to the corresponding next-to-leading order correction in the isospin zero $\pi\pi$ scattering amplitude [4] and in the scalar form factor of the pion [23].

6. The couplings $h^r_{\pm}$ and h^r_s contribute with a *negligible* amount below $E = 400$ MeV [30]-[35]. Above this energy, the inherent uncertainty in h^r_s becomes more important. The influence of $h^r_{\pm}$ is quite small also in the region 400 MeV$\le E \le 600$ MeV.

7. The quark mass expansion of the pion polarizabilities is very similar in structure to the chiral expansion of the threshold parameters in $\pi\pi$ scattering [22]. The effect of the low-energy constants $h^r_{\pm}$ on the value of $\bar{\alpha}_{\pi^0} \mp \bar{\beta}_{\pi^0}$ is large. It will presumably be difficult [14, 19, 24, 45] to extract these couplings from low-energy $\gamma\gamma \rightarrow \pi^0\pi^0$ data alone and to determine in this manner the neutral pion polarizabilities at two-loop order.

8. The DAΦNE facility [18] will have the opportunity to test the chiral predictions at next-to-leading order in much more detail than is possible with present data.

Acknowledgements

It is a pleasure to thank Alexander Bel'kov for drawing my attention to the early work of Ref. [8], Hans Bijnens for discussions and Mikko Sainio for providing me with figure 3.

References

[1] J. Bijnens and F. Cornet, Nucl. Phys. B296 (1988) 557

[2] J.F. Donoghue, B.R. Holstein and Y.C. Lin, Phys. Rev. D37 (1988) 2423

[3] S. Weinberg, Physica 96A (1979) 327

[4] J. Gasser and H. Leutwyler, Ann. Phys. (N.Y.) 158 (1984) 142

[5] J. Gasser and H. Leutwyler, Nucl. Phys. B250 (1985) 465

[6] H. Leutwyler, Bern University preprint BUTP-93/24 (hep-ph/9311274), Ann. Phys., in press

[7] For recent reviews on CHPT see e.g.
H. Leutwyler, Chiral effective lagrangians, Lecture Notes in Physics, vol. 396, Eds. H. Mitter and H. Gausterer, Springer-Verlag (Berlin, 1991);
Nonperturbative methods, Proc. XXVI Int. Conf. on High Energy Physics, Dallas, 1992, Ed. by J.R. Sanford, AIP Conf. Proc. No. 272 (AIP, New York, 1993) p. 185;
G. Ecker, Chiral perturbation theory, in: Quantitative Particle Physics: Cargèse 1992, Eds. M. Lévy et al., Plenum Publ. Co. (New York, 1993);
U.G. Meißner, Rep. Prog. Phys. 56 (1993) 903;
A. Pich, Lectures given at the V Mexican School of Particles and Fields, Guanajuato, México, December 1992, preprint CERN-Th.6978/93 (hep-ph/9308351);

J.F. Donoghue, E. Golowich and B.R. Holstein, "Dynamics of the Standard Model" (1992), Cambridge University Press, Cambridge

[8] M.K. Volkov and V.N. Pervushin, Yad. Fiz. 22 (1975) 346 (Sov. J. Nucl. Phys. 22 (1975) 179), in particular appendix 1

[9] The Mark II Collaboration (J. Boyer et al.), Phys. Rev. D42 (1990) 1350

[10] The Crystal Ball Collaboration (H. Marsiske et al.), Phys. Rev. D41 (1990) 3324

[11] R.L. Goble, R. Rosenfeld and J.L. Rosner, Phys. Rev. D39 (1989) 3264. The literature on earlier work may be traced from this reference

[12] D. Morgan and M.R. Pennington, Phys. Lett. B272 (1991) 134

[13] M.R. Pennington, in [18], p. 379

[14] J.F. Donoghue and B.R. Holstein, Phys. Rev. D48 (1993) 137

[15] A. Dobado and J.R. Peláez, Z. Phys. C57 (1993) 501

[16] T.N. Truong, Phys. Lett. B313 (1993) 221

[17] A.E. Kaloshin and V.V. Serebryakov, preprint ISU-IAP.Th93-03 (hep-ph/9306224)

[18] The DAΦNE Physics Handbook, Eds. L. Maiani, G. Pancheri and N. Paver (INFN, Frascati, 1992)

[19] S. Bellucci, J. Gasser and M.E. Sainio, Nucl. Phys. B423 (1994) 80

[20] T.N. Truong, Phys. Rev. Lett. 61 (1988) 2526

[21] S. Weinberg, Phys. Rev. Lett. 17 (1966) 616; 18 (1967) 188

[22] J. Gasser and H. Leutwyler, Phys. Lett. B125 (1983) 325

[23] J. Gasser and U.G. Meißner, Nucl. Phys. B357 (1991) 90

[24] M. Knecht, B. Moussallam and J. Stern, preprint IPNO/Th 94-08 (hep-ph/9402318), to appear in Nucl. Phys. B; M. Knecht, talk given in the working group on polarizabilities, see Ref. [38]

[25] H.W. Fearing and S. Scherer, preprint TRI-PP-94-68 (hep-ph/9408346)

[26] G. Ecker, J. Gasser, A. Pich and E. de Rafael, Nucl. Phys. B321 (1989) 311;
G. Ecker et al., Phys. Lett. B223 (1989) 425;
J.F. Donoghue, C. Ramirez and G. Valencia, Phys. Rev. D39 (1989) 1947;
M. Praszalowicz and G. Valencia, Nucl. Phys. B341 (1990) 27;
J. Bijnens, S. Dawson and G. Valencia, Phys. Rev. D44 (1991) 3555;
Ll. Ametller, J. Bijnens, A. Bramon and F. Cornet, Phys. Lett. B276 (1992) 185

[27] E. Golowich and J. Kambor, in preparation, and J. Kambor, talk given in the working group on polarizabilities, see Ref. [38]

[28] A. Schenk, Nucl. Phys. B363 (1991) 97

[29] J. Stern, H. Sazdjian and N.H. Fuchs, Phys. Rev. D47 (1993) 3814

[30] P. Ko, Phys. Rev. D41 (1990) 1531

[31] J. Bijnens, S. Dawson and G. Valencia, Ref. [26]

[32] S. Bellucci and D. Babusci, in [36], p. 351

[33] S. Bellucci, in [18], p. 419

[34] P. Ko, Phys. Rev. D47 (1993) 3933

[35] D. Babusci, S. Bellucci, G. Giordano and G. Matone, Phys. Lett. B314 (1993) 112

[36] Proceedings of the Workshop on Physics and Detectors for DAΦNE, Frascati, 1991, Ed. G. Pancheri (INFN, Frascati, 1991)

[37] M.A. Moinester, contribution to this workshop

[38] R. Baldini and S. Bellucci, contribution to this workshop

[39] Yu.M. Antipov et al., Phys. Lett. 121B (1983) 445; Z. Phys. C24 (1984) 39

[40] T.A. Aibergenov et al., Czech. J. Phys. B36 (1986) 948

[41] Yu.M. Antipov et al., Z. Phys. C26 (1985) 495

[42] A.E. Kaloshin and V.V. Serebryakov, Phys. Lett. B278 (1992) 198

[43] A.E. Kaloshin, V.M. Persikov and V.V. Serebryakov, preprint ISU-IAP.Th94-01 (hep-ph/9402220)

[44] The CELLO collaboration, H.-J. Behrend at al., Z. Phys. C56 (1992) 381

[45] J. Portolés and M.R. Pennington, preprint DTP-94/52; hep-ph/9407295; M.R. Pennington, talk given at the working group on polarizabilities, see Ref. [38]

[46] V.B. Golubev et al., Sov. J. Nucl. Phys. 45 (1987) 622

[47] D. Babusci et al., in [36], p. 383; D. Babusci et al., Phys. Lett. B277 (1992) 158

[48] M.V. Terent'ev, Sov. J. Nucl. Phys. 16 (1973) 87

[49] B.R. Holstein, Comments Nucl. Part. Phys. 19 (1990) 221

[50] U. Bürgi, work in progress

[51] G. Höhler, in: Landolt-Börnstein, Vol. 9 b2 (Ed. H. Schopper, Springer, Berlin, 1983)

Pion and Sigma Polarizabilities and Radiative Transitions

Murray A. Moinester [1]

[1]School of Physics and Astronomy
Raymond and Beverly Sackler Faculty of Exact Sciences,
Tel Aviv University, 69978 Ramat Aviv, Israel,
e-mail: murray tauphy.tau.ac.il

Abstract: Fermilab E781 plans measurements of γ-Sigma and γ-pion interactions using a 600 GeV beam of Sigmas and pions, and a virtual photon target. Pion polarizabilities and radiative transitions will be measured in this experiment. The former can test a precise prediction of chiral symmetry; the latter for $a_1(1260) \rightarrow \pi\gamma$ is important for understanding the polarizability. The experiment also measures polarizabilities and radiative transitions for Sigma hyperons. The polarizabilities can test predictions of baryon chiral perturbation theory. The radiative transitions to the $\Sigma^*(1385)$ provide a measure of the magnetic moment of the s-quark. Previous experimental and theoretical results for $\gamma\pi$ and $\gamma\Sigma$ interactions are given. The E781 experiment is described.

1 Introduction

Pion and $\Sigma(1189)$ polarizabilities and associated radiative transitions will be measured in the Fermilab E781 SELEX experiment [1, 2]. Hadron electric (α) and magnetic (β) intrinsic polarizabilities [2 − 6] characterize the induced transient dipole moments of hadrons subjected to external oscillating electric $\boldsymbol{E}$ and magnetic $\boldsymbol{H}$ fields. The dipole moments are given by $\boldsymbol{d} = \alpha\boldsymbol{E}$ and by $\boldsymbol{\mu} = \beta\boldsymbol{H}$. The polarizabilities can be obtained from precise measurements of the gamma-hadron Compton scattering differential cross section. They probe the rigidity of the internal structure of baryons and mesons, since they are induced by the rearrangement of the hadron constituents via action of the photon electromagnetic fields during scattering. For the light charged pion, chiral symmetry leads to a precise prediction for the polarizabilities [7, 8]. The experimental polarizabilities therefore subject the chiral perturbation techniques of QCD to new and serious tests. The availability of high energy Σ beams raises the possibility of investigating their Compton scattering polarizabilities and radiative transitions. The Σ polarizabilities can test predictions of baryon chiral perturbation theory [9].

The radiative transitions to the $\Sigma^*(1385)$ can provide a valuable measure of the magnetic moment of the s-quark [10, 11].

2 Sigma Polarizabilities and Radiative Transitions

Electromagnetic interactions of Σ's can be studied [2, 11] in FNAL experiment E781 with high energy Σ beams via the Primakoff interaction of an incident Σ with a virtual photon in the Coulomb field of a target nucleus of atomic number Z. The $\gamma\Sigma$ Compton scattering and associated polarizabilities are studied [2, 11] by detecting the final state γ-ray and Σ in coincidence. The final state Sigma or other charged particles will be measured [1, 2] in magnetic spectrometers, while final state γ-rays will be measured with lead glass electromagnetic calorimeters. The laboratory decay length ($L = \gamma\beta c\tau$) for very high energy Σ's exceeds ten meters, so that magnetic detection is possible. Transition radiation detectors upstream of the target and ring Cerenkov detectors downstream will provide particle identification [1]. The decay signature for $\Sigma^*(1385)$ $3/2^+$ production is clean as it decays to $\Lambda\pi$, and the Λ to π^-p. Such radiative transition studies were first suggested by Lipkin [10] and considered in more detail afterwards [2, 11 − 15]. The $\Sigma^0 \rightarrow \Lambda\gamma$ radiative transition (lifetime) was measured in this way [15] using a Λ beam, and this to date is the only precision hyperon radiative transition measured. Because radiative transitions involve the well understood electromagnetic process, they provide precision tests of the quark wave functions characterizing the configurations of the low lying excited states of hadrons.

The Σ^+ (uus) and Σ^- (dds) baryons are of interest because electromagnetic observables are particularly sensitive to their underlying quark structure [11]. The Σ^+ differs from the proton only by replacing the d quark by a strange quark which also has charge -1/3. Thus, any difference between the electromagnetic properties of the Σ^+ and proton can only arise from flavor SU(3) symmetry breaking. The two Σ states are isospin mirrors of one another, symmetrically placed in the same flavor SU(3) octet and have very similar strong interactions. However, their electromagnetic interactions are completely different because the Σ^+, like the nucleon and Ξ^o, have valence quarks of two flavors having the opposite sign of electric charge. External electromagnetic fields therefore exert forces in opposite directions on the two flavors and rotate spins in opposite directions, thereby producing internal excitation. The Σ^- on the other hand, and also the Ξ^-, have three valence quarks all with charge -1/3, and the principle effect of an external electromagnetic field is to exert forces in the same direction on all three valence quarks and rotate their spins in the same direction. This produces no internal excitation. We discuss below experimental implications of this effect, first noted [10] by a selection rule involving the U-spin SU(2) subgroup of SU(3). U-spin is conserved to all orders in any combination of electromagnetic interactions and strong interactions invariant under SU(3). The resulting SU(3) prediction is that the Primakoff excitation $\Sigma^- \rightarrow \Sigma^{*-}$ is forbidden, while excitation of $\Sigma^+ \rightarrow \Sigma^{*+}$ is allowed. The U-spin values of interest are U=1/2 for Σ^-, Σ^+, $\Sigma^+(1385)$ and U=3/2 for $\Sigma^-(1385)$.

The E1 and M1 Σ transitions are related to the intrinsic electric and magnetic Σ polarizabilities. We will elucidate below this connection, which can be tested experimentally. Hadron electric (α) and magnetic (β) intrinsic polarizabilities characterize the induced transient dipole moments. During γ-hadron Compton scattering, the lowest order scattering (Thomson) is determined by the charge and magnetic moment. The next order scattering (Rayleigh) is determined by the induced dipole moments. The Compton cross section data determine the Compton polarizabilities $\bar{\alpha}$ and $\bar{\beta}$, expressed here in Gaussian units of 10^{-43} cm^3. The angular distribution formulae for low γ-ray energies in terms of $\bar{\alpha}$ and $\bar{\beta}$ for γp scattering are given by Petrunkin and Lvov [4,5], and similar expressions apply for the Σ hyperons. Perturbation theory for the Σ^+, Σ^-, and proton polarizabilities can be applied, leading to the expressions [4] of Petrunkin:

$$\bar{\alpha} = \frac{2 \sum |\langle 0|d_z|n\rangle|^2}{E_n - E_0} + \frac{1}{3}\frac{e^2}{M}\langle r^2 \rangle = \alpha + \Delta\alpha, \tag{1}$$

$$\bar{\beta} = 2 \sum \frac{|\langle 0|\mu_z|n\rangle|^2}{E_n - E_0} - k\frac{e^2}{M}\langle r^2 \rangle = \beta + \Delta\beta. \tag{2}$$

Here $d = \sum e_k r_k$ and μ are the electric and magnetic dipole operator respectively. We obtained [11] k = 0.68, 0.60, 0.26, and we use rms charge radii from lattice QCD calculations [11, 16] of R = 0.86, 0.96, 0.76 fm, for the proton, Σ^+, and Σ^- respectively. The first and second terms in these equations give the intrinsic and center of charge oscillation contributions, respectively. The sums are over all E1 and M1 excitations. The intrinsic polarizabilities probe the internal structure of baryons and mesons.

Many theoretical and experimental polarizability studies have been made for the proton and neutron [17, 18]. In the proton polarizability calculation of Weiner and Weise [18], the intrinsic part is mainly due to the charged pion or kaon cloud surrounding the proton core. Only two calculations were reported for the Σ. One is a simple non-relativistic quark bag calculation [4, 11] with no cloud. The second is via heavy baryon chiral perturbation theory [9]. Calculations are in progress by Lvov [5] for the Sigma Compton scattering with a dispersion relationship approach. Such an approach is needed to account for the effects of inelastic reactions such as $\gamma\Sigma \rightarrow \pi\Lambda$.

The odd parity $\Sigma^*1/2^-$ and $\Sigma^*3/2^-$ states near 520 MeV excitation above the nucleon have the s orbit strange quark promoted to the p orbit; and also excitations of the nonstrange quarks. The intrinsic contribution of eqn. 1 can be evaluated using closure by saturating the sum over these states, giving [2, 11]:

$$\alpha_{\Sigma^+} \sim \frac{2}{\Delta E}\langle \Sigma^+|(\sum q_i z_i)^2|\Sigma^+ \rangle^2 \sim \frac{2}{3}\frac{\langle r^2 \rangle}{\Delta E}e^2(\frac{4}{9} + \frac{4}{9} + \frac{1}{9}) \approx 17.1. \tag{3}$$

The intrinsic magnetic polarizability can be evaluated by saturating the magnetic dipole excitations with the Σ^+ to $\Sigma(1385)$ transition, giving [4, 11]:

$$\beta_{\Sigma^+} \sim \frac{2}{M_{\Sigma^{*+}} - M_\Sigma}|\langle \Sigma^+|\mu_z|\Sigma^* \rangle|^2 \sim \frac{2}{M_{\Sigma^{*+}} - M_\Sigma}(\frac{2\sqrt{2}}{3}\mu_{\Sigma^+})^2 \sim 8.5, \tag{4}$$

where μ_{Σ^+} is the Σ^+ magnetic moment. The Σ^+ to Σ^* magnetic dipole transition matrix element is written in terms of the Σ^+ magnetic moment, following the non-relativistic quark model. For β_{Σ^+}, the magnetic dipole transition to the Σ^* $3/2^+$ resonance saturates the magnetic dipole excitations, in analogy to the nucleon to Δ transition; the proton and Σ^+ have the same matrix element expression as members of the same U-spin doublet. For the proton intrinsic magnetic polarizability β_p, the analog of eqn. 4 gives [4] $\beta_p \approx 7.4$.

Consider matrix elements for magnetic moment operators μ between states having the quark constituents of nucleons and Σ's. In the SU(3) symmetry limit, the contributions to the magnetic moment of the odd d quark to μ_p and of the odd s quark to μ_Σ are equal. Consider experimental tests of SU(3) breaking mechanisms based on the assumptions that the u-flavor contributions to μ_p and μ_{Σ^+} are equal, and that $SU(3)$ breaking only occurs in μ_s. One can then show [11] that the strange contribution to the magnetic moment is suppressed by an order of magnitude:

$$\frac{\mu_{\Sigma^+} + 2\mu_{\Sigma^-}}{\mu_p + 2\mu_n} \approx \frac{\langle \Sigma^+ | \mu_s | \Sigma^+ \rangle}{\langle p | \mu_d | p \rangle} \approx 0.11 \pm 0.04. \tag{5}$$

The main physical difficulty is in the small value of the d quark contribution to the proton moment and the large contribution 2.45 n.m. of the u quarks. If the u quark contribution to μ_{Σ^+} is the same 2.45 n.m. as in the proton, then the experimental value $\mu_{\Sigma^+} = 2.42 \pm 0.02$ n.m. can be fit only by requiring a nearly vanishing s quark contribution. That the strange quark contribution to μ_Σ is suppressed can also be seen by comparing the strange contributions to μ_Σ and μ_Λ. The contribution to μ_Σ is only 6% of the strange contribution to μ_Λ [11].

Experimental data on related processes may give additional insight. One such process involves octet-decuplet transitions for nucleons and Σ's. It will be interesting to measure these transitions and compare their systematics with those of the magnetic moments. In a description of SU(3) symmetry breaking in the Σ^- to Σ^{*-} transition, the ratio of M1 decay widths or of intrinsic magnetic polarizabilities for the Σ^- and proton can be estimated as [10 − 12]:

$$\frac{\Gamma(\Sigma \rightarrow \Sigma^{*-})}{\Gamma(p \rightarrow \Delta^+)} = \frac{(M_{\Sigma^{*-}} - M_{\Sigma^-})\beta_{\Sigma^-}}{(M_{\Delta^+} - M_p)\beta_p} \approx \frac{1}{9}(1 - \frac{\mu_s}{\mu_d})^2 \approx 0.011. \tag{6}$$

Here μ_s and μ_d are empirical magnetic moments of s and d quarks (estimated in a simple additive quark model [19] using proton, neutron, and Lambda magnetic moments, as $\mu_u = 1.852\mu_N$, $\mu_d = -0.972\mu_N$, $\mu_s = -0.613\mu_N$), the s and d values unequal due to SU(3) symmetry breaking. As shown in eqn. 5, μ_s may be significantly lower than the value fixed by the Lambda magnetic moment; which would significantly increase the predicted decay width of eqn. 6. Following the assumptions underlying eqns. 5-6, the Σ^- M1 transition width directly determines the relative magnitudes of the s and d-quark magnetic moments. Explicit quark model calculations [20] give a larger value of 0.034 for the ratio of eqn. 6, with $\Gamma(\Sigma^{*+}) = 104$ keV [20] and 117 keV [21]; and $\Gamma(\Sigma^{*-}) = 2.5$ keV [20]. These calculations correspond to $\Gamma(\Delta^+) = 74.3$ keV using eqn. 15 of Ref. 11.

The Δ radiative width of the model is smaller than the experimental value, but this should not affect the reliability of the ratio calculation of eqn. 6. One expects from eqns. (1-4,6) to observe very large and very small values of the intrinsic magnetic polarizability for the Σ^+ and Σ^-, respectively; and similarly for the Σ^* radiative decay width. The predictions [11] $\overline{\alpha_{\Sigma^+}} =20.8$, $\overline{\alpha_{\Sigma^-}} =5.7$, $\beta_{\Sigma^+} = 8.5$, $\beta_{\Sigma^-} = 0.12$, $\overline{\beta_{\Sigma^+}} = 1.7$, $\overline{\beta_{\Sigma^-}} = -1.7$ satisfy this expectation. Bernard et al. [9] predict quite different values.

The Σ beam at FNAL is polarized, so that asymmetry measurements are possible; as in the recent study [22] of the $\Sigma \to p\gamma$ weak decay. Asymmetry data can also be of value for the Σ polarizability determination. For radiative transitions, they are sensitive to the M1 or E2 nature of the exchanged photon, and therefore to the L=2 admixtures. The ratio E2/M1 of these amplitudes is a subject of considerable current interest [23, 24] for the nucleon to Δ transition; so that measuring this ratio for the Σ^* (and Ξ^*) will be extremely valuable. It was suggested [2, 11] that the ratios may be different for Σ^*, Ξ^*, and Δ, given that the s-quark mass is significantly heavier than the d quark mass. Recent calculations [25, 26] give predictions for such transitions.

We consider now the event rate expectations and backgrounds for the Σ polarizability and radiative transition experiments. We give the signal and background rates for a C^{12} target. For higher Z targets, the Z^2 dependence of the Primakoff cross section compared to the $A^{2/3}$ dependence of the strong cross section backgrounds, will improve the situation. We take a t-interval up to $6.0 \times 10^{-3} \ GeV^2$, where t is the four momentum transfer to the target nucleus, as discussed in section 3. We compare the Σ expected rates to data and calculations for the $\pi \to \rho\gamma$ radiative transition measurement [27] at 156 GeV on a C^{12} target. For the ρ to π transition, for this t-interval, the Primakoff cross section was 2.4 μb. and the strong background cross section was 0.75 μb; which gives a signal to background ratio of 3. The hadronic cross section falls [12, 28] as 1/E (factor=1/3.8), while the Primakoff cross section rises as ln(E) (factor = 1.3). In addition, the Primakoff cross section has a coefficient K, given by:

$$K = \frac{2J_a^* + 1}{2J_b^* + 1} \left(\frac{M_a^*}{M_{a*}^2 - M_a^2} \right)^3 \Gamma_{a^* \to a\gamma}. \tag{7}$$

Considering mass and spin values, the coefficient K is 7 times larger for the $\Sigma^* \to \Sigma$ transition, compared to the $\rho \to \pi$ transition. We take [11] also the widths $\Gamma(\Sigma^+) = 1000$ KeV, and $\Gamma(\Sigma^-)= 25$ KeV. These are fixed by the widths given above, normalized to the experimental width of the $\Delta \to N$ radiative transition. The theoretical Δ width calculation is roughly 10 times lower than the experimental value. We assume here that the model calculation of width ratios is more accurate than the absolute values. We also assume that the strong backgrounds for the ρ transition are of the same order as those of the Σ transition. We obtain therefore the rough estimate of signal cross sections of 314 microbarns (Σ^+), 7.8 microbarns (Σ^-), with a background cross section of 0.2 microbarns. The count rates expected are very good, via comparison to the rates explicitly given later for pion polarizability, with a roughly 4.0 microbarn cross section. Furthermore, the width of the mass distribution for diffractive production of

$\Sigma \rightarrow \Lambda\pi^-$ is roughly 100 MeV, as shown in previous data, and calculations [29 − 31] for this process at BNL. Since the Σ^* width is 36 MeV, and our mass resolution is 10 MeV, we will gain another factor of 2.5 from a mass cut.

We consider also the background from the small Ξ^- component in the beam. This decays 100% to $\Lambda\pi^-$, and is therefore a background for the Σ^{*-} decay. The mass difference of 64 MeV between Ξ^- and Σ^{*-} is large compared to the expected mass resolution of 10 MeV. This background can therefore be distinguished in the off-line analysis. The on-line trigger may reduce this background, by requiring the vertex of the detected pion to be centered at the target position. Other background reactions have been considered, but will not be discussed here in detail. These include the high cross section process of $\Sigma^- \rightarrow \Sigma^-\pi^0$ diffractive dissociation; and the $\Sigma^- \rightarrow n\pi^-$ decay of beam particles.

3 Pion Polarizabilities and Radiative Transitions

For the γ-π interaction at low energy, chiral perturbation theory (χPT) provides a rigorous way to make predictions; because it stems directly from QCD and relies only on the solid assumptions of spontaneously broken $SU(3)_L \times SU(3)_R$ chiral symmetry, Lorentz invariance and low momentum transfer. Unitarity is achieved by adding pion loop corrections to lowest order, and the resulting infinite divergences are absorbed into physical (renormalized) coupling constants L_i^r (tree-level coefficients in $L^{(4)}$, see Refs. ([32, 33])). With a perturbative expansion of the effective Lagrangian limited to terms quartic in the momenta and masses ($O(p^4)$), the method establishes relationships between different processes in terms of the L_i^r. For example, the radiative pion beta decay and electric pion polarizability are expressed as: [7, 8, 32, 33]:

$$h_A/h_V = 32\pi^2(L_9^r + L_{10}^r); \bar{\alpha}_\pi = \frac{4\alpha_f}{m_\pi f_\pi^2}(L_9^r + L_{10}^r); \tag{8}$$

where $f_\pi = 92.4$ MeV [34] is the pion decay constant, h_A and h_V are the axial vector and vector coupling constants in the decay, and α_f is the fine structure constant. The experimental ratio $h_A/h_V = 0.45 \pm 0.06$, leads to $\bar{\alpha}_\pi = -\bar{\beta}_\pi = 2.7 \pm 0.4$, where the error shown is due to the uncertainty in the h_A/h_V measurement.

Other pion polarizability calculations [35 − 37] find values for $\bar{\alpha}_\pi$ ranging from $\sim 4 - 14$. Holstein [7] showed that meson exchange via a pole diagram involving the $a_1(1260)$ resonance provides the main contribution ($\bar{\alpha}_\pi = 2.6$) to the polarizability. The E781 high energy pion experiments can obtain new high statistics data for radiative transitions leading from the pion to the $a_1(1260)$, and to other meson resonances. For $a_1(1260) \rightarrow \pi\gamma$, the experimental width [38] is $\Gamma = 0.64 \pm 0.25$ MeV. Xiong, Shuryak, and Brown (XSB) [39] estimate this radiative width to be $\Gamma = 1.4$ MeV, more than two times higher than the experimental value [38]. With this estimated width, they calculate the pion polarizability to be $\bar{\alpha}_\pi = 1.8$. A remeasurement of the $a_1(1260)$ width and of the pion polarizability will allow checking the consistency of their expected relationship.

For the pion polarizability, Antipov et al. [40, 41] measured the $\gamma\pi$ scattering with 40 GeV pions at Serpukhov via radiative pion scattering in the nuclear Coulomb field ($\pi^- + Z \rightarrow \pi^- + Z + $ gamma). The final state gamma ray and pion were detected in coincidence. This reaction corresponded to $\gamma + \pi^- \rightarrow \gamma + \pi^-$ scattering for laboratory gamma-ray energies in the range 60–600 MeV on a target π^- at rest. The data selection criteria at Serpukhov and E781 requires one photon and one charged particle in the final state, their total energy consistent with the beam energy, small t, and other position, angle, and energy conditions. Only the angular distribution in the backward hemisphere were measured at 40 GeV. The angular acceptance and detector threshold of the γ-ray calorimeter in E781 allows measuring more complete angular distributions than was achievable in the 40 GeV experiment.

The pion electric polarizability $\bar{\alpha}_\pi$ was deduced [40, 41] in this low statistics experiment ($\sim$ 7000 events) to be $\bar{\alpha}_\pi = -\bar{\beta}_\pi = 6.8 \pm 1.4_{stat} \pm 1.2_{syst}$, where it was assumed in the analysis that $\bar{\alpha}_\pi + \bar{\beta}_\pi = 0$, as expected theoretically [7]. An important result of s-channel ($\gamma + \pi \rightarrow \gamma + \pi$) dispersion sum rules [4] for charged pions is: $\bar{\alpha}_\pi + \bar{\beta}_\pi = 0.39 \pm 0.04$. This result can be used for high statistics data as a constraint in the data analysis. Charged pion polarizabilities were also deduced [42] from $\gamma\gamma \rightarrow \pi^+\pi^-$ data which is related to the Compton scattering by crossing symmetry. Pennington [43] claims that such determinations are very insensitive to the polarizability value. This is so, since two-photon cross section data at low $\pi\pi$ invariant mass agree well with calculations for a large range of choices of the (undetermined) position of the chiral zero. However, very different values of polarizability are associated with the different choices.

In the radiative pion scattering experiments, it was shown experimentally [40, 41] and theoretically [44] that the Coulomb cross section scales as Z^2 and yields sharp peaks in t-distributions at very small four momentum transfers to the target nucleus $t \leq 6 \times 10^{-4}(GeV/c)^2$. Background from other processes could easily be estimated and subtracted by extrapolating in t from events in the region of flat t-distribution of 3–8 $\times 10^{-3}(GeV/c)^2$. The sources of these backgrounds are the coherent process of pion elastic scattering accompanied by gamma emission [44], contributions of pion (or rho) rescattering [42, 45], and higher cross section inelastic channels [27, 46]. Available calculations and data show that these backgrounds are manageable, and that the signal to background improves [12, 28, 44] with increasing incident energy.

One must also evaluate electromagnetic corrections to radiative pion scattering, where the requirement is to measure only single photon bremsstrahlung emission. Here the detailed properties of the gamma detector are important, such as the photon detector threshold, t-resolution, and the two-photon angular resolution. Such calculations were carried out [47, 48] for the conditions of the planned 600 GeV FNAL experiment. The corrections are at the level of 4%, and can be easily made.

The polarizability Compton scattering and chiral anomaly processes in E781 represents a difficult triggering challenge. They have one negative charged track in the final state, and one or two γ-rays. This trigger rate must match that of the charm component of the experiment. The trigger problems are deadtime problems generated in the initial trigger levels when information from the γ-ray detector is not available yet for trigger rate suppresssion. The first level of hardware trigger T_0 is meant to identify a beam particle and an interaction in the target. However, a T_0 trigger with multiplicity 1 will generate a trigger for every beam particle, in contrast to other trigger types which will fire at the beam interaction rate. To differentiate between a single negatively-charged Primakoff interaction product coming from the target and a non-interacting beam track, it is planned to have three hodoscopes to measure a vertical deflection angle of the outgoing track after the Primakoff target. We need to form a coincidence between the projective elements of the first two hodoscopes to form the track road assuming a beam particle, and then veto in a window around this road in the third hodoscope. The required electronics must give the decision in about 140 nsec, in order to enter the next trigger level on time. A simpler beam prescale trigger will also be used; which will help control systematic uncertainties.

To specifically illustrate some of the kinematics germane to a FNAL experiment, the reaction:

$$\pi + Z \rightarrow \pi' + Z' + \gamma' \tag{9}$$

is considered for a 600 GeV incident pion, where Z is the nuclear charge. The 4-momentum of each particle is P_π, P_Z, $P_{\pi'}$, $P_{Z'}$, k', respectively. In the one photon exchange domain, eqn. 9 is equivalent to:

$$\gamma + \pi \rightarrow \gamma' + \pi', \tag{10}$$

and the 4-momentum of the incident virtual photon is $k = P_Z$-$P_{Z'}$. The cross section for the reaction of eqn. 9 is described by the the well tested Primakoff formalism [27, 49] that relates processes involving real photon interactions to production cross sections involving the exchange of virtual photons. We have:

$$\frac{d\sigma}{dt ds d\Omega} = \frac{Z^2 \alpha_f}{\pi} \frac{|F(t)|^2}{s - m_\pi^2} \frac{t - t_0}{t^2} \frac{d\sigma_{\gamma\pi}}{d\Omega}, \tag{11}$$

where $d\sigma_{\gamma\pi}/d\Omega$ is the unpolarized differential cross section for eqn. 10 (for real photons), t is the square of the four-momentum transfer to the nucleus, F(t) is the nuclear form factor (essentially unity at small t-values), $\sqrt{s}$ is the mass of the $\gamma\pi$ final state, and t_0 is the minimum value of t to produce a mass $\sqrt{s}$. The analysis to determine polarizabilities by fitting the data use the known [40 − 42, 50] formula for $d\sigma_{\gamma\pi}/d\Omega$. This cross section depends on the polarizabilities and on s and on t_1, the square of the 4-momentum transfer between initial and final state γ's. We have:

$$t = k^2 \equiv -M(V)^2, \tag{12}$$

where k is the 4-momentum transferred to the nucleus, and M(V) is the virtual photon mass. Since $t = 2M_Z(E_{Z',lab} - M_Z) > 0$, the virtual photon mass is imaginary. To approximate real pion Compton scattering, the virtual photon must be almost real; M(V)<0.077 GeV corresponding to $t < 6.0 \times 10^{-3}(GeV/c)^2$ can be required in E781. In addition,

$$s = (P_{\pi'} + k')^2 \equiv M(\gamma\pi)^2, \tag{13}$$

where $M(\gamma\pi)$ is the $\gamma\pi$ invariant mass. The minimum value for t [51] is given by:

$$t_0 \sim (s - m_\pi^2)^2/4|P_\pi|^2, \tag{14}$$

corresponding to $t_0 \sim 5.4 \times 10^{-8}(GeV/c)^2$ for $\sqrt{s} = 1.75m_\pi$ at 600 GeV incident energy. The maximum of the differential cross section for reaction (11) occurs at $t=2t_0$, and the integral to $100t_0$ gives essentially the entire cross section.

With lead glass detectors [40, 41], a t-resolution $\sigma_t \sim 6.0 \times 10^{-4}(GeV/c)^2$ was achieved at 40 GeV. The t-resolution sets the experimental maximum in t for accepted events. The SELEX t-resolution will be 10 times worse. The strong backgrounds are associated with particle exchange, and such cross sections are known [12, 27, 28, 44] to fall rapidly with increasing energy. From this point of view, the t-resolution is therefore not a critical parameter for 600 GeV experiments. The integrated Compton cross section up to $t = 6.0 \times 10^{-3}$ $(GeV/c)^2$ grows [44, 51] as $\ln(P_\pi)$, where P_π is the laboratory incident pion 3-momentum. With strong backgrounds that decrease roughly as $1/E$, the percent strong background would decrease for a C^{12} target, from the estimated [44] 2.5% at 40 GeV to roughly 0.6% at 600 GeV at FNAL.

The energy of the incident virtual photon in the pion rest frame is:

$$E(V) = (s - m_\pi{}^2 + t)/2m_\pi \sim (s - m_\pi{}^2)/2m_\pi \tag{15}$$

at small t; so that the energy of the photon is determined by s. The elemental cross section at low E(V) is a function of E(V), $cos(\theta)$, $\bar{\alpha}$, $\bar{\beta}$; where θ is the Compton scattering angle in the pion rest frame. In this frame, due to Lorentz contraction, the nucleus Z represents a transverse virtual photon pulse sweeping past the pion. One should require that E(V) be sufficiently low in energy, such that ρ meson production does not occur on-shell for an incident γ-ray on a target pion at rest. The γ-ray energy required to produce a ρ meson, not counting the ρ width, is given by $\omega = E(V) = (M_\rho^2 - m_\pi^2)/(2m_\pi) = 2.0$ GeV. Considering the ρ width, one could limit the incident energy to be lower than 1 GeV, corresponding to $s < 15.3m_\pi^2$. We will analyze the lower energy data for polarizability purposes, and at higher energies to understand the ρ and a_1 meson contributions.

Consider the case of 600 GeV incident laboratory pions. The laboratory outgoing γ-rays are emitted up to 5 mrad, and the corresponding outgoing pions are emitted up to 0.3 mrad. The angular resolution for the pion is roughly 0.04 mrad due to multiple scattering in the Primakoff target and in the in-beam tracking detectors. The gamma ray energies range from 0 - 400 GeV, and the corresponding outgoing pion energies range from 200 - 600 GeV. The corresponding Compton scattering angular range is 0 - 180 degrees in the π rest frame. In practice, the most forward Compton scattering angles are less accessible, as they correspond to the larger γ-ray angles in the laboratory frame which can miss the γ-ray detector, and where the γ-ray energies are also possibly below the detector threshold. But these forward angles are anyhow not sensitive to the polarizabilities, as discussed below.

We consider the uncertainties achievable for the polarizabilities in the FNAL E781 experiment, based on Monte Carlo simulations. An important consideration is the information content [3, 51] of the data versus x and s, where $x=\cos(\theta)$. Consider the fraction of the $\gamma\pi$ Compton cross section arising from the $\bar{\alpha}$ terms in the $\gamma\pi$ center of mass cross section expression, for $\bar{\alpha}=6.8$. High s values and back angles (large t_1) have maximal information content for the polarizabilities. For example, the fraction of the cross section at back angles due to the polarizability term is only 5% at $E(V) = 140$ MeV, and roughly 30% at $E(V) = 600$ MeV.

We consider initially a beam energy of 600 GeV, a C^{12} target, and an s-range of 2. - 10. m_π^2, corresponding to a Primakoff cross section of 4.0μb, and $E(V)$ = 70 - 628 MeV. We take the π-Carbon total cross section at 600 GeV to be 192. mb, eight times the total inelastic π-nucleon cross section at 600 GeV. We assume the simultaneous use of a 5% Carbon interaction target and also a 0.3% Pb interaction target. We calculate below the event rate from the Carbon target, and will in addition have a factor of roughly 2 times higher rate from the Pb target. The probability P per inelastic interaction for a Primakoff interaction is then $P = 4. \times 10^{-6}/192. \times 10^{-3} = 2.1 \times 10^{-5}$. The number of interactions planned [1] for E781 is roughly 3. $\times$ 10^{10}, corresponding to roughly 6.3×10^5 events for 4.0μb. Including the Pb target events, the statistics are roughly 3 times higher. The trigger requirement for a vertical deflection angle of the pion of more than 100 μrad cuts the statistics roughly in half. The losses are mainly events at small s and t_1, which minimally affects polarizability uncertainties. Other experimental efficiencies and acceptances will lower the statistics.

We cite here some Monte Carlo results for 600 GeV Primakoff experiments on Carbon, assuming the dispersion sum rule result $\bar{\alpha}+\bar{\beta} = 0.4$ and also $\bar{\alpha}=6.8$. At 600 GeV, including the Pb target, for 580,000 events in an s interval (2.0 - 10.0)m_π^2, we find by fitting simulated data that $\bar{\alpha} = 7.1 \pm 0.4$, $\bar{\alpha}+\bar{\beta} = 0.3 \pm 0.1$. The overall statistical error for polarizabilities will be improved from the expected higher statistics; and by additional data in the sensitive s-interval 10.-15.3 m_π^2. Data in different s-intervals can be analyzed to give independent values for the polarizabilities, which will help control the systematic uncertainties. These Monte Carlo simulations show that the objective of obtaining pion polarizabilities with significantly smaller statistical and systematic uncertainties is realistic.

4 Conclusions

The beams at FNAL and CERN invite hadron Compton scattering and radiative transition studies for different particle types, such as $\pi^{+,-}$, $K^{+,-}$, p, $\bar{p}$, Σ, Ξ, Λ hyperons, and others. The E781 experiment was described. Because these transitions involve the well understood electromagnetic process, they provide precision tests of the quark wave functions characterizing the configurations of the low lying excited states of hadrons. The 600 GeV beam energy at FNAL is important to get a good yield for low t events in the radiative scattering, and also to reduce backgrounds from the decay of unstable hadrons by significantly boosting their lifetime. We will measure the $\gamma\pi$ and $\gamma\Sigma$ Compton scattering cross sections, thereby enabling determinations of the pion and Sigma polarizabilities. E781 will also measure the formation and decay of the $\Sigma^-(1385)$, $\Sigma^+(1385)$, and $a_1(1260)$ excited states. These various Σ and pion experiments will allow serious tests of chiral symmetry and chiral perturbation theory; and of different available polarizability and radiative decay calculations in QCD. The Σ experiments will shed new light on puzzles associated with the size of the s-quark magnetic moment.

5 Acknowledgements

This work was supported by the U.S.-Israel Binational Science Foundation, Jerusalem, Israel. Thanks are due to S. Bellucci, P. Cooper, T. Ferbel, L. Frankfort, S. Gerzon, G. Giordano, A. Kulyavtsev, J. Lach, H. J. Lipkin, A. Lvov, J. Russ, and N. Terentyev for valuable discussions.

References

1. J. Russ, spokesman, FNAL E781 Collaboration: Carnegie-Mellon U., Fermilab, U. Iowa, U. Rochester, U. Washington, Petersburg Nuclear Physics Institute, ITEP (Moscow), IHEP (Protvino), Moscow State U., U. Sao Paulo, Centro Brasileiro de Pesquisas Fisicas, Universidade Federale de Paraiba, IHEP (Beijing), U. Bristol, Tel Aviv U., Max Planck Institut fur Kernphysik-Heidelberg, Universidad Autonoma de San Luis Potosi;
    J. Russ: Proceedings of the CHARM2000 Workshop, Fermilab, June 1994, Eds. D. M. Kaplan and S. Kwan, Fermilab-Conf-94/190, P. 111, (1994)
2. M. A. Moinester: Proceedings of the Conference on the Intersections Between Particle and Nuclear Physics, Tucson, Arizona, 1991, AIP Conference Proceedings 243, P. 553, 1992, Ed. W. Van Oers.
3. M.A. Moinester: Workshop on Hadron Structure from Photo-Reactions at Intermediate Energies, Brookhaven National Laboratory, May 1992, Eds. A. Nathan, A. Sandorfi, BNL Report 47972, Tel Aviv U. preprint TAUP 1972/92.
4. V. A. Petrunkin: Sov. J. Part. Nucl. **12** 278 (1981)
5. A. I. Lvov: Sov. J. Nucl. Phys. **42** 583 (1985); Int. J. Mod. Phys. A **8** 5267 (1993); Phys. Lett. B **304** 29 (1993) ; private communication.
6. J. L. Friar: Workshop on Electron-Nucleus Scattering (1988, EIPC), Eds. A. Fabrocini et al., World Scientific Publishing Co. (1989)
7. B. R. Holstein: Comments Nucl. Part. Phys. **19** 239 (1990)
8. J. F. Donoghue, B. R. Holstein: Phys. Rev. D **40** 2378 (1989)
9. V. Bernard et al: Phys. Rev. D **46** 2756 (1992); Phys. Lett. B **319** 269 (1993)
10. H. J. Lipkin: Phys. Rev. D **7** 846 (1973)
11. H. J. Lipkin, M. A. Moinester: Phys. Lett. B **287** 179 (1992)
12. A. V. Vanyashin et al: Sov. J. Nucl. Phys. **34** 90 (1981)
13. T. Goldman et al: Physics with LAMPF II, LA-9798-P, P. 319 (1984)

14. M. V. Hynes: Physics with LAMPF II, LA-9798-P, P. 333 (1984)
15. F. Dydak et al: Nucl. Phys. B **118** 1 (1977)
16. D. B. Leinweber, R. M. Woloshyn, T. Draper: Phys. Rev. D **43** 1659 (1991) ;
    D. B. Leinweber, T. D. Cohen: Phys. Rev. D **47** 2147 (1993)
17. B. R. Holstein, A. M. Nathan: Phys. Rev. D **49** 6101 (1994)
18. R. Weiner, W. Weise: Phys. Lett. B **159** 85 (1985)
19. Particle Data Group, G.P.Yost et al: Phys. Lett. B **204** 1 (1988)
20. J. W. Darewych et al: Phys. Rev. D **28** 1125 (1983)
21. E. Kaxiras, E. J. Moniz, M. Soyeur: Phys. Rev. D **32** 695, (1985)
22. M. Foucher et al: Phys. Rev. Lett. **68** 3004 (1992)
23. R. M. Davidson, N. C. Mukhopadhyay, R.S. Wittman: Phys. Rev. D **43** 71 (1991)
24. A. Bernstein, S. Nozawa, M. A. Moinester: Phys. Rev. C **47** 1274 (1993)
25. D. B. Leinweber, T. Draper, R. M. Woloshyn: Phys. Rev. D **48** 2230 (1993)
26. M. N. Butler, M. J. Savage, R. P. Springer: Phys. Lett. B **304** 353 (1993);
    Phys. Lett. **314** 122(E)(1993); Nucl. Phys. B **399** 69 (1993)
27. T. Jensen et al: Phys. Rev. D **27** 26 (1983)
28. G. Berlad et al: Ann. Phys. **75** 461 (1973)
29. R. T. Deck: Phys. Rev. Lett. **13** 169 (1964)
30. L. Stodolsky: Phys. Rev. Lett. **18** 973 (1967)
31. V. Hungerbuhler et al: Phys. Rev. D **10** 2051 (1974)
32. J.Gasser, H.Leutwyler: Nucl. Phys. B **250** 465 (1985)
33. J.Gasser, H.Leutwyler: Ann.Phys. (N.Y.) **158** 142 (1984)
34. B.R.Holstein: Phys.Lett.B **244** 83 (1990)
35. V. Bernard, B. Hiller, W. Weise: Phys. Lett. B **159** 85 (1988)
36. V. Bernard, D. Vautherin: Phys. Rev. D **40** 1615 (1989)
37. M. A. Ivanov, T. Mizutani: Phys. Rev. D **45** 1580 (1992)
38. M. Zielinski et al: Phys. Rev. Lett. **52** 1195 (1984)
39. L. Xiong, E. Shuryak, G. Brown: Phys. Rev. D **46** 3798 (1992)
40. Y. M. Antipov et al: Phys. Lett. B **121** 445 (1983)
41. Y. M. Antipov et al: Z. Phys. C–Particles and Fields **26** 495 (1985)
42. D. Babusci, S. Bellucci, G. Giordano, G. Matone, A. M. Sandorfi, M. A.
    Moinester: Phys. Lett. B **277** 158 (1992)
43. J. Portoles, M. R. Pennington: U. Durham preprint DTP-94/52, 1994, submitted
    to Second DAΦNE Physics Handbook, Eds. G. Pancheri and N. Paver.
44. A. S. Galperin et al: Sov. J. Nucl. Phys. **32** 545 (1980)
45. G. V. Mitselmakher, V. N. Pervushkin: Sov. J. Nucl. Phys. **37** (1983)
46. M. Zielinski et al: Z. Phys. C **16** 197 (1983)
47. A. A. Akhundov, D. Yu. Bardin, G. V. Mitselmakher, A. G. Olshevsky: Sov. J.
    Nucl. Phys. **42** 426 (1984)
48. A. A. Akhundov, S. Gerzon, S. Kananov, M. A. Moinester: I.C.T.P. (Trieste)
    Preprint IC/94/203; Tel Aviv U. Preprint TAUP-2183-94, 1994
49. M. Zielinski et al: Phys. Rev. D **29** 2633 (1984)
50. L. V. Fil'kov, I. Guiasu, E. E. Radescu: Phys. Rev. **26** 3146 (1982)
51. N. I. Starkov, L. V. Fil'kov, V. A. Tsarev: Sov. J. Nucl. Phys. **36** 707 (1982)

Nucleon Polarizabilities

Alan M. Nathan

Nuclear Physics Laboratory and Department of Physics
University of Illinois at Urbana-Champaign, Urbana, IL, USA

1 Introduction

The electric and magnetic polarizabilities, labeled $\bar{\alpha}$ and $\bar{\beta}$ respectively, measure the ease with which an electric or magnetic dipole moment can be induced in a composite system by static external electric or magnetic fields. These structure constants are therefore fundamentally as important as the charge or magnetic radius of the system, although in the case of the nucleon they are considerably less well known. With the high present–day interest in QCD–based theoretical descriptions of the nucleon, it is clear that the additional information represented by an accurate determination of its polarizabilities would be of substantial importance.

In this paper, we review the current experimental status for the nucleon polarizabilities. The primary emphasis will be on the proton (Section 2), since that is where most of the recent experimental activity has been concentrated. The neutron is reviewed in Section 3. In Section 4, the polarizabilities are discussed in the context of dispersion sum rules. Our conclusions are summarized in Section 5.

Lecture delivered at the Workshop on Chiral Dynamics: Theory and Experiments, Massachusetts Institute of Technology, Cambridge, MA, July 25 - 29, 1994

2 Proton Polarizabilities: Compton Scattering Experiments

2.1 The Low Energy Expansion

Measurements of the proton polarizabilities have exclusively come from Compton scattering experiments. These measurements rely on a theorem to establish a unique relation between the low–energy expansion (LEX) of the Compton scattering cross section and the polarizabilities. For photon energies sufficiently low, this expansion reads [1]

$$
\frac{d\sigma}{d\Omega}(\omega,\theta) = \frac{d\sigma}{d\Omega}^{\text{Born}}(\omega,\theta)
$$
$$
- \frac{e^2}{4\pi M}\left(\frac{\omega'}{\omega}\right)^2 (\omega\omega')\left\{\frac{\bar\alpha+\bar\beta}{2}(1+\cos\theta)^2 + \frac{\bar\alpha-\bar\beta}{2}(1-\cos\theta)^2\right\}
\tag{1}
$$

where ω and ω' are the energies of the incident and scattered photon, respectively, and $d\sigma^{\text{Born}}/d\Omega$ is the exact cross section for a proton with an anomalous magnetic moment κ but no other structure [2]. The LEX is an expansion of the cross section to first order in $(\omega\omega')$; besides κ the only structure-dependent terms to that order are the polarizabilities. The equation shows that the forward and backward cross sections are sensitive mainly to $\bar\alpha + \bar\beta$ and $\bar\alpha - \bar\beta$, respectively, whereas the 90° cross section is sensitive only to $\bar\alpha$. The sum $\bar\alpha + \bar\beta$ is independently constrained by a model–independent dispersion sum rule [3]:

$$
\bar\alpha + \bar\beta = \frac{1}{2\pi^2}\int_{m_\pi}^{\infty}\frac{\sigma_\gamma(\omega)d\omega}{\omega^2} = 14.2 \pm 0.5,
\tag{2}
$$

in units of 10^{-4} fm^3, which will be understood hereafter, and where $\sigma_\gamma(\omega)$ is the total photoabsorption cross section on the proton. The numerical value is obtained using both the available experimental data and a reasonable theoretical *ansatz* for continuing the integral to infinite energy [4]. Eq. 2 will henceforth be referred to as the *sum rule constraint*.

A sample plot of the LEX and Born cross sections is shown in Fig. 1. The curves show that the sensitivity of the LEX to the polarizabilities increases with energy. However, if the photon energy becomes too large, the LEX breaks down, thereby introducing theoretical uncertainty into the extraction of the polarizabilities from the measured cross sections. This is demonstrated in Fig. 1, where we show a full dispersion-theoretic calculation of the cross section, valid in principle to all orders in energy, as well as an expansion of the cross section to second order in $(\omega\omega')$. The formalism used to calculate these curves will be discussed more fully in Section 2.3. Since much of the experimental data is at energies outside the range of validity of the LEX, it is necessary to pay particular attention to the model dependence in the determination of the polarizabilities from the cross sections. This issue is discussed at length in Section 2.3.

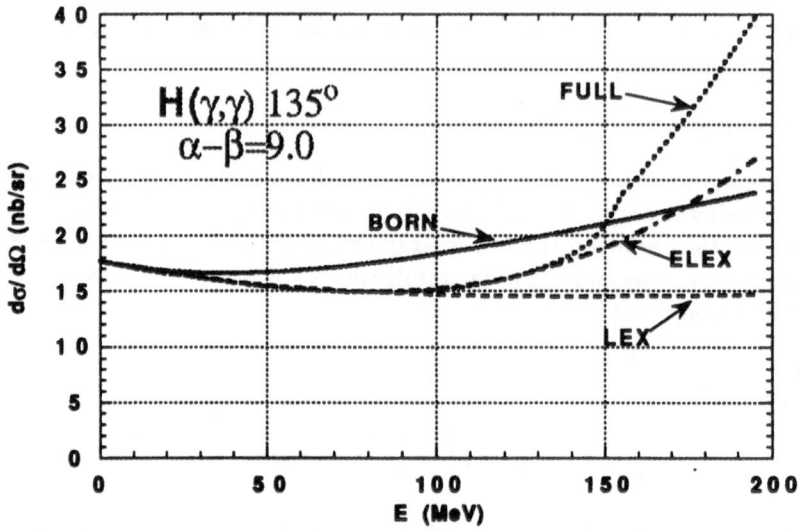

Fig. 1. Calculations of the Compton scattering cross section. The FULL calculation uses the fixed-t dispersion relations discussed in Section 2.3.2, the LEX is the expression of Eq. 1, and the ELEX is the expression of Eq. 3.

2.2 The Experimental Data

The experiments we consider in this review are summarized in Table I. Those reported prior to 1990 were reviewed earlier by Petrun'kin [1], so the present review will concentrate on the more recent ones. One of the principal motivations for the more recent experiments is that the Moscow 1975 experiment, the most precise and accurate of the pre-1990 experiments, reported a result for $\bar{\alpha} + \bar{\beta}$ that was significantly at variance with the dispersion sum rule, Eq. 2. We return to this point later on.

Data Set	Ref.	Energies	Angles	Syst. Err.
Chicago 1958	[5]	60±28 MeV	70°–150°	8.0%
Moscow 1960	[6]	56±25 MeV	75°–150°	6.0%
Moscow 1975	[7]	70-110 MeV	90°, 150°	3.0%
Illinois 1991	[8]	32-72 MeV	60°, 135°	3.0%
Mainz 1992	[9]	98, 132 MeV	180°	4.3%
SAL 1993	[10]	149-286 MeV	24°–135°	4.0%
Illinois/SAL 1994	[11]	70-150 MeV	90°, 135°	4.0%

Table I. Data sets used in the present analysis. SAL refers to the Saskatchewan Accelerator Laboratory. The Moscow 1975 experiment does not quote a systematic error, so we *assign* a systematic error of 3% to this experiment for the purposes of this review.

The Illinois 1991 experiment was the first to use tagged photon beams. The principal advantage of this technique is the ability to measure absolute cross sections with excellent systematic accuracy, since it allows one to measure accurately the incident photon flux by simply counting the associated tagging electrons. The principal drawback is that the beam intensities are inherently low, thereby making it difficult to achieve good statistical precision. Data were taken in the energy range 32–72 MeV, using two NaI detectors positioned at scattering angles of 60° (which is sensitive to $\bar{\alpha} + \bar{\beta}$) and 135° (which is senstive to $\bar{\alpha} - \bar{\beta}$). The LEX is valid over the full energy range. Typical statistical errors are approximately ±8% while the systematic error is estimated to be ±3.0%. The details of the experimental setup, data reduction, various corrections, and systematic errors can be found in Federspiel, *et al.* [8].

The Mainz 1992 experiment utilized an untagged bremsstrahlung beam to measure the 180°-Compton scattering cross section by detecting the recoil proton at 0° in a magnetic spectrometer. This allows a determination of $\bar{\alpha} - \bar{\beta}$ that is completely independent of the value of $\bar{\alpha} + \bar{\beta}$. Good systematic accuracy in the absolute cross sections was achieved by measuring the Compton *proton* yield relative to the Compton *electron* yield at 0°, the cross section for the latter being well-known. Cross sections were measured at energies of 98 and 132 MeV, both of which lie outside the range of validity of the LEX. Typical statistical and systematic uncertainties are each about ±4%. The details of the experimental setup, data reduction including the background subtraction, various corrections, and systematic errors can be found in Zieger, *et al.* [9].

The Saskatchewan 1993 experiment used the bremsstrahlung endpoint technique and a single very large, high resolution NaI detector. Complete angular distributions were measured at four separate energies between the pion threshold and just below the Δ peak, thereby allowing independent measurements of both $\bar{\alpha}$ and $\bar{\beta}$. All the data lie outside the range of validity of the LEX. The statistical precision of the cross sections were quite good, typically a few percent. The systematic accuracy of 4% was largely limited by the calibration of the quantameter used to measure the photon flux. Additional details can be found in Hallin, *et al.* [10].

The Illinois/Saskatchewan 1994 experiment measured the Compton cross section in the energy range 70-100 MeV using tagged photons and simultaneously in the range 100-150 MeV using bremsstrahlung folding. The tagged photon experiment was no different in principle from the Illinois 1991 experiment. The bremsstrahlung experiment consisted of both scattering and calibration runs. The latter were performed with an NaI detector placed directly in the photon beam in order to determine the shape of the upper 50 MeV of the bremsstrahlung spectrum and to calibrate the photon flux. The cross normalization between the scattering and calibration runs was provided by the electron detectors on the focal plane of the tagging spectrometer. Measurements were taken simultaneously at 90° and 135°, thereby allowing independent measurements of $\bar{\alpha}$ and $\bar{\alpha} - \bar{\beta}$, respectively. The LEX is valid for some but certainly not all of the data. Systematic errors are approximately 3%. Further details can be found in the Ph.D. thesis of MacGibbon [11].

2.3 Theoretical Considerations: Beyond the LEX

Since much of the data described above lie outside the range of validity of the LEX, we need to address the issue of how to extract the polarizabilities from the cross section and the model dependence therein. We describe here two separate approaches: the Extended Low Energy Expansion (ELEX) of Guiasu and Radescu [12] and the dispersion relation approach of L'vov [13].

2.3.1 Extended Low Energy Expansion Approach

The most straightforward way to go beyond the LEX is to extend the expansion to order $(\omega\omega')^2$. The result is a formula that we call the ELEX, which was first written down in Ref. [12]:

$$
\frac{d\sigma}{d\Omega} = \frac{d\sigma}{d\Omega}^{\text{Born}} + \frac{d\sigma}{d\Omega}^{\text{pol}} + \frac{d\sigma}{d\Omega}^{\pi^0}
$$
$$
- \frac{e^2}{256\pi^2 M^6} \left(\frac{\omega'}{\omega}\right)^2 (\omega\omega')^2 \{A + Bz + Cz^2 + Dz^3\}
$$
(3)

where $z = \cos\theta$ and $A - D$ are four unknown structure constants. The term labeled $(d\sigma/d\Omega)^{\text{pol}}$ is the polarizability-dependent part of the cross section to order $(\omega\omega')^2$ (that cross section to order $\omega\omega'$ is given by the 2nd line of Eq. 1). The term labeled $(d\sigma/d\Omega)^{\pi^0}$ is the contribution due to the t-channel exchange of a π^0 which is calculated exactly and explicitly separated out in order to extend the validity of the expansion to higher energy. Comparison with the fixed-t dispersion relations, which are discussed in Section 2.3.2, shows that the expansion to this order is accurate for energies below about 140 MeV (see Fig. 1). One way to apply this technique is to fit this expression to the experimental cross section with six adjustable parameters: the two polarizabilities and the four new constants. This allows one to use all of the data below pion threshold to extract the polarizabilities in a completely model independent manner. The obvious drawback is that the statistical precision is reduced since additional parameters need to be determined from the data.

2.3.2 Dispersion Relation Approach

This is the approach described by L'vov [13]. One starts by writing down the six independent invariant amplitudes, $A_i(s, u, t)$, which are free of kinematical singularities and constraints and even functions of $\nu = (s - u)/4M$. If one writes the Compton cross section in terms of the A_i and takes the low-energy limit, one can identify the polarizabilities by comparison with Eq. 1:

$$
\bar{\alpha} + \bar{\beta} = -\left(\frac{1}{2\pi}\right)\left[A_6^{\text{NB}}(\nu = 0, t = 0) + A_3^{\text{NB}}(\nu = 0, t = 0)\right]
$$
$$
\bar{\alpha} - \bar{\beta} = -\left(\frac{1}{2\pi}\right) A_1^{\text{NB}}(\nu = 0, t = 0)
$$
(4)

where NB means the non-Born part of the amplitude. Ignoring for the moment the question of asymptotic behavior, the A_i satisfy an unsubtracted fixed-t dispersion relation:

$$\text{Re}\,[A_i(s,u,t)] = A_i^{\text{Born}}(s,u,t) + \frac{1}{\pi}\int_{M^2}^{\infty}\left(\frac{1}{s'-s}+\frac{1}{s'-u}\right)\text{Im}\,[A_i(s',u',t)]\,ds'.$$

(5)

The Born parts are calculated exactly in terms of the charge and magnetic moment and the usual Feynman rules. Unitarity allows one to relate the $\text{Im}[A_i]$ to the multipole amplitudes for the total photoabsorption cross section on the proton. For Compton scattering in the Δ region and below, the most important intermediate states contributing to the dispersion integrals are the πN states, which have both the largest cross section and the largest energy weighting. For this contribution, the $\text{Im}[A_i]$ are sums of bilinear combinations of single-pion photoproduction multipole amplitudes, the most important of which have been very well measured. Thus the πN contribution to the dispersion integrals can be calculated reliably, unambiguously, and with an accuracy that is limited only by the accuracy with which the multipole amplitudes have been measured. On the other hand the multipole amplitudes for multi-pion photoproduction, which dominates the photoabsorption cross section for $E \geq 600$ MeV, are poorly known experimentally and therefore can only be treated in the context of a model. Fortunately for sufficiently low photon scattering energies, the multi-pion contribution to the dispersion integrals are not expected to be very large, so one can hope that the predicted scattering cross sections will be only weakly dependent on the model assumptions. Approximately 30-40% of the multi-pion cross section proceeds through known resonances, and these amplitudes can be straightforwardly decomposed into multipoles. The non-resonant cross section is partly due to $\pi\Delta$ production. For partial waves $l \geq 1$, the multipole amplitudes for this process are calculated in the Born approximation, assuming a single-pion-exchange mechanism. In L'vov's approach the remaining contributions needed to reproduce the experimental multi-pion cross section are treated crudely: They are ascribed to photoabsorption that is either purely electric dipole (corresponding to $\pi\Delta$ with $l = 0$) or purely magnetic dipole (corresponding to $\pi\pi N$ with everything in a relative $l = 0$ state). These extreme possibilities give us a tool to check the sensitivity of the scattering cross section to the model.

We next consider the asymptotic part of the A_i, since that governs the convergence of the dispersion integrals. Using Regge phenomonology one can show that for i=3-6, the $A_i(s,t)$ drop sufficiently rapidly at high s and fixed t to assure convergence of the dispersion integral. However both A_1 and A_2 are essentially s-independent at high s so the integral does not converge. L'vov's technique is to write the $\text{Re}[A_i]$ as a sum of an integral part, which results from doing the dispersion integral up to the fixed energy ω_M=1.5 GeV, and an asymptotic part A_i^{asymp}, which is a Regge-inspired parametrization of the remaining contribution to the dispersion integral. A_2^{asymp} is expected to be dominated by the t-channel exchange of a π^0:

$$A_2^{\text{asymp}}(t) = \frac{g_{\pi NN} F_{\pi^0 \gamma \gamma}}{t - \mu^2} \tag{6}$$

where both coupling constants are known. A_1 is expected to be similarly dominated by the exchange of the lowest mass scaler meson (the fictitious σ), for which the couplings and the mass are unknown. For sufficiently small t, one instead writes

$$A_1^{\text{asymp}}(t) = C e^{Bt/2} \tag{7}$$

where B and C are essentially unknown constants. The slope parameter B can be estimated from the systematics of the t-dependence of Compton scattering or other cross sections in the vicinity of 1.5 GeV, from which one estimates B $\approx$ 6-10 GeV^{-2} [14]. The remaining constant C is related to the difference of polarizabilities, as can be seen by combining and rearranging Eqs. 4, 5, and 7:

$$\bar{\alpha} - \bar{\beta} = -\left(\frac{1}{\pi^2}\right) \left\{ \int_0^{\omega_M} \text{Im}\left[A_1(\omega, 0)\right] \frac{d\omega}{\omega} + \frac{\pi C}{2} \right\} \tag{8}$$

with ω_M=1.5 GeV.

We thus see that the fixed-t dispersion relations can be used to predict the Compton scattering cross section below about 300 MeV in terms of only one unknown parameter, C (or equivalently $\bar{\alpha} - \bar{\beta}$), which is then adjusted to fit the scattering data. The calculations rely on the experimentally known single-pion photoproduction multipole amplitudes [15,16], a model for the multi-pion photoproduction multipole amplitudes, and the validity of Regge phenomonology, π^0-pole dominance, and the *ansatz* of Eq. 7 to describe the high-energy behavior of $A_3 - A_6$, A_2, and A_1, respectively. The principal sources of model dependence are the calculation of the multi-pion multipole amplitudes and the slope parameter B. In the analysis to follow, we will check the sensitivity of the result derived for $\bar{\alpha} - \bar{\beta}$ to those specific aspects of the calculation.

2.4 Fitting Procedure

In this section we describe the procedure used to derive the polarizabilities from the experimental Compton scattering cross sections, taking account of both the statistical and the systematic errors in the data. We assume that the systematic errors in each data set are mainly errors of normalization and represent a correlated uncertainty in the measured cross sections of that data set. We use the standard technique to take into account the systematic errors from different independent data sets. For each data set, we define χ^2 as follows:

$$\chi^2 = \sum_i \left(\frac{D\sigma_i^{\text{th}} - \sigma_i^{\text{exp}}}{\epsilon_i}\right)^2 + \left(\frac{D-1}{\epsilon_D}\right)^2, \tag{9}$$

where σ_i^{exp} and ϵ_i are the experimental scattering cross section and statistical error, respectively; σ_i^{th} is the corresponding calculated scattering cross section; and D is the normalization constant for that data set. For each data set, the normalization has the experimental value of 1 and uncertainty, ϵ_D, equal to the systematic error in the data set. The second term of Eq. 9 takes into account the contribution of the normalization to χ^2. The summing index i runs over the individual cross section measurements in the data set. The total value of χ^2 is obtained by summing the χ^2 for each data set. Standard least-squares-fitting procedures are used to adjust $\bar{\alpha}$, $\bar{\beta}$, and the D's in order to minimize the total χ^2. The net result of this procedure is that each data set is properly weighted based on its systematic error; the uncertainties in the fitted parameters include contributions from both the statistical and the systematic errors. The goal of the present analysis is to determine a best value for $\bar{\alpha} - \bar{\beta}$ rather than test the dispersion sum rule. Therefore all of the fits described here are subjected to the sum rule constraint of Eq. 2. We achieve this by including an additional data set consisting of a single datum, whose experimental value and uncertainty are given by the RHS of Eq. 2 and whose corresponding calculated value is equal to $\bar{\alpha} + \bar{\beta}$.

2.5 Results

As a first step, we inquire about the consistency of the various data sets with each other. One way to do that is to extract polarizabilities from each data set and ask whether those are consistent with each other. We therefore fit each of the data sets independently, subject to the sum rule constraint and taking into account the systematic errors in the manner described in Section 2.4, with the only other free parameter being $\bar{\alpha} - \bar{\beta}$. We use the fixed-$t$ dispersion relations with the model treatment of the multi-pion photoabsorption, the slope parameter B = 6 GeV^{-2}, and πN photoproduction multipoles of Ref. [15]. The results are shown in Fig. 2, where it is evident that there is excellent overall consistency among all but the Moscow 1975 experiment [7]. The inconsistency of that experiment can be traced to their 150° cross sections which, together with their 90° cross sections, lead to a strong violation of the dispersion sum rule, Eq. 2. If one eliminates the 150° data, the resulting $\bar{\alpha} - \bar{\beta}$ is consistent with the other experiments. Therefore, for the purpose of this review, we eliminate the 150° cross sections of Ref. [7] from further consideration.

We now seek an overall best fit to the data, subject to the sum rule constraint and again taking into account the systematic errors. The results of the fits are summarized in Table II. For the fits that use the fixed-t dispersion relations, we quote a range of values corresponding to varying B between 6 and 10 GeV^{-2} and using both the E1 and M1 option for the poorly known parts of the multi-pion photoproduction. As expected, the sensitivity to the model is reduced (and the fitting error is enlarged) if we use only the data below pion threshold. To proceed further one must make choices. We choose the dispersion calculation for E≤150 MeV as the best compromise between statistical precision and insensitivity to the model, and we take the mean between the two extremes as our result and attach a "theoretical uncertainty" equal to half the spread.

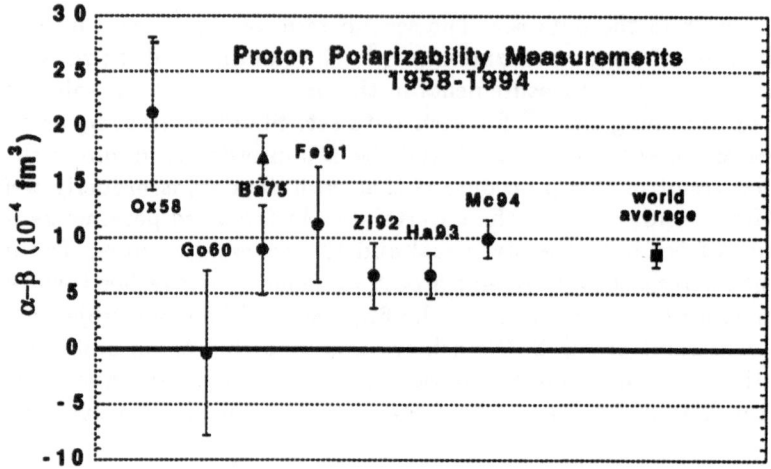

Fig. 2. Values of $\bar{\alpha} - \bar{\beta}$ derived from each experiment, which is denoted by its reference number. For the Moscow 1975 experiment [7], the triangle is derived using all the data and the filled circle omits the 150° data. The "world average" also omits these data.

Technique	E_{max}	$\bar{\alpha} - \bar{\beta}$
LEX	100	8.3 ± 2.3
ELEX	150	10.8 ± 3.2
Dispersion	100	$(8.1 - 8.5) \pm 2.7$
Dispersion	150	$(9.1 - 10.6) \pm 1.5$
Dispersion	290	$(8.2 - 11.4) \pm 1.1$

Table II. Results of fitting the scattering data to obtain $\bar{\alpha} - \bar{\beta}$. E_{max} denotes the maximum energy scattering data used in the fit. The range of results obtained with the dispersion technique corresponds to variations in the theory, as described in the text. The quoted error is the combined statistical and systematic error from the fitting procedure.

We therefore obtain as our final result

$$\bar{\alpha} - \bar{\beta} = 9.8 \pm 1.5 \pm 0.8 \tag{10}$$

Combining this with Eq. 2, we arrive at

$$\begin{aligned}
\bar{\alpha}_p &= 12.0 \pm 0.8 \pm 0.4 \\
\bar{\beta}_p &= 2.2 \mp 0.8 \mp 0.4
\end{aligned} \tag{11}$$

where the errors are anticorrelated. We close this section be remarking that one can reduce considerably the "theoretical uncertainty" on this result by doing a high accuracy, high statistics Compton scattering experiment covering a broad energy and angular range below the pion threshold. Such an experiment is planned for the MAMI facility at Mainz [17].

3 Experiments on the Neutron

Since this topic was reviewed recently by Schmiedmayer [18] and since there have been no new experiments since then, the discussion here will be brief. The only technique to yield anything but an upper limit to the neutron polarizability involves the scattering of a very low energy neutron in the Coulomb field of a heavy nucleus. In effect the Coulomb field polarizes the neutron giving rise to a "tidal force" whose strength is directly proportional to $\bar{\alpha}_n$. The most precise measurement of this type was the Vienna/Oak Ridge experiment of Schmied-mayer, *et al.* [19]. Rather than measure the scattering cross section, they instead measured the total neutron-nucleus cross sections $\sigma_a(k)$ as a function of the momentum of the incoming neutron k, for neutron kinetic energies between 50 eV and 40 keV. If one expands the cross section in powers of k

$$\sigma_a(k) = \sigma_a(0) + ak + bk^2 + ck^4 + ..., \tag{12}$$

then the term *linear* in k depends only on $\bar{\alpha}_n$, whereas the other terms depend mainly on the nuclear part of the neutron-nucleus interaction. By appropriately choosing a target (^{208}Pb), the nuclear contribution can be minimized. Analysis of the momentum dependence of the cross section yields

$$\bar{\alpha}_n = 12.5 \pm 1.5 \pm 2.0, \tag{13}$$

where the first error is statistical and the second is systematic. This can be combined with the dispersion sum rule [1]

$$\bar{\alpha}_n + \bar{\beta}_n = 15.8 \pm 1.0. \tag{14}$$

to yield

$$\bar{\beta}_n = 3.3 \pm 1.8 \pm 2.0 \tag{15}$$

Two comments are in order. First, we have learned at this workshop from S. Kopecky that the Vienna/Oak Ridge collaboration is in the process of redoing the neutron absorption experiment with improved statistical and systematic uncertainties. Second, there are plans at Saskatoon (and perhaps other places) to attempt to measure the neutron polarizabilities by doing either quasi-free or elastic Compton scattering on the deuteron.

4 Dispersion Sum Rules

The same fixed-t dispersion relations that we have used to analyze the Compton scattering data can be used to establish sum rules for the polarizabilities. These sum rules are completely rigorous yet semi-phenomonological, since they relate the polarizabilities to features of the photoabsorption cross section. This is quite appealing physically since it helps one identify the physics giving rise to $\bar{\alpha}$ and $\bar{\beta}$, such as the contribution of a particular resonance. We have already written down two of these sum rules: Eq. 2 for $\bar{\alpha} + \bar{\beta}$ and Eq. 8 for $\bar{\alpha} - \bar{\beta}$. We have also seen that the sum rule for $\bar{\alpha} + \bar{\beta}$ is nearly saturated by 1.5 GeV, whereas that for $\bar{\alpha} - \bar{\beta}$ has an asymptotic contribution that is not constrained by the photoabsorption on the nucleon.

It is illuminating to combine the two sum rules so that one effectively has a sum rule for $\bar{\alpha}$ and $\bar{\beta}$ separately. In turn, each can be written as a sum of an integral part which one can evaluate and an asymptotic part. The integrand for the integral part is shown in Fig. 3 (for the proton). One sees that $\bar{\alpha}$ has a major positive contribution due to nonresonant electric dipole photoproduction near threshold and near 600 MeV and a negative contribution from the Δ, the latter being a relativistic effect. The Δ makes the dominant contribution to $\bar{\beta}$. After integrating one finds

$$\bar{\alpha}_p \approx 6 + \frac{1}{2} \left(\bar{\alpha} - \bar{\beta} \right)_p^{\text{asymp}}$$

$$\bar{\beta}_p \approx 8 - \frac{1}{2} \left(\bar{\alpha} - \bar{\beta} \right)_p^{\text{asymp}}$$

$$\bar{\alpha}_n \approx 8 + \frac{1}{2} \left(\bar{\alpha} - \bar{\beta} \right)_n^{\text{asymp}} \tag{16}$$

$$\bar{\beta}_n \approx 8 - \frac{1}{2} \left(\bar{\alpha} - \bar{\beta} \right)_n^{\text{asymp}}$$

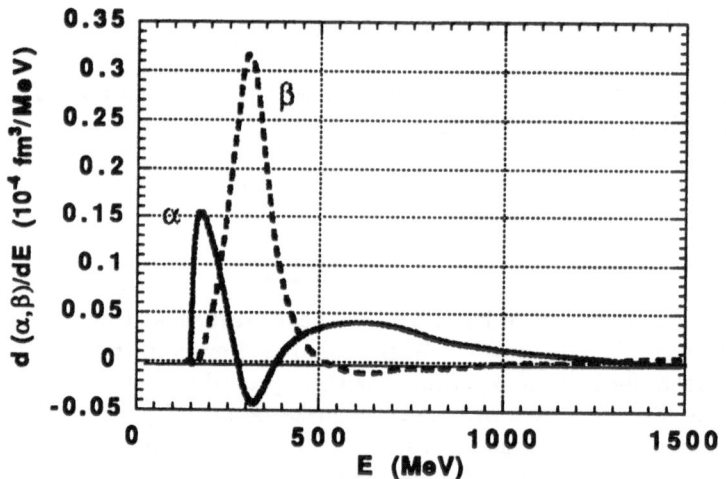

Fig. 3. Integrands to the dispersion sum rules for $\bar{\alpha}_p$ and $\bar{\beta}_p$.

If we compare with the experimental numbers, we find that the asymptotic parts are substantial and approximately equal for the proton and neutron:

$$\frac{1}{2} \left(\bar{\alpha} - \bar{\beta} \right)_p^{\text{asymp}} \approx \frac{1}{2} \left(\bar{\alpha} - \bar{\beta} \right)_n^{\text{asymp}} \approx 3 - 7$$

In fact, the combination $\bar{\alpha} - \bar{\beta}$ is *dominated* by asymptotic contributions. It is interesting to speculate about the physics underlying these asymptotic contributions. As already discussed, this comes from the asymptotic part of the A_1

amplitude, which is expected to be dominated by the t-channel exchange of a scaler meson (for example, correlated two-pion exchange). In effect, it is due to scattering from the pion cloud surrounding the nucleon. However, L'vov points out [20] that the Born part of the scattering from the pion cloud is already contained in the dispersion integral, so it is primarily the non-Born part that enters into the asymptotic amplitude. This suspicion is confirmed by an alternate technique for writing a sum rule based on a backward dispersion relation [21]. In effect the asymptotic part is written as a t-channel integral, which involves the physical amplitudes for the process $\gamma\gamma \rightarrow \pi\pi \rightarrow NN$. One finds that a *substantial* contribution to the integral is due to the non-Born part of $\gamma\gamma \rightarrow \pi\pi$, which is essentially due to pionic structure (including, among other things, the polarizability of the pion). Since the energy needed to excite the pion is large, this is just another indication that much of the physics of the nucleon polarizability is the physics of energies beyond 1 GeV. Any QCD-based structure model that attempts to calculate the polarizabilities will need to address this physics.

5 Concluding Remarks

We have reviewed the existing experimental data for the electric and magnetic polarizabilites of the nucleon and find that reasonably precise values now exist. New experiments are desired to remove the model dependence in the determination of the proton values and to improve the neutron measurements. Dispersion relations indicate that the polarizabilities are only partially constrained by existing photoproduction data and that a substantial part is due to physics beyond 1 GeV, such as pionic structure.

6 Acknowledgements

It is a pleasure to acknowldege many fruitful discussions with Anatoly L'vov and Bruce MacGibbon, as well as a critical reading of this manuscript by Bruce MacGibbon. This research was supported in part by the National Science Foundation under Grant No. NSF-PHY-93-10871.

References

1. V. A. Petrun'kin, Sov. J. Part. Nucl. **12**, 278 (1981).
2. J. L. Powell, Phys. Rev. **75**, 32 (1949).
3. A. M. Baldin, Nucl. Phys. **18**, 318 (1960).
4. M. Damashek and F. J. Gilman, Phys. Rev. D **1**, 1319 (1970).
5. C. L. Oxley, Phys. Rev. **110**, 733 (1958).
6. V. I. Gol'danski, *et al.*, Sov. Phys. JETP 11, 1223 (1960).
7. P. S. Baranov, *et al.*, Sov. J. Nucl. Phys. **21**, 355 (1975).

8.   F. J. Federspiel, *et al.*, Phys. Rev. Lett. **67**, 1511 (1991).
9.   A. Zieger, *et al.*, Phys. Lett. **278B**, 34 (1992).
10. E. Hallin, *et al.*, Phys. Rev. C **48**, 1497 (1993).
11   B. J. MacGibbon, Ph.D. thesis, University of Illinois (1994), unpublished.
12   I. Guiasu and E. E. Radescu, Phys. Rev. D **18**, 651 (1978).
13   A. I. L'vov, Sov. J. Nucl. Phys., **34**, 597 (1981).
14   R. P. Feynman, *Photon-Hadron Interactions*, (Addison-Wesley, Reading, 1989), p. 215.
15   R. G. Moorhouse, *et al.*, Phys. Rev. D **9**, 1 (1974).
16   R. A. Arndt, *et al.*, Phys. Rev. C **42**, 1853 (1990).
17   J. Ahrens, private communication.
18   J. Schmiedmayer, *Baryons '92*, M. Gai ed. (World Scientific, Singapore, 1993), p. 240.
19   J. Schmiedmayer, *et al.*, Phys. Rev. Lett. **66**, 1015 (1991).
20   A. I. L'vov, Int. J. Mod. Phys A, **8**, 5267 (1993).
21   B. R. Holstein and A. M. Nathan, Phys. Rev. D **49**, 6101 (1994).

Pion (Kaon) and Sigma polarizabilities [i]

R. Baldini and S. Bellucci

INFN-Laboratori Nazionali di Frascati,
P.O.Box 13, 00044 Frascati, Italy
e-mail: Baldini@lnf.infn.it; Bellucci@lnf.infn.it

Abstract: We report the results of the working group on "Pion (Kaon) and Sigma Polarizabilities". Interesting possibilities to measure these polarizabilities include the radiative pion photoproduction in the MAMI experiment at Mainz, as well as at the GRAAL facility (actually the latter is being considered for an experimental determination of the pion polarizabilities here for the first time), the experimental plans on Primakoff effect at FNAL, and the measurements at the Frascati Φ-factory DAΦNE.

1   Introduction

We report here about the activity of the working group we have been running on "Pion (Kaon) and Sigma Polarizabilities" at the Workshop on Chiral Dynamics: Theory and Experiment, held July 25-29,1994 at MIT. The goal of this working group was to identify the processes that are more suitable to measure the electric and magnetic polarizabilities of the abovementioned hadrons and probe chiral dynamics in the photon-pion (-kaon) and photon-sigma physics.

The agenda of the group was a mixture of theory and experiment that allowed us to summarize the current status in this field, determine what is to be done in order to improve it, from both the theoretical and the experimental side, and quantify the level of accuracy needed to make the improvement significant.

We considered the following general areas:

1. Theoretical predictions and models for Compton scattering $\gamma\pi(K) \rightarrow \gamma\pi(K)$, as well as for $\gamma\gamma \rightarrow \pi\pi(KK)$, and the relation to the pion (kaon) polarizabilities. 2. Experiments to measure the pion (kaon) and sigma polarizabilities. 3. Phenomenology required in order to extract the polarizabilities from the experimental data.

We have considered the following methods of measurement:

i) Radiative photoproduction of the pion and extrapolating to the pion pole, in order to extract the polarizabilities form the data. ii) Experiments to measure the pseudoscalar meson polarizabilities using the Primakoff effect. iii) The measurements at the Frascati DAΦNE with the KLOE detector.

Murray Moinester illustrated the plans at FNAL and the reaction $\pi\rho \rightarrow \pi\gamma$ in connection with the pion and sigma polarizabilities. Thomas Walcher went over the MAMI project at Mainz. Annalisa D'Angelo discussed the potential of the Graal synchrotron light facility in Grenoble to carry out polarizability measurements. S. Kananov discussed the need for a careful estimate of the radiative corrections for the FNAL experiments where the pion polarizabilities can be measured. A presentation of the DAΦNE capabilities has been done in the workshop.

The related talks had the following titles: A. D'Angelo: "The experimental plans in Grenoble", T. Walcher: "The MAMI experiment at Mainz", M. Moinester: "Pion polarizabilities and quark-gluon plasma signatures", S. Kananov: "Radiative corrections for pion polarizability experiments",

In the second part of the working group, devoted to the theoretical contributions, we had the following talks: J. Kambor: "Determination of a $O(p^6)$ counterterm from sum rules", M. Knecht: "$\gamma\gamma \rightarrow \pi^0\pi^0$ and π^0 polarizabilities in generalized chiral perturbation theory", M. Pennington: "Dispersion relations and pion polarizabilities", S. Bellucci: "Difficulties in extrapolating to the pion pole the data on radiative pion photoproduction".

2 Experimental plans for polarizability measurements

2.1 Radiative pion photoproduction

Let us consider first of all the unexpected and very innovative contribution by Annalisa D'Angelo. She has reported about the possibility to measure the pion polarizability by Graal, the new facility at the electron storage ring ESRF in Grenoble.

The Graal facility consists of a tagged and highly polarized γ-ray beam, produced by the backscattering of Laser light against the high energy electrons circulating in the storage ring ESRF at Grenoble [1]-[3].

If commercial Ion-Argon and Nd-Yag Lasers are used, either linearly or circularly polarized, a γ-ray beam of energy ranging from about 300 MeV to 1.5 GeV with a degree of polarization higher then 70% over almost the entire spectrum may be obtained, by appropriate choice of the Laser line.

A large solid angle multi purpose detector [4] is part of the Graal facility and it will be used to perform experiments on photo-nuclear reactions [5]. It consists of a crystal ball made of 480 BGO crystals (24 cm long) covering all azimuthal angles for polar angles in the interval between 25° and 155°. It may be used as electromagnetic calorimeter, with measured energy resolution of 2% FWHM for 1 GeV photons, or to detect protons of energy up to 300 MeV [6]-[8].

The central hole of the BGO ball ($\phi = 20$ cm) will contain a barrel of 32 plastic scintillators; it will be used to discriminate between charged and neutral particles and to identify the charged particles with the $\frac{\Delta E}{E}$ technique. Inside the barrel two cylindrical wire chambers will provide improved angular resolution for the reconstruction of the trajectories of the charged particles.

In the forward direction two detectors, each consisting of three plane wire chambers rotated of 45°, and a scintillating wall will provide charged particle tracking information and TOF measurements; 10% efficiency is expected for neutron detection in the scintillating wall.

We have started to investigate the possibility of using the Graal facility to study the radiative pion photoproduction from the proton (namely the reaction $\vec{\gamma} + p \rightarrow n + \pi^+ + \gamma$), in order to extrapolate the experimental data to the pion pole and determine the cross-section of the Compton scattering on the pion [9].

The interest of this measure, in order to get information on the pion polarizability, has been pointed out, among other authors, by D. Drechsel and L.V. Fil'kov [9]; they have stressed that in order to obtain a reliable extrapolation it is necessary to have experimental data in kinematical condition as close as possible to the point $t = 0$, being t the momentum transferred between the final neutron and the initial proton.

A measurement performed at 1.5 GeV photon energy, using polarized photons, would fulfill the experimental requirements of Ref. [9], also providing higher sensitivity to the pion polarizability contributions through the polarization structure functions.

A fundamental issue of the experimental set-up is the capability of discriminating the reaction of interest from the background events, like those coming from asymmetric decay of π^0 in the $\pi^+\pi^0 n$ reaction channel.

In principle all these requirements are fulfilled by the Graal facility: the scattered photons may be detected in the BGO ball for laboratory angles between 25° and 155°; low energy neutrons my be detected in the forward direction using the plastic wall and they may be identified using the TOF information; finally the pions may be detected at all angles by the wire chambers with good angular resolution. All these experimental information should allow a complete reconstruction of the interesting events in selected kinematical conditions, with expected good background rejection.

The Graal facility set-up is therefore a promising tool to perform the first experiment with polarized photons on radiative pion photoproduction in order to extract information on the pion polarizability.

Thomas Walcher has reviewed, from his general talk on the experimental activity at MAMI in Mainz, the measurement of the pion polarizability still by means of radiative pion photoproduction.

Walcher has shown some kinematical conditions suitable for the measurement of the charged pion polarizability. The sensitivity in the Chew-Low extrapolation at the pion pole has been stressed [9]. For instance a variation of α_π from 0 up to 7 is equivalent to a 20 % variation in the extrapolated amplitude. Hence the extrapolation has to be done at the 2 % level , if α_π has to be measured at a 10 % level. This sensitivity depends, of course, on the minimum momentum transfer t_{min} achieved. For instance for a 700 MeV incident photon, a 152 MeV final photon, a neutron in the angular range 12^0-32^0 and in the energy range 400 - 350 MeV, it is $t = 0.31$, in pion mass squared units. Nucleon, pion and $\Delta(1236)$ pole diagrams have been evaluated and the extrapolation seems feasible.

Conversely the measurement of the neutral pion polarizability is not realistic at the MAMI energies. The situation may be improved at higher incident photon energies and a polarized beam would be very welcome, just like the Graal facility! By the way M. Moinester has stressed that the maximum energy meaningful for extracting the pion polarizability is about 2 GeV, corresponding to the ρ mass in the photon-pion c.m. system.

The MAMI detector consists of a MWPC system to detect the charged pion, close to the incident photon beam, a segmented BaF_2 to detect the scattered photon and a system of scintillators for detecting the neutron and providing the neutron time of flight. The expected yield of radiative pion photoproduction is $\simeq 2000$ events/day. The background due to double photoproduction, simulating radiative pion photoproduction, is $\simeq 160$ events/day.

2.2 Primakoff effect

The contribution of Murray Moinester has concerned the measurement of π (K) and Σ polarizabilities via π (K) and Σ high energy beams at Fermilab by E781 [10].

This experimental activity has been reported already in detail by M. Moinester in his contribution to the workshop. Therefore only the main topics are emphasized here, first of all the relationship between polarizability and radiative transitions.

The A_1 radiative width is a good illustration of this statement. It has been demostrated [11] according to the current algebra the main contribution to the pion electric polarizability comes from the exchange of the A_1. Xiong, Shuryak and Brown [12] have shown that a radiative width $\Gamma(A_1 \to \pi\gamma)= 1.4$ MeV is needed to get the current algebra value, on the contrary the experimental value is $\Gamma(A_1 \to \pi\gamma)= 0.64 \pm 0.25$ MeV [13]. More data from E781 are welcome to settle this relevant problem.

By the way the unexpected correlation between the reaction $\pi\rho \to \pi\gamma$ and the photon flux in a quark-gluon plasma has been pointed out.

Experimental results from the previous FNAL experiment E272 have been shown to demonstrate the experimental feasibility of the Primakoff effect, even if E272 did not get enough statistics to measure α_π. The main experimental problem in measuring the Primakoff effect by the new FNAL experiment, E781, concerns a suitable fast trigger. It is not implemented at the moment, taking into account the high rate and the difficulties for using any signal from the scattered photon detector, which is 50 meters downstream the target.

From a theoretical point of view an open question remains the fair disagreement between the charged ρ radiative width, as obtained via the Primakoff effect, and the neutral ρ radiative width.

Finally M. Moinester has stressed the role of radiative transitions in the case of a Σ beam, which should also be available in E781. The radiative transitions to the Sigma*(1385) provide a measure of the s-quark magnetic moment of the Sigma[14]. Positive and negative Σ are expected to have very different polarizabilities. In particular the Σ^- magnetic moment should be negligible, both taking into account the 3 quarks have the same charge and according to the U-spin symmetry. Furthermore it has been shown that the event rate for measuring Σ radiative transitions by a 600 GeV Σ beam is higher than the expected rate for measuring the pion Primakoff effect by E781.

S. Kananov has reported about radiative corrections in the scattering of pions by nuclei at high energies [15]. It has been shown that radiative corrections can simulate a variation of the magnetic polarizability $\beta_\pi \simeq$ -0.2 from $\beta_\pi \simeq$ -5, with plausible cuts for the outgoing photon.

2.3 $\gamma\gamma \rightarrow \pi\pi$ at threshold

Another way to get the pion polarizability is by means of the measurement of $\gamma\gamma \rightarrow \pi\pi$ at threshold, performed at the new Frascati Φ-factory DAΦNE [16].

A presentation of this new experimental facility has been done already in the workshop and no further discussion on experimental details has been done in this working group.

In summary the new Frascati e^+e^- storage ring DAΦNE is supposed to deliver a luminosity $\simeq 5 \cdot 10^{32}$ cm^2 s^{-1} at the Φ mass, with the possibility to increase the total energy up to 1.5 GeV. Two detectors are under construction: an all purposes detector, KLOE [17] mainly dedicated to CP violation in K decay, and FINUDA [18] mainly dedicated to hypernuclei physics.

KLOE is expected [19] to detect $10^3 \div 10^4$ times the events collected at present in $e^+e^- \rightarrow e^+e^-\pi^+\pi^-$ and $e^+e^- \rightarrow e^+e^-\pi^0\pi^0$ at threshold. Unfortunately at the Φ energy the decay $\Phi \rightarrow K_S K_L$ is an overwhelming background (in $\sim$ 15 % of the events K_L are not detected) and tagging the outgoing e^+e^- is needed. Two kind of tagging system for the outgoing leptons are foreseen. First of all there are two different rings for electrons and positrons and the splitter magnet after the interaction region is a suitable magnetic analyzer for the outgoing e^+e^-, mostly forward emitted. Furthermore e^+e^- emitted at larger angles are, in part, detected by the central tracking detector in KLOE. The e^+e^- angular distribution depends on m_e/E_e : therefore there are more events at large angles in DAΦNE respect to the high energy e^+e^- storage rings.

By the way correlations in the azimuthal angles between the pions and the outgoing leptons could be performed [20], increasing the possibility to disentangle the D wave contribution. Otherwise $\gamma\gamma$ interactions near threshold are supposed to provide mainly the $\pi\pi$ S wave, which provides only $\alpha_\pi - \beta_\pi$.

The overall double tagging efficiency is $\sim$ 15 % [19]. The background from beam-beam bremmstrhalung is still under study for evaluating the single tagging efficiency.

In $\gamma\gamma$ interactions complications related to any nuclear target are avoided, but it has been demonstrated in the following theoretical discussion that the extrapolation to the pion pole is much more difficult. Therefore the conclusion has been achieved that $\gamma\gamma$ interactions are not the best way to get the pion polarizability. Nevertheless $\gamma\gamma$ interactions near threshold remain a very clean test of any theoretical description of strong interactions at low energies.

Another possibility pointed out for measuring neutral and charged α_π in e^+e^- is by means of $e^+e^- \to \pi\pi\gamma$ increased by the interference both with $\omega \to \pi^0 \rho^0 \to \pi_0\pi^0\gamma$ and, for the charged one, also with radiative ρ production.

3 Theoretical issues in polarizability experiments

We begin with the process $\gamma\gamma \to \pi^0\pi^0$. In this case the Born amplitude vanishes and the one-loop corrections in Chiral Perturbation Theory (CHPT)[21]-[25] are finite [26, 27]. The corresponding cross section is independent of the free parameters of the chiral lagrangian and does not agree with the experimental measurements at Crystal Ball [28], as well as with calculations based on dispersion relations [29]-[35], even at low-energy. The low-energy amplitude recently calculated to two-loops in CHPT [36] agrees with the Crystal Ball data and compares very well with the results of a dispersive calculation by Donoghue and Holstein [32].

The value of the low-energy constants can be obtained in several ways, e.g. by resonance exchange. The resonance saturation method provides empirical values for the scale-dependent renormalized constants of CHPT [22]-[23]

$$L_i^r(\mu) = L_i^r(\mu_0) - \frac{\Gamma_i}{16\pi^2} ln\frac{\mu}{\mu_0} \quad , \; i = 1, .., 10 \tag{1}$$

with a scale μ in the range $0.5 \; GeV - 1 \; GeV$ and a set of constants Γ_i defined in [22]-[23]. This method has been used in Ref. [36] to pin down the couplings in the $\gamma\gamma \to \pi^0\pi^0$ amplitude to order p^6. In his talk J. Kambor discussed how to determine these couplings from sum rules, exploiting the low- and high-energy behaviour.

Let us consider the vector-vector two-point function

$$i \int d^4x e^{iqx} < 0|T(V_\mu^a(x)V_\nu^b(0)|0 >= \delta^{ab}(q_\mu q_\nu - g_{\mu\nu}q^2)\Pi^a(q^2). \tag{2}$$

Following [21]-[23] and [32] we write a dispersion relation for $\Pi^a(q^2)$

$$\Pi^a(q^2) = \frac{1}{\pi} \int ds \frac{Im\Pi^a(s)}{s - q^2 - i\epsilon} + subtractions \; , \tag{3}$$

in order to make contact with the high-energy behaviour of the theory desumed from the perturbative QCD sum rules

$$\begin{aligned} \rho_V^a(s) &= \frac{1}{\pi} Im\Pi^a(s) \; , \\ lim_{s\to\infty}\rho_V^a(s) &= \frac{1}{8\pi^2} + O(\frac{1}{s}) \; , \; a = 3, 8 \end{aligned} \tag{4}$$

showing that the spectral function $\rho_V^a(s)$ at high energy goes like a constant plus higher-order terms that are suppressed at least as $\frac{1}{s}$. Hence, the difference between two spectral functions goes like s at large s, and the integral

$$\int ds \frac{\rho_V^3 - \rho_V^8}{s} \tag{5}$$

converges. Also the once-subtracted dispersion relation for $\rho_V^3(s)$ converges

$$\int ds \frac{\rho_V^3}{s^2} . \tag{6}$$

From the dispersion relation

$$\Pi^3(q^2) - \Pi^8(q^2) = \frac{1}{\pi} \int ds \frac{Im\Pi^3(s) - Im\Pi^8(s)}{s - q^2 - i\epsilon} \tag{7}$$

the sum rule is readily obtained

$$\Pi^3(0) - \Pi^8(0) = \int ds \frac{\rho_V^3 - \rho_V^8}{s} . \tag{8}$$

Taking the q^2 derivative of the dispersion relations (7) and (3) yields the sum rules

$$\frac{d}{dq^2}(\Pi^3(0) - \Pi^8(0)) = \int ds \frac{\rho_V^3 - \rho_V^8}{s^2} \tag{9}$$

and, respectively,

$$\frac{d}{dq^2}\Pi^3(0) = \int ds \frac{\rho_V^3}{s^2} . \tag{10}$$

Notice that if one takes too many q^2-derivatives, then the integrals become dominated by the threshold region.

As for the low-energy behaviour, J. Kambor showed how to use CHPT to calculate $\Pi^a(q^2)$ for small q^2 values. This calculation is carried out in $SU(3) \times SU(3)$ to the two-loop order, and the result depends on the $O(p^4)$ and $O(p^6)$ low-energy constants L_i and, respectively, d_j [37]. The integral on the r.h.s. of Eq. (8) can be evaluated from the e^+e^- data. In the narrow width approximation one gets the following estimate [38]:

$$\int ds \frac{\rho_V^3 - \rho_V^8}{s} = \frac{3}{4\pi\alpha^2}\left(\frac{\Gamma_{\rho \to e^+e^-}}{M_\rho} - 3\frac{\Gamma_{\omega \to e^+e^-}}{M_\omega} - 3\frac{\Gamma_{\phi \to e^+e^-}}{M_\phi}\right) = (11.1 \pm 2.0) \cdot 10^{-3} . \tag{11}$$

Thus, the integration region becomes divided into three pieces, i.e. $4M_\pi^2 \leq s \leq \Lambda_1$, $\Lambda_1 \leq s \leq \Lambda_2$, and $s \geq \Lambda_2$. Here we denote by $\Lambda_{1,2}$ two cutoff values of about 0.4 GeV and 2 GeV, respectively, and M_π is the pion mass. In the first region the shape of the spectral function is obtained from the two-loop CHPT calculation, i.e. $\rho_V = \rho_V^{1-loop} + \rho_V^{2-loop}$. In the second (third) region the e^+e^- data (the perturbative QCD calculation) can be used to obtain $\rho_V(s)$.

The result of this calculation (once it is completed) yields a scale independent determination of d_3 contributing to $\gamma\gamma \to \pi^0\pi^0$. J. Kambor expects an accuracy for this estimate between 10 and 20 percent. This would be more precise than the estimate carried out in [38] using a similar procedure (excluding, however, the two-loop contribution), within the framework of the Generalized Chiral Perturbation Theory (GCHPT)

$$d_3 = (9.4 \pm 4.7) \cdot 10^{-6} \ . \tag{12}$$

There are two more sum rules to consider. In particular Eq. (9) is effectively a sum rule for L_9 (the d_i contributions to Π^3 and Π^8 drop out of the sum rule), whereas Eq. (10) is a sum rule for d_5, d_6 contributing to $\gamma\gamma \to \pi^+\pi^-$ and $\gamma\pi^\pm \to \gamma\pi^\pm$. A similar treatment can be applied to the sum rule for the axial-axial two-point function. This gives an estimate of the low-energy constants $L_{1,2,3}$ contributing to $\pi\pi$-scattering.

The GCHPT approach is described in [39] (see also M. Knecht's talk in the $\pi\pi$-scattering Working Group Section of these Proceedings). Within this approach the cross section for $\gamma\gamma \to \pi^0\pi^0$ and the pion polarizabilities have been calculated up to and including $O(p^5)$ [38]. M. Knecht discussed the result of this calculation. He showed that the cross section depends on the quark mass ratio $r = \frac{2m_s}{m_u+m_d}$ and is consistent with the data from Crystal Ball [28], provided the value of the ratio is at least a factor of two or three smaller than its standard value in CHPT [38]. M. Knecht showed that a low-energy theorem analogous to the one outlined above, but without taking into account the 2-loop contribution, yields, through the evaluation of the dispersive integral via resonance saturation, the following value for the $O(p^5)$ constant c defined in [38]:

$$c = -\frac{1}{r-1}(4.6 \pm 2.3) \cdot 10^{-3} \ . \tag{13}$$

In the standard CHPT case, i.e. for

$$r = 2\frac{M_K^2}{M_\pi^2} - 1 = 25.9 \ , \tag{14}$$

the expression for c given in Eq. (13) corresponds to the value reported in Eq. (12). This is to be compared with the value calculated from Appendix D of Ref. [36], using resonance exchange

$$d_3 = \pm 3.9 \cdot 10^{-6} \ . \tag{15}$$

M. Pennington reviewed the dispersion relation treatment of the $\gamma\gamma \to \pi^0\pi^0$ amplitude that represents the data quite well [30], [31]. He showed how to calculate the cross section from first principles, using a relativistic and causal description based on the unitarity of the scattering matrix. The prediction for low-energy $\gamma\gamma \to \pi^0\pi^0$ and $\gamma\gamma \to \pi^+\pi^-$ is based on the present knowledge of the $\pi\pi$ phases and $PCAC$. M. Pennington argued that precision measurement of $\gamma\gamma \to \pi^0\pi^0, \pi^+\pi^-$ at the Frascati Φ-factory DAΦNE will restrict further the $\pi\pi$ phases and determine where the chiral zero appears on-shell.

The electric and magnetic polarizabilities enter the low-energy limit of the coupling with the photon in the Compton amplitude for any composite system. The dynamics of hadronic systems can be probed by measuring the hadron polarizabilities [40]. In particular, the pion Compton scattering can be investigated in this connection. The charged pion Compton amplitude

$$\gamma(q_1)\pi^+(p_1) \to \gamma(q_2)\pi^+(p_2) \ , \tag{16}$$

admits an expansion near threshold

$$T^C = 2\left[\vec{\epsilon}_1 \cdot \vec{\epsilon}_2{}^* \left(\frac{\alpha}{M_\pi} - \bar{\alpha}_\pi \omega_1 \omega_2\right) - \bar{\beta}_\pi (\vec{q}_1 \times \vec{\epsilon}_1) \cdot (\vec{q}_2 \times \vec{\epsilon}_2{}^*) + \cdots\right] \tag{17}$$

with $q_i^\mu = (\omega_i, \vec{q}_i)$. Below we denote

$$\begin{aligned}
(\alpha \pm \beta)^C &= \bar{\alpha}_\pi \pm \bar{\beta}_\pi \ , \\
(\alpha \pm \beta)^N &= \bar{\alpha}_{\pi^0} \pm \bar{\beta}_{\pi^0} \ ,
\end{aligned} \tag{18}$$

for charged and neutral pions, respectively.

The charged pion polarizabilities have been determined in an experiment on the radiative pion-nucleus scattering $\pi^- A \to \pi^- \gamma A$ [41] and in the pion photoproduction process $\gamma p \to \gamma \pi^+ n$ [42]. Assuming the constraint $(\alpha+\beta)^C = 0$ the two experiments yield[1]

$$(\alpha - \beta)^C = \begin{cases} 13.6 \pm 2.8 & [41] \\ 40 \pm 24 & [42] \ . \end{cases} \tag{19}$$

Relaxing the constraint $(\alpha + \beta)^C = 0$, one obtains from the Serpukhov data

$$\begin{aligned}
(\alpha + \beta)^C &= 1.4 \pm 3.1(\text{stat.}) \pm 2.5(\text{sys.}) \ [43] \ , \\
(\alpha - \beta)^C &= 15.6 \pm 6.4(\text{stat.}) \pm 4.4(\text{sys.}) \ [43] \ .
\end{aligned} \tag{20}$$

At one-loop in CHPT one has [44, 11, 45]

$$\bar{\alpha}_{\pi^0} = -\bar{\beta}_{\pi^0} = -\frac{\alpha}{96\pi^2 M_\pi F^2} = -0.50 \ . \tag{21}$$

At order $O(p^6)$ it was calculated in Ref. [36]

$$\begin{aligned}
\bar{\alpha}_{\pi^0} &= -0.35 \pm 0.10 \ , \\
\bar{\beta}_{\pi^0} &= 1.50 \pm 0.20.
\end{aligned} \tag{22}$$

The low-energy $\gamma\gamma \to \pi^+\pi^-$ data [46] have been used in Ref. [45] to obtain information on $\bar{\alpha}_\pi$ and $\bar{\beta}_\pi$. The result in [45] yields the numerical value for the

---

[1] Throughout the following, we express the values of the polarizabilities in units of $10^{-4} fm^3$

leading-order $\bar{\alpha}_\pi = 2.7 \pm 0.4$, plus systematic uncertainties due to the $O(p^6)$ corrections. The latter are not yet available. A part of the corrections to the charged pion polarizabilities beyond the one-loop order has been obtained in Refs. [47, 48] including the meson resonance contribution.

M. Knecht analyzed the $O(p^5)$ calculation from Ref. [38] of the π^0 polarizabilities in GCHPT. The result $(\alpha + \beta)^N = 0$ remains valid to this order, whereas the remaining combination depends on the quark mass ratio r. Hence this combination can have positive values for r much less than its standard CHPT value (14), e.g. $(\alpha - \beta)^N = 1.04 \pm 0.60$ for $r = 10$ [38]. For comparison we recall the standard CHPT prediction for both combinations to the $O(p^6)$ order [36]

$$
\begin{aligned}
(\alpha + \beta)^N &= 1.15 \pm 0.30 \ , \\
(\alpha - \beta)^N &= -1.90 \pm 0.20.
\end{aligned}
\tag{23}
$$

The reason for the sign difference of $(\alpha - \beta)^N$ in GCHPT with respect to the standard CHPT value has been traced back by M. Knecht to a dominance by the positive $O(p^5)$ contribution over the strongly suppressed negative $O(p^4)$ contribution in GCHPT. This suppression is related to a shift in the position of the chiral zero, as r becomes much smaller than the standard CHPT value (14) [38].

Starting from the unitarized S-wave amplitudes for neutral pions, M. Pennington displayed a proportionality relation between $(\alpha - \beta)^N$ and the position of the chiral zero s_N, showing that the former assumes values between -0.6 and -2.7 as the latter runs from $\frac{1}{2}M_\pi^2$ and $2M_\pi^2$. He also discussed the validity of the errors quoted in a recent estimate of $(\alpha + \beta)^{C,N}$ by Kaloshin and collaborators [49]. Here the polarizabilities appear as adjustable parameters in the unitarized D-wave amplitudes, hence the values of $(\alpha + \beta)^{C,N}$ can be determined from the data with the result [49]

$$
\begin{aligned}
(\alpha + \beta)^C &= 0.22 \pm 0.06 \ [46] \ , \\
(\alpha + \beta)^N &= 1.00 \pm 0.05 \ [28] \ .
\end{aligned}
\tag{24}
$$

M. Pennington, arguing on the partial wave analysis of the data that shows large uncertainties even at the $f_2(1270)$ mass, concluded that the errors quoted in (24) for $(\alpha + \beta)^N$ are unbelievably small. His final conclusion, that one must measure $\gamma\pi \to \gamma\pi$ can be wholeheartedly shared, in view of future measurements of the pion polarizabilities. In this respect, it is very important to devise fully reliable methods that allow to extract the pion Compton scattering amplitude from the measurement of the radiative pion photoproduction, as discussed by one of us (S.B.).

Acknowledgements

It is a pleasure to thank the organizers of this Workshop for a very stimulating working environment and the participants to this working group for their precious help in preparing this report. One of us (SB) acknowledges partial financial support from the EEC Human Capital and Mobility Program.

References

[1] D. Babusci, L. Casano, A. D'Angelo, P. Picozza, C. Schaerf and B. Girolami, Internal report LNF-90/060(P) - Nuovo Cimento 103 A (1990) 1555.

[2] D. Babusci, L. Casano, A. D'Angelo, B. Girolami, D. Moricciani, P. Picozza, C. Schaerf, Proc. Workshop "Future of Nuclear Physics in Europe with Polarized Electrons and Photons" Orsay July 4-6 (1990) 149.

[3] J.P. Bocquet, M. Capogni, L. Casano, A. D'Angelo, F. Ghio, B. Girolami, D. Moricciani, C. Perrin, P. Picozza, C. Schaerf, Book of abstr. PANIC XIII Perugia (1993) 835.

[4] M. Anghinolfi, V. Bellini, G. Berrier, N. Bianchi, J. P. Bocquet, H. Bougnet, P. Calvat, M. Capogni, L. Casano, P. Corvisiero, A. D'Angelo, E. De Sanctis, J.P. Didelez, Ch. Djalali, M.A. Duval, A. Ebolese, A. Elayi, A. Fantoni, R. Frascaria, G. Gervino, F. Ghio, B. Girolami, E. Hourani, P. Levi Sandri, L. Mazzaschi, V. Mokeev, D. Moricciani, M. Morlet, V. Muccifora, C. Perrin, P. Picozza, E. Polli, D. Rebreyend, A.R. Reolon, G. Ricco, M. Rigney, M. Ripani, R. Rosier, P. Rossi, Th. Russew, M. Sanzone, C. Schaerf, M. Taiuti, J. Van de Wiele, A. Zucchiatti, Book of abstr. PANIC XIII Perugia (1993) 850.

[5] The BGO Collaboration, Internal report LNF-90/084(1990).

[6] L. Mazzaschi, P. Levi Sandri, A. Zucchiatti, M. Anghinolfi, D. Babusci, N. Bianchi, P. Corvisiero, A. D'Angelo, E. De Sanctis, G. Gervino, B. Girolami, V. Lucherini, V. Mokeev, V. Muccifora, P. Picozza, E. Polli, A.R. Reolon, G. Ricco, M. Ripani, P. Rossi, M. Sanzone, C. Schaerf, M. Taiuti, Nucl. Instr. and Meth. A305 (1991) 391-394.

[7] A. Zucchiatti, M. Castoldi, G. Gervino, F. Terzi, M. Anghinolfi, N. Bianchi, L. Casano, P. Corvisiero, A. D'Angelo, E. De Sanctis, B. Girolami, P. Levi Sandri, V. Lucherini, L. Mazzaschi, V.I. Mokeev, V. Muccifora, P. Picozza, E. Polli, A.R. Reolon, G. Ricco, M. Ripani, P. Rossi, M. Sanzone, C. Schaerf, M. Taiuti, Nucl. Instr. and Meth. A317 (1992) 492.

[8] A. Zucchiatti, P. Levi Sandri, C. Schaerf, A.A. Cowley, J.V. Pilcher, L. Mazzaschi, M. Anghinolfi, N. Bianchi, L. Casano, P. Corvisiero, A. D'Angelo, E. De Sanctis, B. Girolami, V. Lucherini, V.I. Mokeev, V. Muccifora, P. Picozza, E. Polli, A.R. Reolon, G. Ricco, M. Ripani, P. Rossi, M. Sanzone, M. Taiuti, Nucl. Instr. and Meth. A321 (1992) 219.

[9] D. Drechsel and L.V. Fil'kov, Internal Report MKPH-T-94-4.

[10] M. Moinester, Proceedings of the Conference on the Intersection between Particle and Nuclear Physics, Tucson(1991) 553, Ed.W.Van Oers.

[11] B.R. Holstein, Comments Nucl. Part. Phys. 19 (1990) 221.

[12] L. Xiong, E. Shuryak, G. Brown, Phys. Rev. 46D (1992) 3798.

[13] M. Zieliski et al., Phys. Rev. Lett. 52 (1984) 1195.

[14] H.J. Lipkin, M.A. Moinester, Phys. Lett. B287 (1992) 179.

[15] A. A. Akhundov, S. Gerzon, S. Kananov, M. A. Moinester, TAUP-2183-94 (1994).

[16] M. E. Biagini, PANIC Proceedings, Perugia (1993) 763.

[17] KLOE Collaboration Lab. Naz. di Frascati LNF-92/019 (1993), KLOE Collaboration (addendum), LNF-94/028 (1994).

[18] FINUDA Collaboration Lab. Naz. di Frascati LNF-93/021 (1993).

[19] G. Alexander et al., Il Nuovo Cimento 107A (1994) 837.

[20] S. Ong, P. Kessler and A. Courau, Mod. Phys. Lett. A 4 (1989) 909.

[21] S. Weinberg, Physica 96A (1979) 327.

[22] J. Gasser and H. Leutwyler, Ann. Phys. (N.Y.) 158 (1984) 142.

[23] J. Gasser and H. Leutwyler, Nucl. Phys. B250 (1985) 465.

[24] H. Leutwyler, Bern University preprint BUTP-93/24 (hep-ph/9311274).

[25] For recent reviews on CHPT see e.g.
H. Leutwyler, in: Proc. XXVI Int. Conf. on High Energy Physics, Dallas, 1992, edited by J.R. Sanford, AIP Conf. Proc. No. 272 (AIP, New York, 1993) p. 185;
U.G. Meißner, Rep. Prog. Phys. 56 (1993) 903;
A. Pich, Lectures given at the V Mexican School of Particles and Fields, Guanajuato, México, December 1992, preprint CERN-Th.6978/93 (hep-ph/9308351);
G. Ecker, Lectures given at the 6th Indian–Summer School on Intermediate Energy Physics Interaction in Hadronic Systems Prague, August 25 - 31, 1993, to appear in the Proceedings (Czech. J. Phys.), preprint UWThPh -1993-31 (hep-ph/9309268).

[26] J. Bijnens and F. Cornet, Nucl. Phys. B296 (1988) 557.

[27] J.F. Donoghue, B.R. Holstein and Y.C. Lin, Phys. Rev. D37 (1988) 2423.

[28] The Crystal Ball Collaboration (H. Marsiske et al.), Phys. Rev. D41 (1990) 3324.

[29] R.L. Goble, R. Rosenfeld and J.L. Rosner, Phys. Rev. D39 (1989) 3264. The literature on earlier work may be traced from this reference.

[30] D. Morgan and M.R. Pennington, Phys. Lett. B272 (1991) 134.

[31] M.R. Pennington, in The DAΦNE Physics Handbook, edited by L. Maiani, G. Pancheri and N. Paver (INFN, Frascati, 1992), p. 379.

[32] J.F. Donoghue and B.R. Holstein, Phys. Rev. D48 (1993) 137.

[33] A. Dobado and J.R. Peláez, Z. Phys. C57 (1993) 501.

[34] T.N. Truong, Phys. Lett. B313 (1993) 221.

[35] A.E. Kaloshin and V.V. Serebryakov, Irkutsk preprint ISU-IAP.Th93-03 (hep-ph/9306224).

[36] S. Bellucci, J. Gasser and M.E. Sainio, Nucl. Phys. B423 (1994) 80.

[37] E. Golowich and J. Kambor, in preparation.

[38] M. Knecht, B. Moussallam and J. Stern, Orsay preprint IPNO/TH 94-08, to appear in Nuc. Phys. B.

[39] N.H. Fuchs, H. Sazdjian and J. Stern, Phys. Lett. B269 (1991) 183.

[40] V.A. Petrunkin, Sov. J. Part. Nucl. 12 (1981) 278;
J.L. Friar, in: Proc. of the Workshop on Electron-Nucleus Scattering, Marciana Marina, 7-15 June 1988, eds. A. Fabricini et al., (World Scientific, Singapore, 1989) p. 3;
M.A. Moinester, in: Proc. 4th Conf. on the Intersections Between Particle and Nuclear Physics, Tucson, Arizona, May 24-29, 1991, AIP Conf. Proc. No. 243, ed. W.T.H. Van Oers (AIP, New York, 1992), p. 553.

[41] Yu.M. Antipov et al., Phys. Lett. 121B (1983) 445; Z. Phys. C24 (1984) 39.

[42] T.A. Aibergenov et al., Czech. J. Phys. B36 (1986) 948.

[43] Yu.M. Antipov et al., Z. Phys. C26 (1985) 495.

[44] J.F. Donoghue and B.R. Holstein, Phys. Rev. D40 (1989) 2378.

[45] D. Babusci et al., in The DAΦNE Physics Handbook, edited by L. Maiani, G. Pancheri and N. Paver (INFN, Frascati, 1992), p. 383;
D. Babusci et al., Phys. Lett. B277 (1992) 158.

[46] The Mark II Collaboration (J. Boyer et al.), Phys. Rev. D42 (1990) 1350.

[47] A.E. Kaloshin and V.V. Serebryakov, Z. Phys. C32 (1986) 279.

[48] D. Babusci, S. Bellucci, G. Giordano and G. Matone, Phys. Lett. B314 (1993) 112.

[49] A.E. Kaloshin, V.M. Persikov and V.V. Serebryakov, "First estimates of the $(\alpha+\beta)^\pi$ from two-photon experiments", Irkutsk preprint ISU-IAP.Th94-01.

Report of the Working Group on Nucleon Polarizabilities

Malcolm N. Butler [1], Alan M. Nathan [2]

[1]Department of Astronomy and Physics, Saint Mary's University
Halifax, NS, Canada B3H 3C3
[2]Nuclear Physics Laboratory and Department of Physics
University of Illinois at Urbana-Champaign, Urbana, IL, USA 61801

1 Introduction

The format for our working group emphasized critical discussion and analysis of the issues surrounding nucleon polarizabilities. Formal presentations were made only when they were useful for clarifying concerns and criticisms of specific work. The discussion was quite lively and many issues that were covered complement concerns raised in the plenary sessions.

2 Status of Experimental Studies

2.1 Proton

The proton polarizabilities can be measured via Compton scattering at relatively low energy. In recent years, there have been a number of new experiments providing rather reliable cross sections from which one can attempt to extract the electric and magnetic polarizabilities, usually denoted by $\bar{\alpha}_p$ and $\bar{\beta}_p$, respectively. In fact the sum $\bar{\alpha}_p + \bar{\beta}_p$ is already well constrained by a sum rule derived from a forward dispersion relation,

$$\bar{\alpha}_p + \bar{\beta}_p = 14.2 \pm 0.5 \tag{1}$$

in units of 10^{-4} fm,[3] which are hereafter understood. Therefore the only remaining experimental issue is a measurement of one other linear combination such as $\bar{\alpha}_p - \bar{\beta}_p$. The new data as well as the details of how to extract $\bar{\alpha}_p - \bar{\beta}_p$ from the cross sections were discussed in the plenary talk of Nathan [1]. The principal issue is one of model dependence. For sufficiently low energy, one relies on a low energy theorem (LET) which states that to order E^2 in the expansion of the scattering cross section in powers of energy, the only structure-dependent constants that enter, aside from the anomalous magnetic moment, are $\bar{\alpha}$ and $\bar{\beta}$. At energies for which the LET is valid one can determine the polarizabilities from the cross sections in a completely model independent way. Unfortunately, only a small subset of the existing data were taken at sufficiently low energy. For higher energy, one needs to rely on dispersion relations or other techniques to relate the cross sections to the polarizabilities, with varying degrees of model dependence. The reader is referred to the contribution of Nathan [1] for an extensive discussion of this issue. Keeping only the data below the pion threshold, one arrives at the result

$$\bar{\alpha} - \bar{\beta} = 10.4 \pm 1.5 \pm 1.3, \tag{2}$$

where the first error is the combined statistical and systematic error of several experiments and the second error is an estimate of the theoretical uncertainty due to the model dependence of the extraction process. In order to improve on that situation, a new experiment is planned at Mainz [2], in which the scattering cross section will be measured using the photon tagging technique over a broad angular range from very low energy up to the pion threshold.

2.2 Neutron

We know less about the polarizabilities of the neutron than those of the neutron. As with the proton, the sum of polarizabilities is constrained by the dispersion sum rule:

$$\bar{\alpha}_p + \bar{\beta}_p = 15.8 \pm 1.0. \tag{3}$$

However, since the photoabsorption data that was used to evaluate the sum rule was obtained from experiments on the deuteron, it is quite possible that the uncertainty in that sum rule could be significantly larger than the value given here. There has been only one direct measurement of the polarizabilities with enough precision to be more than an upper limit. This comes from measurements of the total interaction cross section of very low energy (50eV–50keV) neutrons from a Pb target [3]. If one expands the cross section in powers of neutron momentum, then for energies well below the resonance region the linear term in the expansion depends *only* on $\bar{\alpha}_n$. The result of that experiment is

$$\bar{\alpha}_n = 12.5 \pm 2.5, \tag{4}$$

where again the stated error is a combined statistical and systematic error. Stefan Kopecky told us about a Vienna/Oak Ridge collaboration that plans to repeat this experiment, with improved statistical and systematic uncertainties.

Nevertheless, there seems to be a consensus in the community that an alternate technique to measure the neutron polarizabilities is highly desirable. Compton scattering, the technique of choice for the proton, presents difficulties for two reasons. First is the obvious problem with a target. Second is the fact that, for the uncharged neutron, there is no LET that allows one to extract the polarizabilities in a completely model independent manner. Two possible techniques for attacking these problems were discussed.

First is the reaction $D(\gamma, \gamma' n)p$, where one effectively scatters from a (moving) neutron in quasi-free kinematics. Considerable theoretical work has been done by Levchuk et al.[4] to relate the quasi-free cross section to the free cross section. It suffices to say that, if one chooses ones kinematics wisely and if one is also willing to measure simultaneously the quasi-free scattering on the proton, then one can quite reliably extract the free cross section from the quasi-free cross section. This still does not solve the problem of interpreting the free cross section in terms of polarizabilities, which can only be done in the context of a model. The experiment has already been investigated in the closing days of the MAMI A accelerator at Mainz [5], but the poor statistical quality of the data only allowed an upper limit on the polarizabilities. There are plans to redo this experiment at SAL within the next year. It was agreed that this experiment is worth doing even if one does not gain additional information about the polarizabilities, since the Compton scattering cross section on the neutron is essentially unknown.

Second is the reaction $D(\gamma, \gamma)D$, i.e., elastic scattering on the deuteron, at energies below about 100 MeV. After much discussion it was concluded that, although this cross section is sensitive to neutron polarizabilities, our ability to extract them from the data depends on having all of the more "conventional physics" under control, such as the NN interaction, meson exchange currents, etc. Once again, it was agreed that this experiment is probably worth doing *because* of what one might learn about this conventional physics.

3 Theory (ChPT)

It seems to be generally agreed upon now that the best approach for studying meson-baryon physics in ChPT is through the use of a heavy baryon lagrangian [6]. There are two approaches to this problem (ignoring for the moment the question of $SU(2)$ vs. $SU(3)$, which is a detail compared to other concerns): one that includes the Δ resonance as an explicit degree of freedom [7]; and one that integrates it out so that it appears through local counterterms in the lagrangian [8,9].

3.1 Explicit Δ's

In this case we have an additional energy scale, namely the mass splitting

$$\Delta m = m_\Delta - m_N .\qquad (5)$$

The Δ contributes formally at $\mathcal{O}(q^3)$ in this approach, through graphs such as the one shown in Figure 1.

Such graphs yield contributions to the scattering matrix element of the form

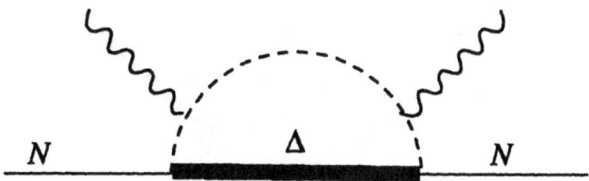

Fig. 1. A typical diagram showing explicit Δ contributions to the nucleon polarizability.

$$\frac{1}{(4\pi f_\pi)^2} \frac{1}{\sqrt{\Delta m^2 - m_\pi^2}} \log\left(\frac{\Delta m - \sqrt{\Delta m^2 - m_\pi^2}}{\Delta m + \sqrt{\Delta m^2 - m_\pi^2}}\right). \tag{6}$$

In the real world $\Delta m \sim 2m_\pi$. Nonetheless, if we were to perform an expansion for $m_\pi << \Delta m$ the expression reduces to

$$\frac{1}{(4\pi f_\pi)^2} \frac{1}{\Delta m} \log\left(\frac{m_\pi^2}{4\Delta m^2}\right). \tag{7}$$

3.2 Δ integrated out

In this case the effects of the Δ appear through local counterterms. The leading contribution of such terms appears formally at $\mathcal{O}(q^4)$ (note the higher order compared to the case with explicit Δ's) in diagrams like those of Figure 2.

If the coefficient of the counterterm (denoted by the black square) is given by c_2, then the chiral behaviour of the graph is given by

$$\frac{c_2}{(4\pi f_\pi)^2} \log m_\pi. \tag{8}$$

Resonance saturation [8] gives

$$c_2 \sim \frac{1}{\Delta m} \tag{9}$$

and we see that this result is similar in form to to Eq. 7.

It is clear that the two approaches amount to selective resummings of the chiral expansion, but it should be possible to determine which one is the best approach. If m_π were much smaller than Δm, the situation might be simpler. However, the simple fact that Δm is itself much smaller than the chiral scale of 1 GeV makes the argument one of practical phenomenology. A significant effort is needed to clear up this issue.

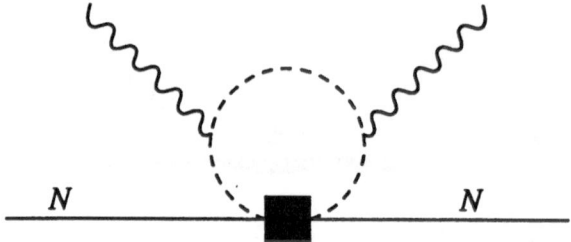

Fig. 2. A typical diagram showing the effect of a local counterterm arising from integrating out the Δ. The black square represents the counterterm.

4 Dispersion Relations

This topic was reviewed by Nathan in his plenary talk [1]. The same dispersion relations that are helpful in extracting the polarizabilities from the Compton scattering data also can be used to establish sum rules for the polarizabilities. These sum rules are completely rigorous yet semi-phenomenological, since they relate the polarizabilities to other measurable quantities. This latter point is appealing, since it helps one relate the polarizabilities to physics we already know. The best studied of these sum rules is the one based on the forward dispersion relation for the spin- independent scattering amplitude, which leads to the sum rule for $\bar{\alpha} + \bar{\beta}$. Explicit numerical computation (see Eq. 1,3) shows that this sum rule is essentially saturated below about 1 GeV, with the principal contributions coming from the Δ resonance region and the pion threshold region, and considerably lesser contributions coming from the two-pion threshold region and higher resonances. It is also possible to write sum rules for $\bar{\alpha} - \bar{\beta}$ based either on a forward [10] or a backward [11] dispersion relation. The forward sum rule calculation, based on an s-channel dispersion relation, would seem to indicate that the contribution below 1 GeV is about -2, whereas the experimental value is about $+10$. Thus, most of $\bar{\alpha} - \bar{\beta}$ would come from physics beyond 1 GeV [10]. L'vov interprets the physics giving rise to this large asymptotic contribution as due to the t-channel exchange of a scaler meson. On the other hand, the backward sum rule is composed of both an s-channel and a t-channel integral. The latter involves amplitudes for the process $\gamma\gamma \rightarrow \pi\pi \rightarrow NN$ and is directly sensitive to pionic structure. In fact, these dispersion relations indicate that the non-Born contributions to the $\gamma\gamma \rightarrow \pi\pi$ (for example, the pion polarizability) make a *significant* contribution to the nucleon polarizability. Since the energy needed to excite a pion is quite large compared to the energy needed to excite a nucleon, this is another indication that the nucleon polarizability has significant contributions from energies beyond 1 GeV.

What is the relationship between the dispersion relation approach and ChPT? In ChPT, counterterms represent physics that has been integrated out of the lagrangian at some energy scale. This is the short distance physics that cannot be described through the interactions of the Goldstone bosons alone. Typically, this scale is taken to be that of chiral symmetry breaking, approximately 1 GeV. On the other hand, the dispersion relations indicate that there may be significant contributions to the polarizabilities above 1 GeV. Such a result would indicate that the physics of counterterms dominates, thereby diminishing greatly the predictive power of ChPT for the polarizabilities. The clear consensus among the working group is that there needs to be more communication between those using ChPT and those using dispersion relations to see if they can really help one another understand the essential physics underlying the polarizabilities.

5 New Physics

To finish, we want to mention some aspects of nucleon polarizabilities that remain unknown, or unmeasured.

The first is the spin–dependent polarizability of the nucleon. This was discussed by Ulf Meißner in the plenary session [8], and it has been calculated in SU(2). It has not yet been measured. The experiment would require a polarized beam and target, with all the associated difficulties.

The second is virtual Compton scattering. The amplitude of interest is shown in Figure 3.

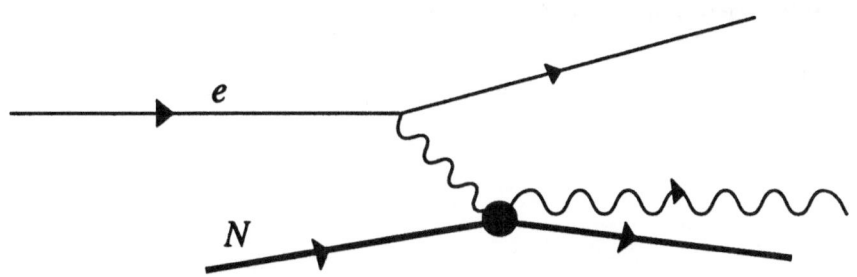

Fig. 3. The amplitude of interest in virtual Compton scattering off the nucleon. It is a probe of the off-shell behaviour of the Compton amplitude and the polarizabilities.

This process studies the off–shell behaviour of the Compton amplitude, and thus of the polarizabilities. Such a measurement at low Q^2 could play an important role in determining the relevant degrees of freedom in meson–baryon ChPT through the Q^2 dependence. Virtual Compton scattering has not been considered in great detail as yet, but will be the subject of a workshop in Seattle in late September 1994.

6 Acknowledgements

We would like to thank the organizers of the workshop, Barry Holstein and Aron Bernstein, for inviting us to chair this working group. We would also like to thank Joanne Gregory for her assistance and the many participants who attended our sessions and contributed to the lively discussion.

MNB would like to acknowledge the support of the Natural Sciences and Engineering Research Council of Canada. AMN's research is supported by the US National Science Foundation.

References

1. A.M. Nathan, these proceedings.
2. J. Ahrens, private communication.
3. J. Schmiedmayer and P. Riehs, Phys. Rev. Lett. **66**, 1015 (1991).
4. M.I. Levchuk, A.I. L'vov and V.A. Petrun'kin, Preprint MKPH-T-93, Mainz (1993).
5. K.W. Rose *et al.*, Phys. Lett. B **234**, 460 (1990).
6. E. Jenkins and A.V. Manohar, Phys. Lett. B **255**, 558 (1991).
7. M.N. Butler and M.J. Savage, Phys. Lett. B **294**, 369 (1992).
8. V. Bernard, N. Kaiser, U.G. Meißner and A. Schmidt, Phys. Lett. B **319**, 269 (1993); Z. Phys. **A348**, 317 (1994).
9. U.G. Meißner, these proceedings.
10. A.I. L'vov, Phys. Lett. B **304**, 29 (1993).
11. B.R. Holstein and A.M. Nathan, preprint 1994.

Part V: πN Interaction

Tests of Predictions from Chiral Perturbation Theory for πN Scattering*

G. Höhler

Universität Karlsruhe, Institut für Theoretische Teilchenphysik
D-76128 Karlsruhe, Germany

Abstract: Problems are discussed which arise in determinations of πN scattering amplitudes from data and in analytic continuations to the unphysical subthreshold region, which are needed for tests of predictions from Chiral Perturbation Theory.

1 Introduction

Since πN scattering amplitudes have been determined with a better accuracy than all other hadron-hadron scattering amplitudes, they have frequently been used in the sixties and seventies for tests of predictions derived from Current Algebra and PCAC. Nowadays, Chiral Perturbation Theory represents a solid mathematical basis for improved versions of these predictions. In this talk, only the first two steps for these tests will be treated: the determination of πN amplitudes from experimental data and the analytic continuation from the physical region to the unphysical subthreshold region. In particular, we shall consider the continuation to the 'Cheng-Dashen point' which is needed for a test of the prediction for the πN sigma term. In his talk, M. Sainio will explain how the result can be used in order to obtain information on the strange quark content of the nucleon ground state.

In the sixties, predictions were made for cases where one or both pions have the mass zero. 'On-shell versions' were given in 1971 [12]. We shall consider only amplitudes for physical pions.

Since the continuation is based on dispersion relations, one could ask if these relations are valid within the framework of QCD. In a recent paper, R. Oehme [46] treated this question and concluded that the original methods of proofs remain applicable.

In Sect. 2 we shall describe the determination of πN scattering amplitudes from experimental data. Sect. 3 is devoted to various methods for the analytic continuation to the unphysical subthreshold region, in particular to the Cheng-Dashen point. In Sect. 4 we shall discuss how the accuracy of the result can be improved by new theoretical and experimental investigations. Details of calculations done before 1983, a collection of formulas and a (hopefully) complete list of references up to this time can be found in [36].

---

* I am grateful for support by the Bundesministerium für Forschung und Technologie and for hospitality at the MIT.

2 Determination of πN Scattering Amplitudes from Data

It is well known that only very little information on the πN system can be obtained directly from πN scattering data. So the first step must be a determination of the scattering amplitudes. Unfortunately, it is not possible to determine the amplitudes from data alone, even if all measurable quantities were known with a high accuracy (see Sect. 2.2.1 in [36]). In practice, there are large gaps in the data base. So one can find a unique solution only if rather strong theoretical constraints are used in addition to Lorentz invariance, unitarity and isospin invariance (which is not exactly valid). It turns out to be necessary to add a further constraint. The only available general principle is the Mandelstam hypothesis. Of course, one can apply only some one-dimensional dispersion relations derived from this hypothesis.

The unitarity constraint can be employed in a suitable way only for partial waves. Therefore the determination of amplitudes from data is called *partial wave analysis*. This leads to the problem that one has to neglect an infinite 'tail of high partial waves'. The pioneers used a sharp cutoff, but then there are always some partial waves which are too small to be determined from the data but too large to be neglected.

Finally, we are interested in *hadronic amplitudes* which fulfil the above mentioned theoretical constraints. But the data are described by *experimental amplitudes* which contain effects of the electromagnetic interactions between the nucleons and pions. Therefore, a first step is to find the relation between hadronic and experimental amplitudes.

2.1 Determination of Hadronic Amplitudes from Experimental Amplitudes

Of course, one has to subtract from the data the contribution of the direct electromagnetic interaction between pion and nucleon, which is described by the 2nd order Feynman graph, modified by form factors and a phase which takes into account interactions of higher order [52]. Then, there is an interference term between electromagnetic and strong interactions. The electromagnetic part is known, but the strong interaction part must be taken from an earlier partial wave analysis. At the end, one has to repeat the calculation with the result of the new analysis.

The amplitudes determined with this procedure are called *nuclear amplitudes*. They still contain electromagnetic effects which occur, in contrast to Coulomb interference, at all angles. One contribution comes from photons exchanged between the proton and the charged pion before or after the strong interaction (*Coulomb barrier effect*). Then, one has to correct for the mass differences (mainly of the pions) and for the $n\gamma$ final state in $\pi^- p$ scattering. A relativistic dispersion relation method for the treatment of these corrections was developed by the NORDITA group (J. Hamilton et al., [52]). This method predicts in addition a 'short range contribution', which has a characteristic energy dependence

$$\delta^{sr}(s) = h_\pm(s) \sin^2 \delta^h(P33), \tag{1}$$

where s is the total c.m energy squared, δ^h the hadronic P33 phase shift and $h_\pm(s)$ a slowly variable function whose magnitude is not known. It is different for $\pi^\pm p$ scattering. This term is usually neglected.

If these corrections are applied to the nuclear phase shifts, one obtains the *hadronic phase shifts* from which the hadronic amplitudes are calculated. These play the role of *standard amplitudes*. It is assumed that they obey unitarity, Mandelstam analyticity and SU(2). However, there is a problem which will be discussed in the next subsection.

2.2 Violation of Isospin Invariance for the Hadronic Amplitudes

Evidence for a violation of isospin invariance in πN scattering was reported many times in the literature, based on violations of isospin bounds (see Sect. A10 in [36]). In the Karlsruhe-Helsinki partial wave analysis for πN scattering, evidence for a violation of SU(2) was found only from the accurate total cross section data in Refs. [13, 47]. An evaluation of the data in [47] based on the NORDITA method [52] by Koch and Pietarinen [39] led to

$$M^0 - M^{++} = (2.7 \pm 0.6) MeV, \qquad \Gamma^0 - \Gamma^{++} = (2 \pm 2) MeV. \qquad (2)$$

The mass is defined as a parameter of the usual Breit-Wigner formula with an energy-dependent width. Bugg [11] and Abaev [1] found almost the same result for the mass splitting. Their values for the difference of the widths are larger, but if two corrections are applied [11], the difference becomes comparable with its error. The results in [1] above the resonance depend on charge-exchange data which were not published, because the final analysis was not carried out [9].

The difference between the P33 phase shifts for $\pi^{\pm} p$ scattering given in [39] has semi-quantitatively the shape expected from (1) [38]. According to Tromborg, the effects of the u- and d-quark mass differences would show up as short range left hand cut contributions to the phase shift corrections in his method [53].

From a theoretical point of view, the approximate validity of SU(2) in πN scattering is due to the fact that the u- and d-quark masses are small in comparison with a characteristic mass parameter [22, 54]. In general, isospin breaking is not dominated by the electromagnetic fields but by a symmetry breaking term which is proportional to the very small quark mass difference $(M_u - M_d) \approx -2MeV$ [44].

Cutkosky obtained recently $M^0 - M^{++} \approx 1.5 MeV$ [19] as a prediction for the mass splitting. (This value corresponds to about thrice the error of each beam momentum at the resonance position [13]). In an earlier paper [16], Cutkosky investigated isospin violation from $\pi^0 - \eta$ mixing. He concluded that significant effects are possible in the neighborhood of the resonance S11(1535). In general, the best chance to find evidence for isospin violations is to consider data in the neighborhood of lines in an energy vs. $\cos(\theta)$ plot along which an isospin bound is saturated (see Sect. 2.4.4 in [36]).

Violations of isospin in πN scattering at low energies and in calculations of the sigma term were discussed by Weinberg [54]. It will be of interest to analyze the new $\pi^- p$ charge-exchange experiments at 10 and 20 MeV by Sadler et al. [51] from this point of view. Experiments of the same group in the energy range of the $\Delta(1232)$ resonance will be important for a better understanding of the above mentioned mass splitting of the isobar states.

2.3 Partial Wave Analysis

In the early seventies, the pioneers in partial wave analysis had a remarkable success with their attempts to determine the quantum numbers of the strongest resonances, in particular Refs.[2, 5]. But they had not found a successful method to impose analyticity constraints as part of the analysis. So our first analytic continuations to the subthreshold region were a separate second step [31].

It was a crucial progress that E. Pietarinen developed an expansion method which made it possible to impose analyticity constraints as part of the procedure of partial wave analysis without calculating Cauchy principal value integrals, which are not well suited for numerical computations [48]. I invited Pietarinen to continue his work in collaboration with our group. This led to the main part of the results to be described in Sects. 2 and 3.

R.E. Cutkosky [15] and others had developed expansion methods for scattering amplitudes and for data several years earlier. But Cutkosky applied his method to πN scattering only some time after he had seen Pietarinen's first papers. Cutkosky's expansions are more sophisticated from a mathematical point of view, but Pietarinen's Taylor expansions are easier to handle. Since Cutkosky knew that we are using fixed-t dispersion relations and studying continuations to the subthreshold region, he decided to concentrate on the parameters of nucleon resonances, the multichannel aspects and the error matrix. The CMB solution covers only the momentum range from 0.43 to 2.5 GeV/c, i.e. it starts on the right wing of the first resonance. So it is, unfortunately, not suited for a continuation to the subthreshold region.

After the publication of the Karlsruhe-Helsinki and CMB solutions [34, 17], the VPI group developed a much simpler method which was based on an *empirical expansion of the partial wave amplitudes*. The price for the simplicity was that they ignored the well-known l.h. cut singularities of the partial wave amplitudes, even the projections of the nucleon pole term. Since the predictions from chiral perturbation theory were made for points in a close neighborhood to these singularities, it is clear that the VPI solutions are not suitable for our applications. Therefore, we shall describe in the following mainly the Karlsruhe-Helsinki method, adding some remarks on the CMB method. Only recently, the VPI group started to impose fixed-t analyticity for the invariant amplitudes [3], but this does not introduce the singularities of the partial waves.

2.4 The Karlsruhe-Helsinki Partial Wave Analysis

In a first step, 'hadronic data' were produced by subtracting from the experimental data the corrections of the NORDITA method [52], using the result of an earlier partial wave analysis. 88 momentum bands were chosen and the data within each band were shifted to its center, using again the result of the earlier analysis. Finally, this was repeated with the new partial waves.

In Pietarinen's method, a unique solution was determined directly by imposing strong analyticity constraints in an iteration procedure together with partial wave analysis. Constraints for analyticity at fixed-t were used in a range $0 > t > -1 \ GeV^2$ and for analyticity at fixed $\cos(\theta)$ for 19 cms angles θ between the backward direction and $\cos(\theta)=0.8$. Details are given in [49, 34].

The expansion methods have a *smooth cutoff*. It is determined by the parameters of the convergence test function which suppresses the higher coefficients. The strength of the smoothing was chosen in such a way that evidence for weak resonances remained visible (in addition to some structures due to experimental errors). In contrast, the VPI solutions do not even show some 3-star resonances due to the strong smoothing.

The solution KH80 differs from the solution in [34] by the improved low-energy part [39]. KH80 ignored the mass splitting of the isobar states, which was seen at that time only in the total cross sections. But almost all readers of the paper [39] did not notice that KH80 can easily be modified in such a way that the splitting is taken into account. The last section in [39] gives two different P33 partial waves derived from the data in [47] which can be used instead of the single P33 in KH80.

Uniqueness of the Partial Wave Solution. The KH solution is in reasonable agreement with the CMB solution [36], although the authors used quite different methods and their own data bases. Furthermore, there are no large discrepancies with the results in Refs. [2, 5]. Therefore, further work can be restricted to a search in the neighborhood of the KH and CMB solutions.

Nevertheless, it is of interest to ask for a mathematical treatment of the remaining ambiguities. For a truncated partial wave expansion, Barrelet [8, 36] has shown that a calculation of the zeros of transversity amplitudes can be used for the construction of other solutions with exactly the same values of differential cross sections and polarization parameters. The role of continuum ambiguities is discussed in [36, 4]. Further work on the uniqueness and the stability of the solution can be found in [50].

2.5 Compatibility of the KH Solution with Consequences of the Mandelstam Hypothesis

Since we intended to perform analytic continuations, it was of interest to check the compatibility of our solution with various one-dimensional dispersion relations and other consequences of Mandelstam's two-variable analyticity.

S-channel Partial Wave Dispersion Relations and Fixed-s Dispersion Relations. πN partial wave dispersion relations [21] were studied in great detail by J. Hamilton et al. [28] in the sixties and early seventies. R. Koch [41] improved the method and used KH80 as input. Furthermore, he employed fixed-s dispersion relations and our evaluation of the Mandelstam double spectral function in a strip near the physical region [35]. He obtained a solution KA84 which differs from KH80 mainly by a smoothing of fluctuations in the momentum dependence of the real and imaginary parts. This procedure differs from the complete smoothing of the partial waves in the second paper [17], since Koch used additional dispersion relation constraints and does not include model-dependent assumptions. Unfortunately, Koch did not find a way to apply the method in a unique way to S-waves, because not calculable short range l.h. cuts are much stronger than for the other partial waves.

Interior Dispersion Relations. It is of interest to evaluate dispersion relations along hyperbolas in the Mandelstam plane which belong to a fixed angle in the lab.system [29]. The r.h. cuts lie in the physical region. The l.h. cuts can be calculated from a convergent partial wave expansion, using our $\pi\pi N\bar{N}$ partial waves (see Sects. 2.4.6 and A6.7 in [36]), if the angle is larger than $96°$. At smaller angles, a discrepancy function method can be applied. The possibility to determine the πN coupling constant f^2 for each of the θ_{lab}-values gives a check of the internal consistency. Applications confirm the compatibility with our partial wave solution [10].

Zero Trajectories. A study of zero trajectories is a sensitive tool for localizing energies where partial wave solutions have problems. The trajectories are curves in a 3-dimensional real s, complex t space along which one of the transversity amplitudes is zero. The plots usually show the projections onto a s, Re t-plane or a s,cos(θ)-plane. These curves can be calculated from partial wave solutions and, for charge-exchange, also from differential cross section and analyzing power data [33, 36].

It follows from two-variable analyticity that in every neighborhood of a zero there exists another zero. Isolated zeros cannot occur and zero trajectories cannot start or end at finite values of Im t. Plots of trajectories calculated from data or from partial wave solutions show at certain energies more or less serious violations of these conditions. This indicates that corrections are necessary. The zero trajectories calculated from KH80 and CMB80 show much more violations than those calculated from KA84.

Intersections of zero trajectories with resonance surfaces can occur only at singularities of the type 0/0 which occur only at certain angles. See [8, 36, 32] for further properties. Abaev et al. [1] have used zero trajectories in their partial wave analysis. Many results on zero trajectories derived by Stefanescu [50] have not yet been exploited.

3 Analytic Continuations to the Subthreshold Region

A first step for a test of the prediction for the πN sigma term is the determination of the invariant πN amplitude $\bar{D}^+$ at the *Cheng-Dashen point* $\nu=0$, $t = 2m_\pi^2$ [12]. s,t,u are the Mandelstam variables, $\nu = (s - u)/4m$ is the crossing variable and m the nucleon mass. $\bar{D}^+(\nu, t)$ is one of the isospin even invariant amplitudes which agrees at $t = 0$ with the forward amplitude. The bar indicates that the PV Born term has been subtracted [36]. The relation to the notation in the talk by M. Sainio is given by

$$\Sigma = F_\pi^2 \bar{D}^+(0, 2m_\pi^2), \tag{3}$$

where F_π^2=92.4 MeV is the pion decay constant. In the following, we shall describe several methods for a determination of Σ.

3.1 The Subthreshold Expansion

The ν-dependence of the invariant amplitudes in the range between the s- and u-channel thresholds follows from an evaluation of fixed-t dispersion relations. It is dominated by the contributions of the nucleon pole terms. If one considers the amplitude $\bar{D}^+$, a slow variation in ν and t is expected as long as one does not approach the s- and u-channel thresholds or $t = 4m_\pi^2$. Therefore, we introduce an expansion at $\nu=0$, $t=0$ (Fig. 1)

$$\bar{D}^+(\nu, t) = d_{00}^+ + d_{01}^+ t + d_{02}^+ t^2 + d_{10}^+ \nu^2 + ..., \tag{4}$$

whose coefficients can be calculated from the dispersion relation for the forward amplitudes and its derivatives, taken at t=0. Since a subtraction is required, the coefficients d_{0n}^+ have the character of subtraction constants. We define at $\nu=0$

$$\Sigma = \Sigma_d + \Delta_D, \qquad \Sigma_d = F_\pi^2(d_{00}^+ + d_{01}^+ 2m_\pi^2). \tag{5}$$

The Σ and Δ are t-dependent if used together with (4). If no argument is written, we refer to the values at $t = 2m_\pi^2$.

Our first table of coefficients [31] was calculated from preliminary CERN partial waves [2]. The result for Σ_d was about 40 MeV in contrast to the much larger value of about 110 MeV found by Cheng and Dashen [12]. The reason for the discrepancy was that the latter authors used a fixed-t dispersion relation in the unphysical region at $t = 2m_\pi^2$ and applied a 'broad area subtraction method' which was not suitable for this purpose [30].

When the new data of Carter et al. [13] became available, Nielsen and Oades [45] found $\Sigma_d \approx 57 MeV$ and $\Sigma \approx 66 MeV$, in contrast to other authors. Our final result for the linear part is (Table 2.4.7.1 in [36], coefficients in natural units)

$$\Sigma_d = 61.2(d_{00}^+ + 2d_{01}^+ m_\pi^2) = [(-89 \pm 6) + (139 \pm 2)]MeV = (50 \pm 7)MeV. \quad (6)$$

It is remarkable that d_{01}^+ is almost the same as in 1971, except for the error. I think that our error estimate in (6) was too optimistic.

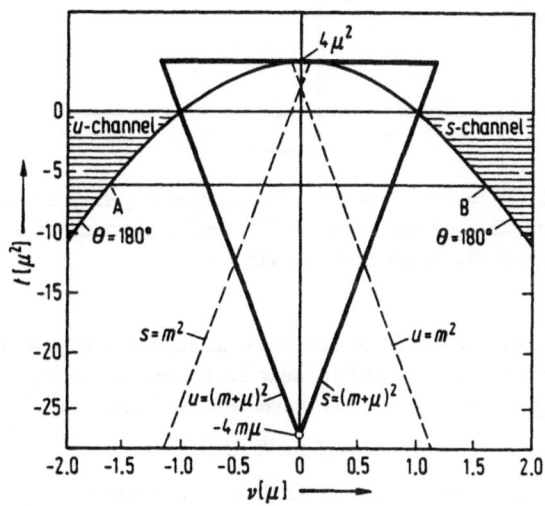

Fig. 1. Mandelstam plot in the ν, t-plane. The Cheng-Dashen point lies at the intersection of the dashed lines. The invariant amplitudes are real within the triangle.

3.2 Evaluation of the Fixed-ν Dispersion Relation at $\nu=0$

In order to determine Δ_D we used a fixed-ν dispersion relation at $\nu=0$. It requires as additional input the S-wave $\pi\pi N\bar{N}$-amplitude which produces a cusp at $t = 4m_\pi^2$ and therefore the t-dependence of Δ_D in (5). The evaluation is described in Sect. 2.5.1 [36]. We obtained

$$\Sigma = (63 \pm 6)MeV, \qquad \Delta_D \approx 13MeV. \quad (7)$$

As in other cases, the error is not the total error but the error of this part of the calculation.

$\Delta_D(t)$ contains not only the tail of the cusp but also a tail of the effect of $\Delta(1232)$ exchange which leads to a *turning point* at $t \approx -4m_\pi^2$.

3.3 Continuation Along 50 Hyperbolas

If Mandelstam's two-variable analyticity is fulfilled, continuations to the Cheng-Dashen point along different paths must lead to the same result. R. Koch calculated continuations along 50 hyperbolas chosen in such a way that, above the s-channel threshold,

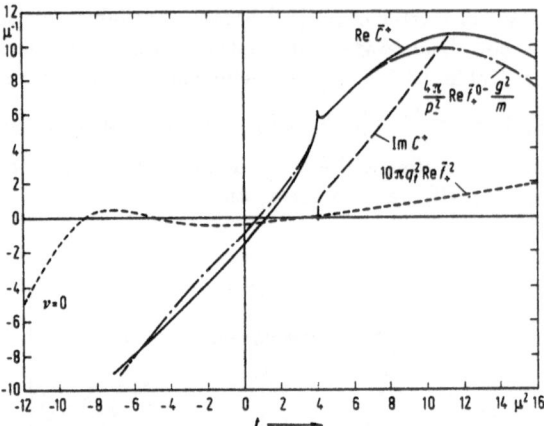

Fig. 2. Solid line: $Re\,\bar{C}^+ = Re\,\bar{D}^+$ at $\nu=0$. The cusp at $t = 4m_\pi^2$ is about 6 times higher than the value at the Cheng-Dashen point ($t = 2m_\pi^2$). Turning point: near $t = -4m_\pi^2$. The t-channel S-wave is dominant near the cusp.

the curves remain almost entirely in the physical region and pass a line s=const at angles distributed from near-forward to near backward scattering [40]. The l.h. cuts were evaluated using our $\pi\pi N\bar{N}$ partial waves and a discrepancy function. The result

$$\Sigma = (64 \pm 8)MeV \tag{8}$$

shows a remarkably good consistency of KH80 with two-variable analyticity. 8 MeV is the uncertainty due to the small inconsistency. It is not possible to give a well-founded estimate of the total error.

The numerical value as well as all others derived from our S-wave $\pi\pi N\bar{N}$ amplitude have to be corrected because our old values for the $\pi\pi$ phases should be replaced by the prediction from chiral symmetry [24]. The magnitude of the correction is discussed in [26].

3.4 Use of Scattering Lengths as Subtraction Constants

Some authors preferred to work with isospin even S and P-wave scattering lengths and the S-wave effective range instead of d_{00}^+ and d_{01}^+. If the scattering lengths are calculated from KH80, this gives exactly the same result. The final data of the experiment with pionic hydrogen [6] and for inverse π^- photoproduction near threshold [43] will lead to an improved accuracy of the S-wave scattering length a^+ and therefore also of d_{00}^+. The preliminary value gives $d_{00}^+ = -1.39$, i.e. a correction to Σ_d and Σ by about $+4$ MeV. The analysis of the new low energy scattering experiments will improve the accuracy of the P-wave scattering lengths. But nevertheless, it will be important to check the compatibility with the usual determination of the subtraction constant d_{01}^+ from new data for the slopes of $d\sigma/dt$ at t=0 in a large momentum range (Sect. 4).

Gasser's discussion of the use of scattering lengths led to an interesting sum rule [25] which expresses the combination of scattering lengths and the S-wave effective range b^+

$$L = m_\pi^2(b^+ + 2a_{1+}^+ + a_{1-}^+) + (m_\pi/2m)a^+ \tag{9}$$

by the result of the one-loop evaluation plus an integral over total πN cross sections. It was noted only several years later that *an exact sum rule* was derived in 1958 from the forward dispersion relation [27] which expresses L by a different integral over total cross sections. The numerical agreement is good, but it would be of interest to discuss the relation between the one-loop calculation and integrals over total cross sections in detail.

3.5 Determination of Σ from S-channel πN Partial Waves

Ericson pointed out that, in terms of πN partial waves, the S-wave gives the dominant contribution to Σ [20]. He attempted to calculate the continuation into the unphysical region, using the values given in [36] for the scattering length and the effective range. Unfortunately, this method cannot lead to a reliable result, because the contribution to Σ follows from the S-wave at an unphysical energy very close to the front of the 'circle cut' of the partial wave amplitude. Its discontinuity was calculated in [28, 41] and it turned out to give an appreciable contribution. It is included in our fixed-ν dispersion relation (Sect. 3.2). See also p. 8ff in [37].

In [42] R. Koch has evaluated the exact formulas for the partial wave projection of fixed-t dispersion relations, using KH80 for the imaginary parts under the dispersion integrals. This work is a continuation of the fundamental paper by 'CGLN' [14]. The results give smoothed real parts of the partial waves not only up to about 500 MeV/c but also in the subthreshold region. The contribution of the circle singularity is included in the subtraction function.

In a simple approximation, the contributions of the S, P and D waves to Σ follow from a direct projection of our subthreshold expansion:

$$\Sigma = (60.9 + 0.9 - 2.8) MeV = 59\ MeV. \tag{10}$$

Corrections are needed due to the fact that the known terms of the expansion do not give a good description of the region near $\theta = 180°$ which lies extremely near to the singularity at $t=4m_\pi^2$.

4 How Can the Accuracy of Σ Be Improved ?

4.1 Experiments

The subtraction constant for the slope d_{01}^+ in Σ_d gives a large contribution, but it is not well determined because there are almost no data for $d\sigma/dt$ at small $|t|$ (Fig. 3). Furthermore, there is a strong discrepancy between the data of Baillon et al. [7] for small $|t|$, which are available only at a few energies, and many other data which cover a large angular range (Fig. 4).

One should consider the possibility to detect a charge-dependence of the πN coupling constant by measuring accurate data for $\pi^- p$ charge-exchange forward scattering and using the forward dispersion relation (see p.223 in [36]).

4.2 Theory

It is necessary to update the KH partial wave analysis as soon as the final results of the new experiments at the meson factories are available. In my opinion, it would be

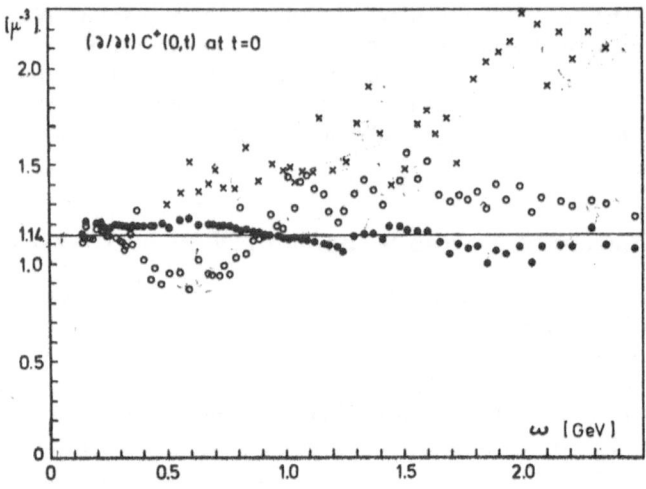

Fig. 3. Determination of the subtraction constant d_{01}^+. From KH80: o, CMB80: ×, KA84: •. All points should approximately lie on a horizontal line. The solid straight line shows the value obtained from the fixed-ν dispersion relation (Sect. 3.2).

preferable to use interior dispersion relations (fixed θ_{lab}) instead of dispersion relations at fixed θ_{cm}, because one has a much better information on the l.h. cut [10]. Furthermore, one obtains for each value of θ_{lab} in part of the range a result for the πNN coupling constant f^2, which gives a check of the internal consistency (J. Stahov et al. in [10]).

The result of the new analysis should be used for an improved determination of the $\pi\pi N\bar{N}$ partial waves. Then one could obtain an updated version of the improvements made by Koch [41] and other calculations.

As long as a new analysis is not available, the $\pi\pi N\bar{N}$ S-wave near threshold should be corrected as described in Sect. 3.3. The πN S-wave scattering lengths and f^2 should be adjusted as soon as the final results of Refs. [6, 43] are available. It is known for some time from single energy analyses and from [1] that in KH80 S11 is somewhat too high, mainly in the range of the new LAMPF data.

One should try to improve the NORDITA method [52] by including the effects of the quark mass differences. In connection with the experiment proposed above, it would be useful to try to estimate the differences between the various πNN coupling constants. It was pointed out a long time ago that in the relation between the PV and PS coupling constants:

$$\frac{g^2}{4\pi} = \frac{4m^2}{m_\pi^2} f^2 \tag{11}$$

usually the charged pion mass is inserted. If the neutral pion mass is used for one of the factors in f^2, the result changes by 3.4%. But this is only a trivial part of the expected splitting.

One should search for isospin breaking effects in charge-exchange scattering at very low momenta [54, 51] and continue to study effects of this type in determinations of the sigma term (Appendix D in [23] and [54, 18]).

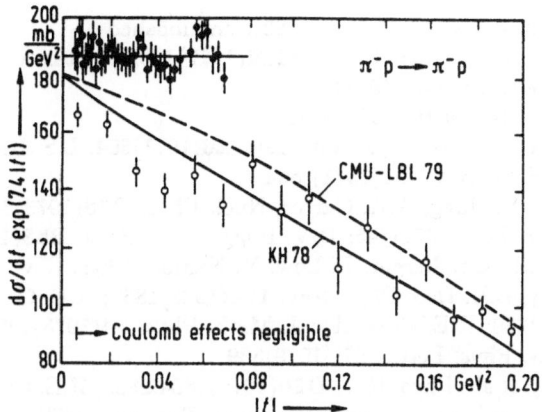

Fig. 4. Discrepancy between scattering data for the slope at t=0. Baillon et al.: •, Bardsley et al.: o. The cross sections have been multiplied by a factor such that the fit to the data of Baillon et al. is horizontal. Lab. momentum: 1.0 GeV/c. KH78 agrees with KH80 above 0.5 GeV/c.

5 Conclusions

Various tests confirm that the solution KH80 is compatible with consequences of the Mandelstam hypothesis. Therefore, quite different methods can be applied for the analytic continuation to the Cheng-Dashen point. The results lead to $\Sigma \approx 60 MeV$. A reliable estimate for the total error cannot be given.

In order to reach an improved accuracy, the points mentioned in Sect. 4 should be considered.

Acknowledgments

I am very grateful to J. Gasser, H. Leutwyler and M.E. Sainio for information on the problems of Chiral Perturbation Theory and for many discussions on the relation to methods based on analyticity.

References

1. V.V. Abaev, S.P. Kruglov, St. Petersburg preprint No. 1794, May 1992
2. S. Almehed, C. Lovelace, Nucl. Phys. **40B**(1972)157
3. R.A. Arndt, R.L. Workman, M.M. Pavan, Phys. Rev. **C49**(1994)2729
4. D. Atkinson, I.S. Stefanescu, Commun. Math. Phys. **101**(1985)291
5. R. Ayed, thesis and report CEA-N-1921, Centre d'Etudes Nucléaire de Saclay, 1976
6. A. Badertscher, these Proceedings
7. P. Baillon et al., Nucl. Phys. **B105**(1976)365
8. E. Barrelet, Nuovo Cim. **8A**(1972)331

9. F.D. Borcherding, UCLA Thesis (1982), unpublished
10. E. Borie, F. Kaiser, Nucl. Phys. **B126**(1977)173; J. Stahov, M.E. Sadler V.V. Abaev, πN Newsletter **6**(1992)91
11. D.V. Bugg, πN Newsletter **3**(1991)1
12. T.P. Cheng, R. Dashen, Phys. Rev. Lett. **26**(1971)594; L.S. Brown, W.J. Pardee, R.D. Peccei, Phys. Rev. **D4**(1971) 594
13. J.R, Carter, D.V. Bugg, A.A. Carter, Nucl. Phys. **B58**(1973)378; A.A. Carter et al., *Nucl. Phys.* **B26**(1971)445; D.V. Bugg, *Nucl. Phys.* **B58**(1973)397
14. G.F. Chew, M.L. Goldberger, F. Low, Y. Nambu, Phys. Rev. **106**(1957)1337
15. R.E. Cutkosky, B.B. Deo, Phys. Rev. **174**(1968)1859; R.E. Cutkosky, Ann. Phys. N.Y. **54**(1969)350; R.E. Cutkosky, J. Math. Phys. **14**(1973)1231
16. R.E. Cutkosky, Phys. Lett. **88B**(1979)339
17. R.E. Cutkosky et al., Phys. Rev. **D20**(1979)2804,2839; R.E. Cutkosky, Baryon 80, Proc. IV Int'l Conf. on Baryon Resonances, Toronto (1980), ed. N. Isgur, p. 19
18. R.E. Cutkosky, Proc. Hadron 91, eds. S. Oneda and D.C. Peasley, World Scientific, Singapore, 1992, p. 13
19. R.E. Cutkosky, πN Newsletter **6**(1992)108
20. T.E.O. Ericson, Phys. Lett. **B195**(1987)116; CERN preprint TH 5621/90 (1990)
21. W.R. Frazer, J.R. Fulco, Phys. Rev. **119**(1960)1420; S.C. Frautschi, J.D. Walecka, Phys. Rev. **120**(1960)1487
22. H. Fritzsch, Proc. Enrico Fermi School, Course LXXI, Varenna 1977, ed. M. Baldo Ceolin, North Holland 1979
23. J. Gasser and H. Leutwyler, Physics Reports **87**(1982)77
24. J. Gasser, H. Leutwyler, Phys. Lett. **B125**(1983)321; Ann. Phys. N.Y.**158**(1984)142
25. J. Gasser. Proc. 2nd Int'l Workshop on πN Physics, Los Alamos 1987, eds. W.R. Gibbs and B.M.K. Nefkens, p. 266
26. J. Gasser, H. Leutwyler, M.E. Sainio, Phys. Lett. **B253**(1991)260
27. D.A. Geffen, Phys. Rev. **112**(1958)1370
28. J. Hamilton in High Energy Physics, Vol.I, ed. E. Burhop, Academic Press 1967; Springer Tracts Mod. Phys. **57**(1971)41
29. G.E. Hite, F. Steiner, Nuovo Cim. **18A**(1973)237
30. G. Höhler, H.P. Jakob, R. Strauss, Phys. Lett. **35B**(1971)445; G. Höhler, H.P. Jakob, Lett. Nuovo Cim. **2**(1971)485
31. G. Höhler, H.P. Jakob, R. Strauss, Nucl. Phys. **B39**(1972)237
32. G.Höhler, I.S. Stefanescu, Z. Phys. **C2**(1979)253; E. Borie et al., Z. Phys. **C4**(1980)333
33. G. Höhler et al., Kernforschungszentrum Karlsruhe, KfK 2735 (1979)
34. G. Höhler, F. Kaiser, R. Koch, E. Pietarinen, Handbook of Pion-Nucleon Scattering, Physics Data **12-1** 1979
35. G. Höhler et al., Karlsruhe report TKP 83-24,1983; P.A. Klumpp, Diplomarbeit, Univ. of Karlsruhe, TKP 84-2,1984, unpublished; D. Grether, Thesis, Univ. of Karlsruhe, 1985
36. G. Höhler: Pion-Nucleon Scattering, Landolt-Börnstein Vol.I/9b2, ed. H. Schopper, Springer, 1983
37. G. Höhler, πN Newsletter **2**(1990)1
38. G. Höhler, Karlsruhe report TTP 92-8, 1992
39. R. Koch, E. Pietarinen, Nucl. Phys. **A336**(1980)331
40. R. Koch, Z. Phys. **C15**(1982)161
41. R. Koch, Z. Phys. **C29**(1985)597

42. R. Koch, Nucl. Phys. **A448**(1986)707
43. M.A. Kovash, these Proceedings
44. H. Leutwyler, Lectures given at the Workshop 'Hadrons 1994', Gramado, RS, Brasil, May 1994 and University of Berne preprint BUTP - 94/13
45. H. Nielsen, G.C. Oades, Nucl. Phys. **B72**(1974)310
46. R. Oehme, πN Newsletter **7**(1992)1
47. E. Pedroni et al., Nucl. Phys. **A300**(1978)321
48. E. Pietarinen, Nucl. Phys. **B49**(1972)110; Nuovo Cim. **12A**(1972)522; Nucl. Phys. **B55**(1973)541
49. E. Pietarinen, Nucl. Phys. **B107**(1976)21; Helsinki reports **HU-TFT**78-13 and 78-23
50. E. Pietarinen, Phys. Scripta **14**(1976)11; I.S. Stefanescu, Phys. Rev. **D21**(1980)3325; J. Math. Phys. **23**(1982)1190; Fortschr. Phys. **35**(1987)573; Z. Phys. **C41**(1988)453
51. M.E. Sadler, preprint of talk at the Dubna Symposium "Mesons and Nuclei at Intermediate Energies", May 1994 and private communication
52. B. Tromborg et al. Phys. Rev. **D15**(1978)725; Helv. Phys. Acta **51**(1978)584
53. Private communication (1979)
54. St. Weinberg, Transactions of the New York Academy of Sciences **38**(1977)185 and these Proceedings

Pion-Nucleon Sigma Term

M.E. Sainio

Department of Theoretical Physics, University of Helsinki,
P.O. Box 9 (Siltavuorenpenger 20 C), FIN-00014 Helsinki, Finland

Abstract: A brief status report on the σ-term is given. The extrapolation of the isoscalar D-amplitude to the Cheng-Dashen point is analyzed in light of the recent low-energy πN data, and the Σ-term is evaluated. The strange quark content of the proton is discussed.

1 Introduction

An empirical measure of the chiral asymmetry generated by the u and d quark masses in QCD is the πN σ-term. It is defined as [1]

$$\sigma = \frac{\hat{m}}{2m} \langle p|\bar{u}u + \bar{d}d|p\rangle , \qquad (1)$$

where $\hat{m} = \frac{1}{2}(m_u + m_d)$, m is the proton mass and $|p\rangle$ denotes a physical one-proton state normalized as $\langle p'|p\rangle = (2\pi)^3\, 2p_0\, \delta(\mathbf{p}' - \mathbf{p})$. Algebraically σ can be written in the form

$$\sigma = \frac{\hat{m}}{2m}\, \frac{\langle p|\bar{u}u + \bar{d}d - 2\bar{s}s|p\rangle}{1 - y}, \qquad (2)$$

where the parameter y, the strange quark content of the proton, is defined as

$$y = \frac{2\langle p|\bar{s}s|p\rangle}{\langle p|\bar{u}u + \bar{d}d|p\rangle}. \qquad (3)$$

The numerator in (2) is proportional to the octet breaking piece in the hamiltonian. To first order in SU(3) breaking, the sigma term can therefore be written as

$$\sigma \simeq \frac{\hat{m}}{m_s - \hat{m}}\, \frac{m_\Xi + m_\Sigma - 2m_N}{1 - y} \simeq \frac{26\ \text{MeV}}{1 - y}, \qquad (4)$$

where the quark mass ratio

$$\frac{m_s}{\hat{m}} = 2\frac{\mu_K^2}{\mu^2} - 1 \simeq 25 \qquad (5)$$

derived from the kaon and pion masses, was used. The Zweig rule would now lead us to expect that $\langle p|\bar{s}s|p\rangle \simeq 0$, i.e. $y = 0$, and, therefore, in this limit $\sigma = 26$ MeV.

On the other hand, chiral symmetry allows one to relate σ to the πN scattering amplitude $(T_{\pi N})$ at the unphysical (but on-shell) Cheng-Dashen point $\nu = (s - u)/4m = 0$, $t = 2\mu^2$ [2-4]. Therefore, elastic πN scattering can be used to measure σ and thus to determine the strange quark content of the proton. The building blocks needed for such a determination have been around for quite some time [4-6], and the apparent result that half of the proton mass is due to the contribution from strange quarks, has led to the discussion of the sigma term puzzle [7].

To facilitate the analysis of the problem, the pion-nucleon Σ-term is defined as the isospin even D-amplitude $(T_{\pi N} = \bar{u}[A(\nu,t) + \frac{1}{2}(\not{q} + \not{q}')B(\nu,t)]u, D = A + \nu B$, see Ref. [8]) at the Cheng-Dashen point

$$\Sigma = F_\pi^2 \bar{D}^+(\nu = 0, t = 2\mu^2), \tag{6}$$

where $F_\pi = 93.3$ MeV is the pion decay constant. The pseudovector Born term has been subtracted from the amplitude, i.e. $\bar{D}^+ = D^+ - D_{pv}^+$. The extrapolation of the pion-nucleon amplitudes to the unphysical region has been extensively studied e.g. by the Karlsruhe group. For a comprehensive summary and a review of the literature until 1982 see the handbook article by Höhler [8].' A recent evaluation [6] based on hyperbolic dispersion relations gives the result $\Sigma = 64 \pm 8$ MeV, where the uncertainty of 8 MeV reflects the internal consistency of the method used. The difference between σ and Σ is of the order $(m_q)^{3/2}$, and numerically the one-loop correction amounts to $\Sigma - \sigma \simeq 5$ MeV [1,9,10]. The value of σ would then be 59±8 MeV, which is off by about a factor of two from the value extracted from the baryon spectrum, Eq. (4) with $y = 0$. Of course, by choosing a value for the parameter y around 0.5 the puzzle would be solved. However, such a large value for the strange quark content would imply that the proton mass in the chiral limit $(m_q=0)$ is nowhere near the value in the physical world. This is the sigma puzzle to be addressed in the following.

The actual evaluation of $\langle p|\bar{s}s|p\rangle$ and σ from the pion-nucleon data involves four nontrivial steps [11]:

i. σ has to be related to baryon masses and y,

ii. the t-dependence of the sigma form factor $\sigma(t)$ has to be determined,

iii. the connection of σ to Σ has to be established, and

iv. Σ has to be constructed from πN data.

It turns out that there are significant corrections to the above discussion from different items in the list, and, as a consequence, the sigma term puzzle finds a solution which implies a much more reasonable value for the proton mass in the limit $m_s=0$ [12]. However, there are still open questions that remain starting from the experimental input required for such an analysis. Also, eventually one would like to reduce the error bar which has to be assigned for σ and y, currently about 20 % and 100 % respectively.

2 The σ-term and the scalar form factor

It has been shown [9] that there are significant corrections to the leading order octet formula (4) from meson cloud effects corresponding to higher order terms, $\mathcal{O}(m_q^{3/2})$ in the quark mass expansion [1,9]. These yield

$$\sigma = \frac{\hat{\sigma}}{1 - y}; \qquad \hat{\sigma} = 35 \pm 5 \text{ MeV}, \tag{7}$$

where the 5 MeV error is due to the theoretical uncertainty in the terms of order $\mathcal{O}(m_q^2)$. The result of Ref. [9] was obtained using an ultraviolet cut-off. Therefore, it is of interest to see the results of a calculation employing modern techniques of chiral perturbation theory. Also, the question of the intermediate decuplet contributions to the baryon masses arises.

One-loop calculations in chiral perturbation theory yield [13,14] results in the range $\hat{\sigma} = 40 - 50$ MeV, i.e. slightly larger numbers than the result with cut-off [9]. If this would be the final word, the prediction $\hat{\sigma} = 26$ MeV, based on first order SU(3), would receive very large corrections – contrary to what one expects from a result based on SU(3). In order to decide the issue, a full calculation to order $\mathcal{O}(m_q^2)$ is needed (including an estimate of the relevant counterterms which enter at this order). These contributions presumably tame the large SU(3) corrections which are generated by the one-loop diagrams. The same effect is expected to be provided by the ultraviolet cut-off used in the derivation of (7). Since a full calculation at order m_q^2 is not yet in sight, we will use in the following the result (7). See also the discussion in Refs. [15,16].

Another issue of importance is to check the relation between the σ-term and the Σ. For this the scalar form factor $\sigma(t)$ is defined as

$$\bar{u}'\sigma(t)u = \langle p'|\hat{m}(\bar{u}u + \bar{d}d)|p\rangle; \qquad t = (p' - p)^2, \tag{8}$$

and the low-energy theorem of chiral symmetry can now be written in the form proposed by Brown, Pardee and Peccei [3]

$$\Sigma = \sigma(2\mu^2) + \Delta_R, \tag{9}$$

where Δ_R is of order $\mu^4 \log \mu^2$. Numerically the result is $\Delta_R = 0.35$ MeV to one-loop in chiral perturbation theory [10] . The crucial piece of information is then the difference

$$\Delta_\sigma = \sigma(2\mu^2) - \sigma(0), \tag{10}$$

where $\sigma(0) = \sigma$. The t-dependence of $\sigma(t)$ has to do with the $\pi\pi$ interaction for $t > 4\mu^2$, and it has to be constrained to be consistent with chiral symmetry [17,18]. The result emerging from a dispersion analysis is [18]

$$\Delta_\sigma = 15.2 \pm 0.4 \text{ MeV}, \tag{11}$$

where the error bar reflects the estimated uncertainty in the $\pi\pi$ phase shifts. This is larger by about 10 MeV than the result from chiral perturbation theory at the one-loop level quoted in the previous section. It turns out that the s-wave t-channel πN amplitude $f^0_+(t)$, which drives the magnitude of Δ_σ, is poorly described by the corresponding Born term away from $t = \mu^2$, and that is what the one-loop calculation is, in fact, using for $t > 4\mu^2$.

The leading nonanalytic contribution to Δ_σ proportional to μ^3 is about 7 MeV [1,4]. A similar contribution is found [16] from the analogous diagram with a Δ intermediate state. The contribution from the K-and η-loops being small yields $\Delta_\sigma \simeq 15$ MeV. The cloudy bag model analysis of Ref. [19] is in line with this result. A similar value for Δ_σ is obtained also from a meson exchange model calculation [20] with input from the Karlsruhe analysis [8].

If the list of steps in the introduction is checked, Eq. (7) provides the relation asked by the step i). As discussed above, there is need for additional work here, especially by extending the calculation to $\mathcal{O}(m_q^2)$. Item ii) is addressed by Eq. (11), and here also model calculations seem to lead to a coherent picture. There is an interesting suggestion [21] to investigate directly the nucleon scalar form factor with neutrino induced pion production. The experimental feasibility still has to be clarified. Eq. (9) provides the answer for item iii). That result has been obtained in a one-loop calculation [10]. At the two-loop level there will be additional contributions of the same order, $\mathcal{O}(m_q^2)$, so the result just gives an indication of the size of the terms appearing.

It now remains to check item iv), the extrapolation of the πN amplitudes to the Cheng-Dashen point and the importance of the experimental uncertainties on the error bar for Σ. The 8 MeV cited in the introduction shows the scatter of values corresponding to different hyperbolae in the dispersion analysis of Koch [6].

3 Σ and dispersion relations

A method, based on forward dispersion relations, was proposed some time ago for extracting the pion-nucleon Σ-term from low-energy data [22]. The purpose was to simplify the analysis of the propagation of the experimental uncertainties to the uncertainty of the Σ-term. The experimental error in Σ is assumed to be dominated by the errors of the low-energy data. The uncertainties in the data sets at higher energies will be taken into account separately and included in the systematic error bar. At low energies, πN scattering is dominated by the six partial waves S_{11}, S_{31}, P_{11}, P_{31}, P_{13} and P_{33}, which have to be extrapolated from the experimentally well known region down to threshold. To determine the energy dependence of these partial waves, six dispersion relations were written. These include the standard [8] equations for the invariant amplitudes $D^\pm$ and $B^\pm$ together with two further amplitudes $E^\pm$ defined by [22]

$$E^\pm = \frac{\partial}{\partial t}(A^\pm + \omega B^\pm)|_{t=0}, \tag{12}$$

where the partial derivative is taken at fixed s, and ω is the pion laboratory energy.

The standard forward dispersion relations ($t = 0$; $\nu = \omega + t/4m$, $\nu = \omega$ for $t=0$) for D and B are [8]

$$Re\bar{D}^+(\omega) = \bar{D}^+(\mu) + \frac{2(\omega^2 - \mu^2)}{\pi} \int_\mu^\infty \frac{d\omega'\omega'}{\omega'^2 - \omega^2} \frac{ImD^+(\omega')}{\omega'^2 - \mu^2}$$

$$Re\bar{D}^-(\omega) = \frac{g^2\omega}{2m^2} + \frac{2\omega}{\pi} \int_\mu^\infty \frac{d\omega'}{\omega'^2 - \omega^2} ImD^-(\omega')$$

$$Re\bar{B}^+(\omega) = \frac{2\omega}{\pi} \int_\mu^\infty \frac{d\omega'}{\omega'^2 - \omega^2} ImB^+(\omega') \tag{13}$$

$$Re\bar{B}^-(\omega) = \frac{g^2}{2m^2} + \frac{2}{\pi} \int_\mu^\infty \frac{d\omega'\omega'}{\omega'^2 - \omega^2} ImB^-(\omega'),$$

where g is the πN coupling constant ($g^2/4\pi = 14.28$; [8]). Here, as before, the bar indicates that the pseudovector Born term has been subtracted from the amplitude. The dispersion relations for the amplitudes $E^\pm$ are [22]

$$Re\bar{E}^+(\omega) = \bar{E}^+(\mu) + \frac{2(\omega^2 - \mu^2)}{\pi} \int_\mu^\infty \frac{d\omega'\omega'}{\omega'^2 - \omega^2} \frac{ImE^+(\omega')}{\omega'^2 - \mu^2}$$

$$- \int_\mu^\infty d\omega' \{\tilde{h}_2(\omega',\omega)ImD^+(\omega') - \tilde{h}_1(\omega',\omega)ImB^+(\omega')\}$$

$$Re\bar{E}^-(\omega) = \frac{2\omega}{\pi} \int_\mu^\infty \frac{d\omega'}{\omega'^2 - \omega^2} ImE^-(\omega') \tag{14}$$

$$+ \int_\mu^\infty d\omega' \{h_2(\omega',\omega)ImD^-(\omega') - h_1(\omega',\omega)ImB^-(\omega')\}$$

$$h_n(\omega',\omega) = (\omega' + \omega)^{-n}/2\pi m$$

$$\tilde{h}_n(\omega',\omega) = h_n(\omega',\omega) - h_n(\omega',\mu).$$

The two subtraction constants, $\bar{D}^+(\mu)$ and $\bar{E}^+(\mu)$, remain to be fixed and they are related to the isoscalar scattering lengths a_{0+}^+ and a_{1+}^+ ($x = \mu/m$)

$$\bar{D}^+(\mu) = 4\pi(1 + x)a_{0+}^+ + \frac{g^2x^3}{\mu(4 - x^2)}$$

$$\bar{E}^+(\mu) = 6\pi(1 + x)a_{1+}^+ - \frac{g^2x^2}{\mu^3(2 - x)^2}. \tag{15}$$

In the present approach, it is assumed that $Im\,D^\pm$, $Im\,B^\pm$ and $Im\,E^\pm$ are known for the momentum range $k_{LAB} \geq k_0$. In this study the imaginary parts are constructed from the partial waves of the KH.80 analysis [23]. In addition, the d-and f-waves are assumed to be known at low energies, $k_{LAB} < k_0$, and they are also taken from the Karlsruhe analysis [24]. By choosing k_0 small enough d-and f-waves can, of course, be neglected. The matching momentum k_0 was initially the lower limit of the experimentally well known momentum range. What is more important, however, is that below the momentum k_0 πN scattering can essentially be described by the above six partial waves. The d-and f-waves are just small corrections to the invariant amplitudes. All in all, with the above mentioned input, if the scattering lengths a_{0+}^+ and a_{1+}^+ are known, the low-energy ($k_{LAB} < k_0$) behaviour of the amplitudes is completely fixed including the threshold parameters for both isoscalar-and isovector channels.

From the first estimates for the real parts of the amplitudes $\bar{D}^\pm$, $\bar{B}^\pm$ and $\bar{E}^\pm$ the real parts of the s-and p-wave partial wave amplitudes can be solved using the expansions [22]

$$D(\omega) = \frac{4\pi\sqrt{s}}{m} \sum_{l=0}^{\infty} \{(l+1)f_{l+}(q) + lf_{l-}(q)\}$$

$$B(\omega) = \frac{4\pi m}{q^2} \sum_{l=0}^{\infty} \{(l+1)(\varepsilon - l - 1)f_{l+}(q) + l(\varepsilon + l)f_{l-}(q)\} \qquad (16)$$

$$E(\omega) = \frac{\pi\sqrt{s}}{mq^2} \sum_{l=0}^{\infty} l(l+1)(l+2)\{f_{l+}(q) + f_{(l+1)-}(q)\},$$

because the higher partial waves are assumed to be known at the low momenta considered here. In (16) the partial wave amplitude is

$$f_l = \frac{1}{q} \exp(i\delta_l) \sin \delta_l$$

and ε stands for $(1 + q^2/m^2)^{\frac{1}{2}}$, where q is the centre-of-mass momentum. Unitarity gives then the imaginary parts of the invariant amplitudes and they can be used in (13) and (14) to compute the next estimate for the real parts. Usually only 4 to 7 iterations of this sort are sufficient to reach stable partial wave amplitudes starting from s-and p-waves identical to zero for $k_{LAB} < k_0$.

To calculate the Σ-term we note that in the Taylor expansion of $\bar{D}^+$ in powers of ν^2 and t

$$\bar{D}^+ = d_{00}^+ + d_{10}^+\nu^2 + d_{01}^+t + d_{20}^+\nu^4 + d_{11}^+\nu^2 t + d_{02}^+t^2 + ... \qquad (17)$$

the coefficients d_{00}^+ and d_{01}^+ have a simple connection to the forward amplitudes at $\omega = 0$, i.e.

$$d_{00}^+ = \bar{D}^+(0), \qquad d_{01}^+ = \bar{E}^+(0). \qquad (18)$$

For given values of a_{0+}^+ and a_{1+}^+, the solutions of the dispersion relations fix then d_{00}^+ and d_{01}^+, and the Σ-term is according to (6)

$$\Sigma = F_\pi^2(d_{00}^+ + 2\mu^2 d_{01}^+) + \Delta_D \equiv \Sigma_d + \Delta_D, \qquad (19)$$

where Δ_D represents terms proportional to t^2 or higher in (17). However, the curvature part requires information from $\pi\pi$ scattering in a manner analogous to the curvature of the scalar form factor $\sigma(t)$. The result obtained in [18] is

$$\Delta_D = 11.9 \pm 0.6 \text{ MeV}. \qquad (20)$$

The corresponding Karlsruhe value for Δ_D is about 12 MeV [8].

There remains two parameters, the subtraction constants a_{0+}^+ and a_{1+}^+, to be fixed. It was shown in Ref. [22] that with the KH.80 values for these parameters [8], $a_{0+}^+ = -0.00967\,\mu^{-1}$ and $a_{1+}^+ = 0.1327\,\mu^{-3}$, the rest of the KH.80 scattering lengths are reproduced well by the present method. The task is then to fix a_{0+}^+ and a_{1+}^+ using directly the experimental pion-nucleon information. Consequently the πN amplitudes at low energy are completely determined and the Σ-term is fixed once the value of $\triangle_D$ is known.

4 Data analysis

In the present approach, it is assumed that $Im\,D^\pm$, $Im\,B^\pm$ and $Im\,E^\pm$ are known above the matching momentum k_0, which has been fixed to 185 MeV/c [12]. The d-and f-waves will give small corrections below k_0, and these are assumed to be known. Two subtraction constants associated with the amplitudes D^+ and E^+ remain to be fixed. That is accomplished with a χ^2-fit to the low-energy pion-nucleon data. With a pair of starting values (a_{0+}^+, a_{1+}^+) the iterative procedure described in section 3 produces partial waves at momenta below k_0, and the observables can then be calculated. For the minimization the CERN library programme, MINUIT [25], is employed. For each data set the contribution to the total χ^2 is

$$\chi^2 \equiv \left(\frac{z-1}{\Delta z}\right)^2 + \sum_{\text{DATA}} \left(\frac{z\sigma^*(\theta) - \sigma(\theta)}{\Delta\sigma(\theta)}\right)^2, \tag{21}$$

where $\sigma^*(\theta)$ is the experimental differential cross section with errors $\Delta\sigma(\theta)$. The theoretical angular distribution, which is a function of the two parameters a_{0+}^+ and a_{1+}^+, is denoted by $\sigma(\theta)$. The normalization of the experimental data is adjusted with the parameter z, and Δz is the experimental overall normalization uncertainty. A "data set" means here an angular distribution at a given momentum in one experiment. The total χ^2 is a sum of the χ^2's for each set, and so the number of parameters used for the χ^2 minimization is *number of data sets +* *2*, where two parameters are related to the two subtraction constants, and the rest comes from the z's for each data set in the minimization. In the theoretical differential cross section, $\sigma(\theta)$, the electromagnetic corrections are calculated according to the formalism of Tromborg et al. [26] as implemented in Ref. [23]. In addition, the P_{33} π^-p phase shifts are corrected for the isospin breaking due to the mass difference $\Delta^{++} - \Delta^0$.

Over the last 20 years several groups have published πp data at low energies [27-34]. Only the data of Bertin et al. [27] and Auld et al. [28] were available at the time of the KH.80 analysis [23]. Koch's result [6] for the Σ-term, $\Sigma = 64 \pm 8$ MeV, was essentially based on the KH.80 amplitudes. In the present analysis the elastic low-energy data are considered, but the information [29] on the charge exchange reaction π^-p $\rightarrow$ π^0n is not used. The charge exchange reaction involves the isospin odd channel and, therefore, constrains only weakly the isospin even channel, which is of interest for the question of the Σ-term [22]. Of course, eventually the charge exchange information should be taken into account in the analysis to check the consistency of the π^+p and π^-p elastic input. Some details of the data selection have been discussed in Ref. [35]. It turns out that only the data sets of Bertin et al. [27] and Frank et al. [30] lead to a solution in the present highly constrained approach.

The Bertin et al. data [27] on π^+p extend down to very low energy (the lowest momentum is $k_{LAB} = 79$ MeV/c corresponding to $T_\pi = 21$ MeV), and so they have contributed in an essential way to fixing the amplitudes at low energy. The data at $k_{LAB} = 153$ MeV/c are found to be inconsistent with the rest and are removed from the data basis. This fact was also observed in Ref. [23]. The minimum, χ^2_{MIN}, with data at 79, 97, 112 and 130 MeV/c included, is found at $a_{0+}^+ = -0.0080 \; \mu^{-1}$, $a_{1+}^+ = 0.1317 \; \mu^{-3}$ with χ^2/data point = 1.03. The values for a_{0+}^+ and a_{1+}^+ are within one standard deviation of the values quoted in Table 2.4.7.2 of Ref. [8] for the KH.80 analysis. The remaining low-energy parameters are reproduced with similar precision. The result for the sigma term is $\Sigma_d = 48$ MeV. Fig. 1 of Ref. [12] displays the values of Σ_d, the curves $\chi^2_{MIN} + 1$ and $\chi^2_{MIN} + 4$ are shown in the (a_{0+}^+, a_{1+}^+)-plane with a solid dot denoting the minimum (solution A). To estimate the error for the resulting Σ_d is a difficult task. The statistical error is well defined as $\Delta\Sigma_d = \pm 4$ MeV, but the main problem is the remaining systematic error. To estimate that, a number of minimizations were performed with different values for the matching momentum k_0. Additional data were included to find out the sensitivity to the data selection. The uncertainty of the dispersion integrals was considered. This was done by comparing the KH.80, KA.84 and KA.85 amplitudes and using the difference as a measure for the uncertainty of the imaginary parts $Im\,D^\pm$, $Im\,B^\pm$ and $Im\,E^\pm$, which are input for the analysis. For the $Im\,D^\pm$-amplitudes the experimental errors of the total cross sections can be directly used for estimating the possible range for the dispersion integrals [36]. The d-and f-wave phase shifts below k_0 were allowed to vary and the effect on Σ_d was calculated. The result for the Bertin data is then

$$\Sigma_d = 48 \pm 4 \pm 4 \pm 4 \; \text{MeV},$$

where the errors refer to statistics, to the data base modifications and to the input uncertainties respectively.

A similar analysis was performed for the Frank et al. data [30]. The quoted normalization uncertainties are generally quite large except for the π^+p data at 95 MeV/c. It turns out that this data set is inconsistent with the remainder and has to be dropped. The minimum is found at $a_{0+}^+ = -0.0115 \; \mu^{-1}$, $a_{1+}^+ = 0.1339 \; \mu^{-3}$ with χ^2/data point = 1.20. The corresponding Σ_d-value is

$$\Sigma_d = 50 \pm 3 \pm 7 \pm 4 \; \text{MeV},$$

where the errors were estimated as in the previous case. The solution is denoted by B in Fig. 1 of Ref. [12].

With the indicated input the two constants a_{0+}^+ and a_{1+}^+ fully fix the low-energy behaviour of the amplitudes. Therefore, the combination

$a_{\pi^- p} = a_{0+}^+ + a_{0+}^-$ is also known on each point of the (a_{0+}^+, a_{1+}^+)-plane. The error band in Fig. 1 of Ref. [12] displays the result of the $\pi^- p$ atomic s-wave level shift measurement [37] $\Delta E = -7.12 \pm 0.32$ eV corresponding to the scattering length $a_{\pi^- p} = 0.086 \pm 0.004\ \mu^{-1}$, which result is consistent with the scattering data.

If the more recent low-energy data of Brack et al. [32] are used as input, it turns out that the $\pi^+ p$ and $\pi^- p$ data cannot be accommodated simultaneously (this statement relies, of course, on the assumption that the input amplitudes above k_0 are correct). The $\pi^- p$ data are roughly consistent with solutions A and B discussed above, but the $\pi^+ p$ data lead to a qualitatively different solution. The solution which fits well the Brack $\pi^+ p$ data is clearly off for the $\pi^- p$ data at the same momentum as can be seen from Fig. 3 of Refs. [35,38]. In principle, the measurement [33] of $Re\,D^+(k, t = 0)$ at 135 MeV/c is able to distinguish between the two values. However, as discussed in Ref. [12] these measurements have to be extremely precise to make a clear distinction.

Of course, there is the possibility that part of the input is not quite correct. There are two sorts of input used in the present calculations, i.e. $Im\,D^\pm, Im\,B^\pm$ and $Im\,E^\pm$ for $k_{LAB} \geq k_0$ and d-and f-wave phase shifts for $k_{LAB} < k_0$. Clearly the discrepancies are so large that the high partial waves cannot be the culprit. There remains the question of the imaginary parts of the invariant amplitudes for $k_{LAB} \geq k_0$. For example, the E-amplitudes are sensitive to high partial waves, because the partial wave amplitudes get multiplied by l^3, see Eq. (16). To allow for variation of the imaginary parts of the invariant amplitudes above k_0 six additional parameters were introduced. They multiply the six dispersion integrals from k_0 to infinity. The initial value for these parameters is, of course, unity, but then the parameters are allowed to vary in a small window around unity. The size of the window should reflect the estimated uncertainty of the input, and here one employs a method similar to the one used for estimating the systematic uncertainty in Σ_d due to the input amplitudes. With this additional freedom it is possible to find solutions, however, with rather high χ^2 [45].

5 Discussion

The results discussed in section 4 have been published in Refs. [12,35]. Since then a number questions have arisen. One of them, the role of Δ in σ, $\sigma(t)$ and $\bar{D}^+(0,t)$ has been discussed in section 2. The numbers used here for Δ_σ and Δ_D are based on dispersion relations [18], so there the Δ contributions have been taken into account. The situation concerning σ is still open, because a full calculation to order $\mathcal{O}(m_q^2)$ is missing. The hope is that such a calculation would allow reducing the theoretical uncertainty for $\hat{\sigma}$.

The connection between $\sigma(t)$ and Σ is made at the Cheng-Dashen point $(\nu = 0, t = 2\mu^2)$. In view of the steep t-dependence of $\bar{D}^+(0,t)$ the question arises [39] of the pion mass to be adopted here. If the value of the pion mass, μ, is taken to be that of the neutral pion instead of the charged pion, the resulting Σ_d would be reduced by about 9 MeV. Of course, the cusp at $t = 4\mu^2$ would also move, so it is extremely difficult to estimate the change in Σ. At present, it is considered safest to use only elastic $\pi^\pm p$ data and keep the charged pion mass throughout [40].

Recently there has been quite some discussion on the value of the pion-nucleon coupling constant g [41-44]. The standard value adopted here is $f^2 = 0.079$ [23] ($f^2 = (\mu/2m)^2 g^2/4\pi$), and the error should be about 0.002 [42]. Analyzing pp, np and $\bar{p}p$ data the Nijmegen group [41] ends up favouring much smaller numbers like $f^2 = 0.0745$ with a very small error. Fixed-t dispersion relations have been used in [43] to obtain $f^2 = 0.0771 \pm 0.0014$. The most recent VPI analysis FA93 [44] gives a low figure $f^2 = 0.076$ as well. Change of f^2 could influence the value of Σ in an analysis like the one described here. In Ref. [12] it was not possible to address this issue, because the input amplitudes were taken from the Karlsruhe analysis, and one was bound to use the value of the πN coupling constant which was consistent with other input. However, with the new VPI set of amplitudes, FA93 [44], it has become possible to make an exploratory study of the effect of a smaller coupling constant. The imaginary parts of the invariant amplitudes were constructed from the FA93 partial waves in the range $k_{\text{LAB}} = 0.185 - 2.07$ GeV/c and above that range the KA.84 partial waves were used. If the data of Frank et al. [30] and Brack et al. [32] (excluding 152 MeV/c) is used, then a solution can be found at $a_{0+}^+ \simeq -0.004$ μ^{-1}, $a_{1+}^+ \simeq 0.129$ μ^{-3} with χ^2/data point = 1.15. The corresponding Σ_d-value is 55 MeV, i.e. 5 MeV higher than the result with Frank et al. data alone (solution B). In view of the fact that $f^2 = 0.076$ for the FA93 solution [44], it seems likely that Σ_d is not too sensitive to the value of the pion-nucleon coupling constant (possibly, because Σ involves $\bar{D}^+$ where the Born term has been subtracted). The resulting value for $a_{\pi^- p}$ is roughly 0.082 μ^{-1}, which is consistent with the published value from pionic hydrogen [37]. However, the new preliminary high precision data seem to favour a somewhat larger value [46]. (The consistency of the new experimental value with the solutions A and B with KH.80 input amplitudes can be estimated from Fig. 2 of [47].)

6 Concluding remarks

The results of the present analysis are not far from the Karlsruhe value for $\Sigma_d = 51$ MeV [8]. To get the final comparison for the Σ-term the curvature part Δ_D is needed. At small t the curvature of $\bar{D}^+(0, t)$ is dictated by the $\pi\pi$ cut at $t > 4\mu^2$. The result consistent with the chiral symmetry constraints near the $\pi\pi$ threshold is $\Delta_D = 12$ MeV [18]. The solution A leads then to the value $\Sigma \simeq 60$ MeV with an error of about 7 MeV (which must be an underestimate). This result is in good agreement with the Karlsruhe value 64±8 MeV [6]. Even a slight change of the value of the pion-nucleon coupling constant seems to lead to a result which is well within the error bar for Σ. Various phenomenological models with parameters fixed by low-energy πN input produce results also in the range $\Sigma \simeq 62 - 66$ MeV [48,49].

Combining then the results $\hat{\sigma} = 35\pm5$ MeV, $\Delta_\sigma \simeq 15$ MeV yields an estimate for the strange quark content of the proton $y = 0.2 \pm 0.2$. The corresponding contribution of the term $m_s \bar{s}s$ to the proton mass is about 130 MeV, which is now more reasonable than the value based on leading order results.

There remains conflicts in low-energy πN data. In many cases the discrepancies are apparent and do not need any phase shift analysis to be made visible. It is an experimental question to clarify these issues. A feeling of the typical sensitivity of the Σ-term on low-energy πN data can be obtained from Fig. 2 of [22]. An additional piece of information to settle the problems at low energies would be provided by the analyzing power measurements [36,50]. They will soon become available [51].

Acknowledgements

I wish to thank J. Gasser, G. Höhler, H. Leutwyler and M.P. Locher for numerous exchanges in the field of pion-nucleon physics. In addition, I thank J. Gasser and A.M. Green for useful remarks on the manuscript.

References

1. J. Gasser, H. Leutwyler: Phys. Reports **87** 77 (1982)
2. T.P Cheng, R. Dashen: Phys. Rev. Lett. **26** 594 (1971)
3. L.S. Brown, W.J. Pardee, R.D. Peccei: Phys. Rev. D **4** 2801 (1971)
4. H. Pagels, W.J. Pardee: Phys. Rev. D **4** 3335 (1971)
5. T.P. Cheng: Phys. Rev. D **13** 2161 (1976)
6. R. Koch: Z. Phys. C **15** 161 (1982)
7. R.L. Jaffe, C.L. Korpa: Comments Nucl. Part. Phys. **17** 163 (1987)
8. G. Höhler, in Landolt-Börnstein, Vol. 9 b2, ed. H. Schopper (Springer, Berlin, 1983).
9. J. Gasser: Ann. Phys. (N.Y.) **136** 62 (1981)
10. J. Gasser, M.E. Sainio, A. Švarc: Nucl. Phys. B **307** 779 (1988)
11. J. Gasser: Proc. of the second international workshop on πN physics, 1987, eds. W.R. Gibbs and B.M.K. Nefkens, Los Alamos report LA-11184-C (1987) p. 266
12. J. Gasser, H. Leutwyler, M.E. Sainio: Phys. Lett. B **253** 252 (1991)
13. S. Dentin: Ph.D. thesis, University of Marseilles II (1990)
14. U. Bürgi: Licentiate thesis, University of Bern (1993)
15. E. Jenkins, A.V. Manohar: Phys. Lett. B **281** 336 (1992)
16. V. Bernard, N. Kaiser, Ulf-G. Meißner: Z. Phys. C **60** 111 (1993)
17. J. Gasser, H. Leutwyler: Ann. Phys. (N.Y.) **158** 142 (1984)
18. J. Gasser, H. Leutwyler, M.E. Sainio: Phys. Lett. B **253** 260 (1991)
19. I. Jameson, A.W. Thomas, G. Chanfray: J. Phys. G **18** L159 (1992)
20. B.C. Pearce, K. Holinde, J. Speth: Nucl. Phys. A **541** 663 (1992)
21. V. Bernard, N. Kaiser, Ulf-G. Meißner: Phys. Lett. B **331** 137 (1994)
22. J. Gasser, H. Leutwyler, M.P. Locher, M.E. Sainio: Phys. Lett. B **213** 85 (1988)
23. R. Koch, E. Pietarinen: Nucl. Phys. A **336** 331 (1980)

24. R. Koch: Nucl. Phys. A **448** 707 (1986)
25. F. James, M. Roos: Comput. Phys. Comm. **10** 343 (1975)
26. B. Tromborg, S. Waldenstrøm, I. Øverbø: Phys. Rev. D **15** 725 (1977)
27. P.Y. Bertin et al.: Nucl. Phys. B **106** 341 (1976)
28. E.G. Auld et al.: Can. J. Phys. **57** 73 (1979)
29. J. Duclos et al.: Phys. Lett. B **43** 245 (1973);
    M. Salomon et al.: Nucl. Phys. A **414** 493 (1984);
    D.H. Fitzgerald et al.: Phys. Rev. C **34** 619 (1986);
    A. Bagheri et al.: Phys. Rev. C **38** 885 (1988)
30. J.S. Frank et al.: Phys. Rev. D **28** 1569 (1983)
31. B.G. Ritchie et al.: Phys. Lett. B **125** 128 (1983)
32. J.T. Brack et al.: Phys. Rev. C **34** 1771 (1986); **38** 2427 (1988); **41** 2202 (1990)
33. U. Wiedner et al.: Phys. Rev. Lett. **58** 648 (1987); Phys. Rev. D **40** 3568 (1989)
34. E. Friedman et al.: Nucl. Phys. A **514** 601 (1990)
35. M.E. Sainio: πN Newsletter **4** 58 (1991)
36. R. Koch: "Inconsistencies in low-energy pion-nucleon scattering", Karlsruhe report TKP 85-5 (1985)
37. W. Beer et al.: Phys. Lett. B **261** 16 (1991)
38. M.P. Locher: Nucl. Phys. A **527** 73c (1991)
39. D.V. Bugg: Nucl. Phys. A **527** 419c (1991)
40. R.E. Cutkosky: Proc. of Hadron 91, eds. S. Oneda and D.C. Peaslee (World Scientific, Singapore, 1992) p. 913
41. V. Stoks, R. Timmermans, J.J. de Swart: Phys. Rev. C **47** 512 (1993)
42. G. Höhler: πN Newsletter **3** 66 (1991); **6** 154 (1992)
43. F.G. Markopoulou-Kalamara, D.V. Bugg: Phys. Lett. B **318** 565 (1993)
44. R.A. Arndt, R.L. Workman, M.M. Pavan: Phys. Rev. C **49** 2729 (1994)
45. M.E. Sainio: πN Newsletter **8** 49 (1993)
46. A. Badertscher et al.: these proceedings
47. M.P. Locher, M.E. Sainio: Nucl. Phys. Å **518** 201 (1990)
48. P.F.A. Goudsmit et al.: Nucl. Phys. A **575** 673 (1994)
49. C. Schütz, J.W Durso, K. Holinde, J. Speth: Phys. Rev. C **49** 2671 (1994)
50. M.P. Locher, M.E. Sainio: Czech. J. Phys. B **39** 943 (1989)
51. C. Joram et al.: πN Newsletter **8** 30 (1993)

A Meson–Exchange Model of Pion–Nucleon Interaction

C. Schütz[1], *J.W. Durso*[2], *K. Holinde*[1], *M.B. Johnson*[3], *B.C. Pearce*[4], and *J. Speth*[1]

[1] Forschungszentrum Jülich GmbH, D–52425 Jülich, Germany
[2] Dept. of Physics, Mount Holyoke College, MA 01075
[3] Los Alamos National Laboratory, Los Alamos, NM 87545
[4] Dept. of Physics and Math. Physics, The University of Adelaide, Adelaide 5005, Australia

Abstract. Starting from a dynamical model for the $N\bar{N} \to \pi\pi$ process, which in the pseudophysical region agrees with available quasiempirical information, the scalar (σ) and vector (ρ) piece of correlated two–pion exchange in the pion–nucleon interaction is derived via dispersion integrals over the unitarity cut. Supplemented by conventional (direct and exchange) pole diagrams involving the nucleon and the Δ isobar, our πN interaction model accounts for the πN phase shifts in the elastic region as well as for low energy parameters and the $\pi N \Sigma$ term. A quantitative description of P_{11}–wave πN scattering is obtained without inclusion of a genuine N^* (Roper) resonance. In the pion production region the effect of coupling to the reaction channel σN is investigated.

1 Introduction

The interaction between a pion and a nucleon plays an outstanding role in low and medium energy physics. First it is of topical interest in itself, being a prominent example of a strong hadronic interaction. Second it is an important ingredient in many other hadronic reactions, e.g. pion production in nucleon–nucleon collisions or scattering of a pion on a nucleus.

In the past, the pion–nucleon (πN) interaction has been often parametrized in separable terms in order to simplify its application in related few- and many-body systems. One disadvantage of such an approach is that the underlying parameters have no physical meaning and thus cannot be related to those occurring in other processes. A simultaneous understanding of various hadronic reactions requires a microscopic treatment based on the fundamental theory of strong interactions, i.e. quantum chromodynamics (QCD). At the relatively low energies of interest here, QCD cannot be solved rigorously, neither at present nor in the near future, in terms of explicit quark–gluon dynamics. On the other hand, baryons and mesons, to be considered as the collective degrees of freedom of QCD, have definitely established their role as the proper variables to describe low and medium energy strong interaction phenomena. For example, models based on meson exchange have been very successful in describing the NN empirical data, up to nucleon kinetic energies of about 1 GeV. Therefore it should be expected that the meson exchange concept works comparably well in the πN system. Such a microscopic model is especially needed if one wants to understand the baryon spectrum because most of these data are derived from πN scattering. Here the interesting question is whether the experimentally observed baryon resonances really contain a genuine three–quark 'excited state' contribution or whether they can be understood purely in terms of the underlying meson–baryon dynamics. In this work we will examine more closely the nature of the Roper resonance, the first excited state with the quantum numbers of the nucleon.

Recently, several papers have been published, which present a meson exchange model for πN scattering [1, 2, 3, 4]. All models include the direct and crossed nucleon as well as Δ–isobar pole terms, see Fig. 1 a)–d). (In an approximation applied in Ref.[3] the crossed Δ term drops out; on the other hand, the Roper (N^*, direct and exchange) and D_{13} (direct) pole diagrams are included in addition). Furthermore, they contain σ– and ρ–exchange terms, Fig. 1 e),f). Although it appears that all models contain essentially the same physics they differ however in detail, especially concerning σ and ρ exchange. Different coupling schemes and coupling constants are used in the various approaches. The σ contribution even differs in sign: In Refs.[1, 3, 4] it is attractive whereas in Ref.[2], by a suitable choice of the coupling constants, it is repulsive.

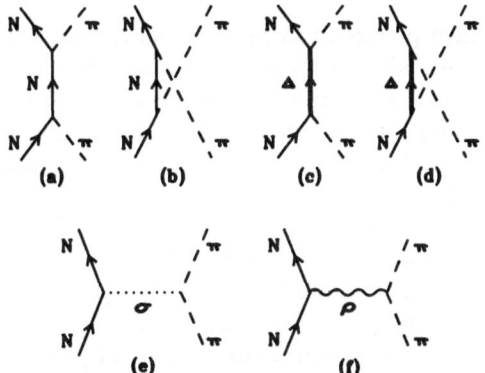

Fig. 1. Direct (a,c) and exchange (b,d) nucleon (N) and delta–isobar (Δ) pole diagrams together with σ, ρ–exchange processes (e,f) used in πN interaction models.

The models of Refs.[1, 3] unitarize the interaction by solving a relativistic scattering equation while Ref.[4] is a tree–level model and is thus restricted to low energies. The model of Ref.[1] provides a satisfactory description up to 400 MeV pion laboratory kinetic energy while the model of Ref.[2] shows some discrepancies in P waves. Due to the inclusion of N^* and D_{13}, the model of Ref.[3] is able to provide a good description up to 600 MeV.

At this point we remind the reader that it is from the NN system that we really learned about the nature of the σ. Extensive dispersion–theoretical calculations have successfully derived the NN intermediate–range attraction from 2π–exchange processes using empirical πN and $\pi\pi$ data. Consequently, the σ

used in meson exchange models of the NN interaction should not be viewed as a genuine particle but as a simple parametrization of (correlated) 2π–exchange processes.

Since this interpretation works so well for the NN system, it should also be correct in the πN system. Thus, also here, σ (as well as ρ) exchange should be likewise viewed as correlated 2π exchanges, in the scalar–isoscalar (vector–isovector) channel as visualized in Fig. 2. This implies that, as in the NN system, this picture should provide constraints for the size of σ and ρ exchange. In a

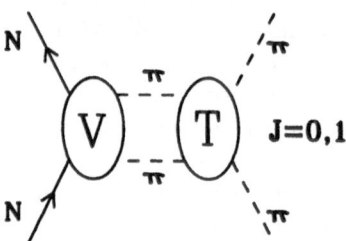

Fig. 2. Correlated 2π exchange in the scalar ($J = 0$) and vector ($J = 1$) t channel.

recent work [5] we have presented such a model for correlated 2π exchange in pion–nucleon interaction which uses as input quasiempirical information about the t channel $N\bar{N} \rightarrow \pi\pi$ amplitudes obtained by Höhler et al. [6]. Although this approach is certainly adequate for free πN scattering, problems arise when this πN interaction is used in other areas of physics. For example, modifications of the interaction in the nuclear medium, which come into play when a pion is scattered by a nucleus, cannot be taken into account. The study of such effects requires an explicit field–theoretic description. Therefore, in this work we present a dynamical model for the correlated 2π–exchange process of Fig. 2, which does not only provide a microscopic understanding of the reaction $N\bar{N} \rightarrow \pi\pi$, but also allows investigation of medium modifications of pion–nucleon interaction.

In a second step this model of correlated 2π–exchange is then supplemented by conventional (direct and exchange) pole diagrams involving the nucleon and Δ isobar, cf. Fig. 1 a)–d), (which are treated as bare particles in case of the direct terms) and the interaction is then unitarized by means of the relativistic Schrödinger equation

$$T(Z) = V + V \frac{1}{Z - H_0} T(Z) \qquad (1)$$

which dresses the direct pole terms by shifting their masses and providing a width (in the case of the Δ) generated by the dynamics of the model.

The following section deals with our meson–exchange models for $\pi\pi$ interaction and the process $N\bar{N} \rightarrow \pi\pi$ which is then transformed to the s (πN) channel. In the third section we present our results for πN scattering before we discuss in more detail πN interaction in the P_{11} partial wave and its implications for the understanding of the Roper resonance. The paper ends with a short summary and outlook.

2 The dynamical model for correlated two–pion exchange

Our investigation of correlated 2π exchange is based on two main ingredients: a dynamical model for $\pi\pi$ scattering which yields the correlation between the exchanged pion pair, and a model for the transition $N\bar{N} \rightarrow \pi\pi$ required to describe the coupling to the nucleon. For the $\pi\pi$ interaction the $\pi\pi - K\bar{K}$ coupled chan-

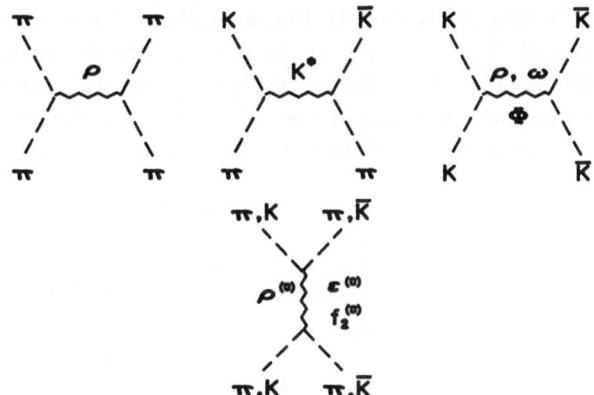

Fig. 3. Born term of the coupled channel $\pi\pi - K\bar{K}$ potential model.

nel model is applied which has been developed by our group [7]. The diagrams building up this model are shown in Fig. 3. Apart from the exchange of physical mesons (ρ, K^*, ω, ...) it also includes pole diagrams involving bare particles (ρ^0, ...) which are renormalized by the iteration procedure. Fig. 4 shows the fit to the $\pi\pi$ phase shifts in the partial waves of relevance here. It is ρ exchange which generates the attraction in the $JI = 00$ phase shift up to 950 MeV leading to a strong correlation between two exchanged pions in a $J = 0$ and $I = 0$ state (σ).

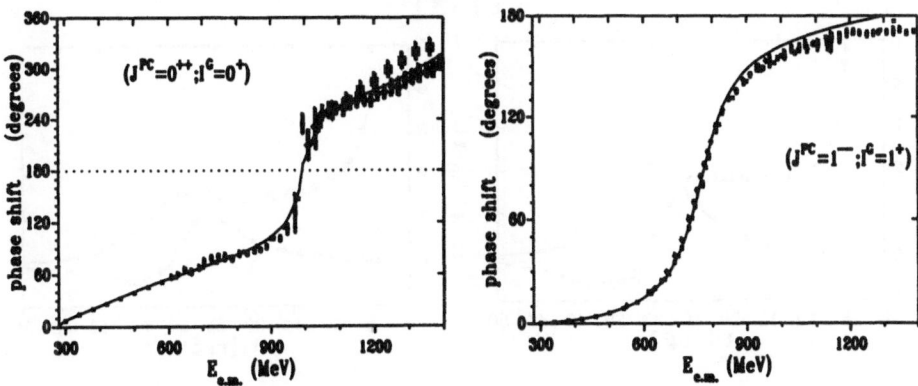

Fig. 4. $\pi\pi$ phase shifts obtained for $J = 0$ and $J = 1$ from our coupled channel $\pi\pi - K\bar{K}$ model. For references to the data, see Ref. [7].

The amplitudes for the process $N\bar{N} \rightarrow \pi\pi$ in the pseudophysical region $(4m_\pi^2 < t < 50m_\pi^2)$ are then generated by solving the scattering equation

$$T_{N\bar{N}\rightarrow\pi\pi} = V_{N\bar{N}\rightarrow\pi\pi} + \sum_{pp=\pi\pi,K\bar{K}} T_{pp\rightarrow\pi\pi} g_{pp} V_{N\bar{N}\rightarrow pp} \qquad (2)$$

Here $V_{N\bar{N}\rightarrow\pi\pi}$ is the transition interaction (cf. Fig. 5) and $T_{pp\rightarrow\pi\pi}$ the transition amplitudes from $\pi\pi$ and $K\bar{K}$ to $\pi\pi$. In Fig. 6 we show the results obtained from our dynamical model. There is one amplitude, f_+^0, for the scalar (σ) channel whereas there are two, f_+^1 and f_-^1, for the vector (ρ) channel. Given that we have only four free parameters (occuring in the $N\bar{N} \rightarrow \pi\pi, K\bar{K}$ transition potential), there is remarkable agreement with the quasiempirical result in all amplitudes.

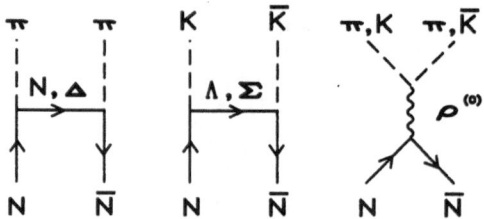

Fig. 5. Ingredients of the $N\bar{N} \rightarrow \pi\pi, K\bar{K}$ transition potentials.

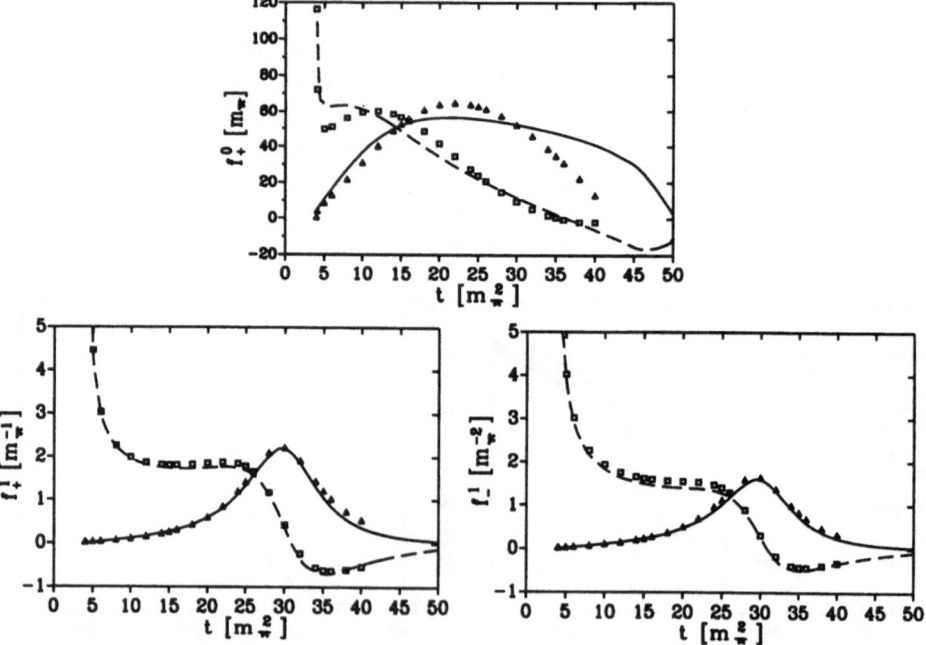

Fig. 6. $N\bar{N} \rightarrow \pi\pi$ helicity amplitudes in the pseudophysical region. The solid lines denote the imaginary parts of the model amplitudes and the dashed lines the real parts. Squares and triangles are the quasiempirical amplitudes taken from Ref. [8].

The small disagreement occuring in the scalar amplitude does not severely affect our final result, the correlated $\pi\pi$ exchange in πN scattering, and moreover it can be compensated by a small readjustment of the cutoff parameter introduced for the σ potential. Here one should keep in mind that the quasiempirical result is anyhow subject to some uncertainties.

In order to define the effective $\sigma-$ and ρ-exchange contribution arising from the correlated $\pi\pi$ exchange in the physical region of πN scattering, dispersion integrals are performed over the unitarity cut as described in Ref. [5]. For the scalar channel a subtraction has been made in the dispersion relation at the Cheng–Dashen point ($t = 2m_\pi^2$). In combination with the pseudovector coupling used at the πNN vertices (Fig. 1 (a), (b)), this subtraction ensures that in Born approximation the isospin even forward scattering amplitude $T^{(+)} = A^{(+)} + \nu B^{(+)}$ vanishes at the Cheng–Dashen point (consistent with chiral symmetry). It is interesting to note that our result has the same analytic structure as a potential for scalar meson exchange in the πN system using derivative coupling at the $\sigma\pi\pi$ vertex, and it can easily be shown that it gives a *repulsive* contribution to the potential $V_{\pi N}^{(\sigma)}$ (in the S waves).

3 Results for πN scattering

3.1 πN phase shifts

After solving the scattering equation the partial wave πN phase shifts are calculated in the standard way. Results are shown in Fig. 7 in comparison to the empirical phase shifts obtained from an analysis of Koch and Pietarinen [9]. We obtain a good agreement in all partial waves.

Fixed parameters of our calculation are the charge–averaged hadron masses and the coupling constants at the πNN and $\pi N\Delta$ vertices appearing in Fig. 1 (b),(d) (cf. Ref. [5]). Free parameters of the model are the cutoff masses ocurring in the form factors of the contributions to our pseudopotential. The bare Δ mass and the bare $N\Delta\pi$ coupling constant appearing in the Δ pole diagram (Fig. 1 (d)) are adjusted to reproduce the P_{33} phase shift. The bare nucleon mass and the bare $NN\pi$ coupling constant for the nucleon pole contribution (Fig. 1 (b)) are determined in a renormalization procedure as function of the other parameters. This renormalization ensures that our full T matrix has a pole at the physical nucleon mass with the residue which reproduces the physical $NN\pi$ coupling constant.

3.2 Low energy parameters

For our πN interaction model, the resulting scattering lengths and volumes, defined as

$$a_L = \lim_{q \to 0} \frac{\sin 2\delta_L}{2q^{2L+1}} \tag{3}$$

are in good agreement with the empirical information, as shown in Table 1.

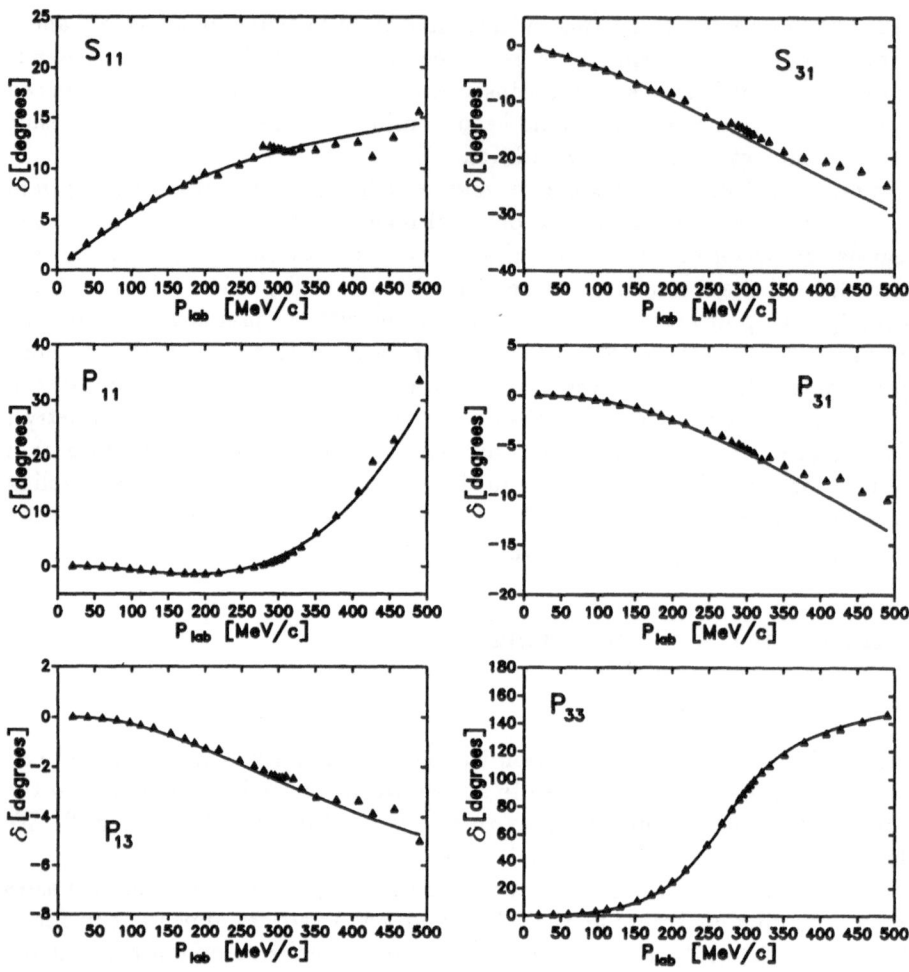

Fig. 7. πN scattering phase shifts in S and P waves, as function of the pion laboratory momentum. The empirical information is taken from Ref. [9].

Table 1. The scattering lengths and volumes. Units are $m_\pi^{-(2L+1)}$.

	model	Koch and Pietarinen [9]
S_{11}	0.165	0.173 ± 0.003
S_{31}	-0.092	-0.101 ± 0.004
P_{11}	-0.080	-0.081 ± 0.002
P_{31}	-0.042	-0.045 ± 0.002
P_{13}	-0.029	-0.030 ± 0.002
P_{33}	0.213	0.214 ± 0.002

3.3 $\pi N \Sigma$ term

According to the Cheng–Dashen theorem [10], the isospin–even forward scattering amplitude at the Cheng–Dashen point is given by

$$T^{(+)}(\nu = 0, \nu_B = 0, q^2 = m_\pi^2, q'^2 = m_\pi^2) = \frac{\Sigma}{f_\pi^2} \quad , \tag{4}$$

where Σ denotes the $\pi N \Sigma$ term and f_π the weak pion decay constant. Using this relation, an extrapolation of our dynamical model below threshold of the πN system enables us to calculate the $\pi N \Sigma$ term. This provides an important test whether a model, which agrees with the empirical situation in the physical region, also satisfies chiral symmetry constraints. The result of our model is 65.5 MeV which is in good agreement with the experimental value of $\Sigma = 60 \pm 10$ MeV [11]. (Note that here $\Sigma(2m_\pi^2)$ has been evaluated. In an earlier work on the scalar form factor of the nucleon [12], we have found within the same framework a 15 MeV contribution to the Σ term arising from the difference $\Sigma(2m_\pi^2) - \Sigma(0)$.)

4 P_{11}–wave πN scattering and the Roper resonance

One of the most interesting partial waves in πN scattering is the P_{11} partial wave. It is a challenge for models of πN interaction to describe quantitatively the repulsion at low energies followed by the attraction leading to the Roper resonance at higher energies. Within the model presented here, the rise of the P_{11} phase shift in the elastic region of the interaction can be reproduced without any contribution from a genuine N^* particle. In Fig. 8 (a) we have plotted the

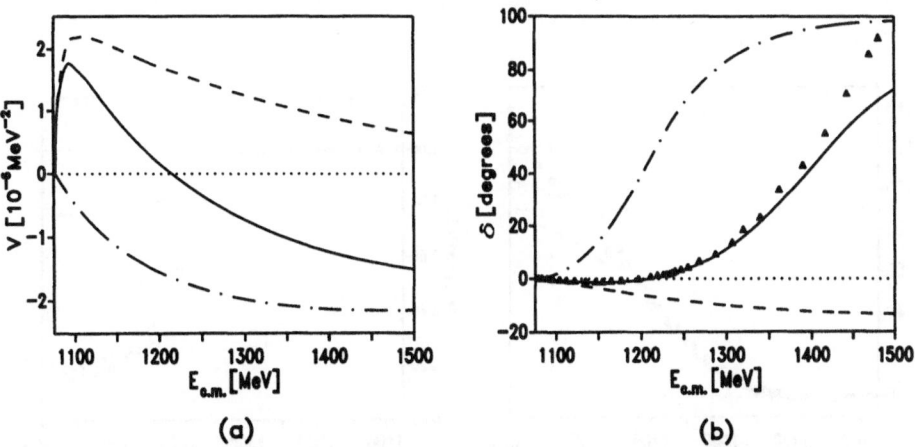

(a) (b)

Fig. 8. (a) On–shell potentials in the P_{11} partial wave as function of the c.m.s. energy. The dashed line arises from the direct nucleon pole; the dash–dotted line belongs to the non–pole part of the potential; the solid line gives the total potential. (b) Resulting phase shifts after iteration of the corresponding contributions to the potential. Empirical data points are taken from Ref. [8].

on–shell potential (i.e. the Born term) in this partial wave split into its pole
part, which is generated by the direct nucleon pole, and and the non–pole part.
This non–pole part of the interaction is dominated by correlated 2π exchange in
the ρ channel. Whereas the pole part of the potential is repulsive the non–pole
part is attractive over the whole energy range investigated here. This attraction
is so strong that the iteration of the non–pole part of the potential even leads
to resonant behavior in this partial wave at relatively low energies ($\simeq 1.3$ GeV,
cf. Fig 7 (b)). If this attraction is counterbalanced by the direct nucleon pole
contribution the data (in the elastic region) can be described quantitatively.

For energies above 1.3 GeV, pion–nucleon interaction gets inelastic. According to the 'Review of Particle Properties' [13], the branching ratio of the Roper
resonance into the $\pi\pi N$ channel is 30–40%. Here the dominant contributions
are the $\pi\Delta$ channel (20–30%) and the σN channel (5–15%), where σ denotes an
isoscalar $\pi\pi$ S–wave state.

As a first step, we have investigated the coupled channel system $\pi N - \sigma N$,
where in our model the additional contributions shown in Fig. 9 are taken into
account. The (effective) σNN coupling is taken from Ref. [14]. Self–energy con-

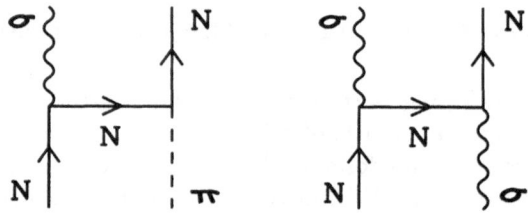

Fig. 9. Additional contributions to our interaction model.

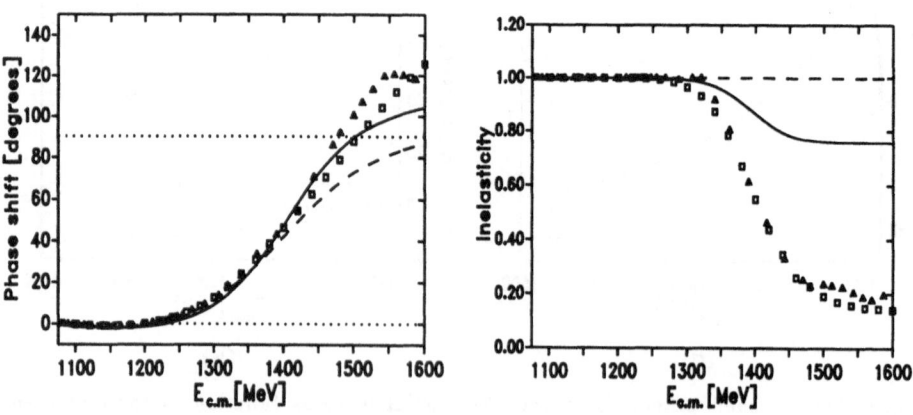

Fig. 10. The $\pi N - \sigma N$ coupled channel calculation (solid lines) compared to the
one–channel calculation (dashed lines). Squares (triangles) denote the empirical information from Ref. [15] (Ref. [8]).

tributions of the σ are implemented in our calculation. As one can see from Fig. 10, inclusion of these additional processes leads to a resonant behavior in the P_{11} partial wave. However, not surprisingly, the model accounts only for part of the inelasticity. Here the additional effect of the reaction channel $\pi\Delta$ remains to be investigated.

5 Summary and outlook

We have presented a meson–exchange model of the πN interaction. It consists, apart from conventional direct and exchange nucleon and delta pole diagrams, of correlated 2π–exchange processes which replace the so far used sharp mass σ– and ρ–exchange contributions. The model provides a quantitative description of πN scattering data in the elastic region. An investigation of coupling to the reaction channel σN indicates that no contribution from a genuine N^* resonance is needed in the P_{11} partial wave to describe the empirical Roper resonance. However, also the effect of the reaction channel $\pi\Delta$ should be taken into account in order to get a correct description of the inelasticities. Work in this direction is under way.

This work was supported in part by the Humboldt senior fellow program, the German Academic Exchange Service (HSP II), and the Science & Technology Cooperation Germany/Canada (KAN-BSC13).

References

1. B.C. Pearce and B.K. Jennings, Nucl. Phys. **A 528** (1991) 655.
2. C. Lee, S.N. Yang, and T.-S. H. Lee, J. Phys. **G 17** (1991) L131.
3. F. Gross and Y. Surya, Phys. Rev. **C 47** (1993) 703.
4. P.F.A. Goudsmit, H.J. Leisi, and E. Matsinos, Phys. Lett. **B 299** (1993) 6; P.F.A. Goudsmit, H.J. Leisi, E. Matsinos, B.L. Birbrair, and A.B. Gridnev, preprint ETHZ–IMP PR/93-2.
5. C. Schütz, J.W. Durso, K. Holinde, and J. Speth, Phys. Rev. **C 49** (1994) 2671.
6. G. Höhler, F. Kaiser, R. Koch, and E. Pietarinen, Handbook of Pion–Nucleon-Scattering, Physics Data 12–1, Fachinformationszentrum Karlsruhe 1979.
7. D. Lohse, J.W. Durso, K. Holinde, and J. Speth, Nucl. Phys. **A 516** (1990) 513.
8. G. Höhler, *"Pion-Nucleon Scattering"*, Landolt–Börnstein Vol.I/9b2, ed. H. Schopper, Springer 1983.
9. R. Koch and E. Pietarinen, Nucl. Phys. **A 336** (1980) 331.
10. T.P. Cheng and R. Dashen, Phys. Rev. Lett. **26** (1971) 594.
11. J. Gasser, H. Leutwyler, and M.E. Sainio, Phys. Lett. **B 253** (1991) 252.
12. B.C. Pearce, K. Holinde, and J. Speth, Nucl. Phys. **A541** (1992) 663.
13. Review of Particle Properties, Phy. Rev. **D 45**, Part 2 (June 1992).
14. J.W. Durso, A.D. Jackson, and B.J. Verwest, Nucl. Phys. **A 345** (1980) 471.
15. R. A. Arndt, VPI analysis FA93, private communication.

This article was processed using the LaTeX macro package with LMAMULT style

πN Working Group Summary

G. Höhler [1], M.E. Sainio [2]

[1]Institut für Theoretische Teilchenphysik, Universität Karlsruhe,
Postfach 6980, D-76021 Karlsruhe, Germany
[2]Department of Theoretical Physics, University of Helsinki,
P.O. Box 9 (Siltavuorenpenger 20 C), FIN-00014 Helsinki, Finland

Abstract: A brief summary of the discussions in the πN working group is given. The individual progress reports will be in separate articles following this summary.

Introduction

The pion-nucleon partial wave analysis is a quite mature field, for a comprehensive summary until about 1982 see [1]. Thereafter e.g. the πN Newsletter has reported progress in the field. The purpose of the analysis is (at least) threefold: (i) At low energies and in the subthreshold region it is important to compare predictions from chiral perturbation theory with results from an analysis of experimental data. After all, the πp system involves the interaction of the lightest baryon with the lightest Goldstone boson, (ii) Spectroscopy of baryon resonances, and analysis in terms of quark models, (iii) Pion-nucleon amplitudes are input for work involving the pion interaction with more complex systems or pion production with other probes. The tools and physical principles involved in the analysis – analyticity, crossing symmetry and unitarity – are well established. There are, however, a number of issues which render the analysis open ended, because of the complexity of the details. Examples of such problems are the question of the isospin breaking effects due to quark mass differences and the contributions of the electromagnetic interaction.

Especially in the context of the πN Σ-term [2] there has been quite some interest in low-energy πN data. Those experiments are, however, difficult, and a certain amount of work will be required to clarify the discrepancies in the existing data. Therefore, much of the discussion in the πN Working Group centred around the analysis of data at low energies and in the Δ-resonance region [3-5]. There are a number of outstanding general issues here: (i) values of the πN threshold parameters (scattering lengths and volumes, effective ranges), (ii) the value of the πN coupling constant, (iii) checking the level of conservation of the isospin symmetry, (*iv*) extrapolation of the amplitudes to the Cheng-Dashen point to determine the Σ-term.

In the Working Group other topics were also discussed: correlated 2π-exchange in πN scattering [6], and the scalar sector in chiral perturbation theory [7]. Some further topics were addressed in a joint session with the Threshold Nucleon, Electro-and Photopion Production Working Group, and they will be discussed in the summary of that session.

Threshold parameters

A very useful way of parametrizing low-energy scattering information is the low-energy expansion in terms of the scattering lengths etc. Of course, there is always the caveat that it is not clear for which range of momenta a given expansion is valid [1]. In any case, dispersion methods provide a well defined and reliable method for extrapolating the amplitudes from the range of energies where there are data down to the physical threshold [1]. In addition to earlier methods for invariant amplitudes [1], dispersion relations for partial wave amplitudes and partial wave projections of fixed-t dispersion relations are important tools [8]. However, with dispersion methods it is quite difficult to determine the uncertainty of the parameters due to the experimental uncertainties. An important check of the validity of the extrapolation is any direct measurement of the threshold parameters. The prime example of such is the determination of the $\pi^- p$ s-wave scattering length from the 1s level shift measurement of pionic hydrogen [3]. The earlier published result [9] had an error of about 4 %, but now the precision has reached the 1 % level $a_{\pi^- p} = 0.0873 \pm 0.0007 \ \mu^{-1}$ [3]. The new experiment has also determined the width of the level and, together with a value for the Panofsky ratio, this can be used to extract the s-wave isovector scattering length $a_{0+}^- = 0.096 \pm 0.008 \ \mu^{-1}$ with the hope of improving the precision by a factor of two due to the analysis of more events. A more accurate result for the s-wave isovector scattering length follows from the Panofsky ratio and a new value given by M.A. Kovash in his talk for the multipole E_{0+} derived from an extrapolation of the data for $\pi^- p \rightarrow \gamma n$ to threshold. The preliminary result for $E_{0+}(\pi^-)$ is $(-34.6 \pm 1.0) \times 10^{-3} \ \mu^{-1}$, where the uncertainty refers to the statistical error only. It is higher by about 5 % than the value used by Spuller et al. [10], so the new isovector s-wave scattering length $a_{0+}^- = 0.092 \ \mu^{-1}$ is also larger than Spuller's value. The new results happen to agree within the errors with the old ones given in [1] and [8]. An interesting recent development is that the lattice calculations start to be able to address questions like the πN scattering lengths [11].

The pion-nucleon coupling constant

The standard value for the πNN coupling constant obtained from the data available in the seventies has been $f^2 = 0.079 \pm 0.002$ [1,12,13]. Then, a number of analyses [14-16] ended up favouring values somewhat lower, in the range $f^2 = 0.075$ to 0.077. Because of the GMO sum rule [17] these authors then arrive at smaller values for the isovector s-wave scattering length. However, if one accepts the preliminary value of Kovash et al., the GMO sum rule gives accidentally the old value $f^2 = 0.079 \pm 0.002$.

Violation of isospin invariance

The only piece of experimental information which has been able to see clearly isospin violation effects at low energies is the splitting of the Δ-resonance parameters between the doubly charged and neutral members of the multiplet. This was first noticed in total cross section data by Carter et al. [18] and confirmed by Pedroni et al. [19]. The result is sensitive to the method applied for the treatment of the electromagnetic interactions. Koch and Pietarinen applied the NORDITA method of [20], which is considered by most authors to be the most reliable one, although there remain some unsolved problems. One can expect to have in the near future in the resonance region new accurate πN elastic scattering data from TRIUMF [5] and charge-exchange data from LAMPF [21], so improved tests of isospin invariance and of the electromagnetic corrections will be possible.

In the low energy region, a valuable piece to the puzzle will be provided by the atomic level shift and width measurements in pionic hydrogen and deuterium. In the case of deuterium the problem is the multiple scattering correction which is about six times larger than the leading term related to the isoscalar s-wave πN scattering length [4]. The theoretical analysis is quite involved and needs dynamical assumptions. Opinions differ at the moment on the precision which can be achieved.

In his talk, S. Weinberg made the remark, that at very low energies reactions which have one neutral pion should show a breaking of the isospin invariance due to the differences of the u and d quark masses. The charge-exchange experiments at 10 and 20 MeV at LAMPF [21] will be of interest for this question.

Conclusion

The results presented at this workshop led to good progress in the determination of πN scattering amplitudes at low energies. Further experiments have been completed and the data will be available in the near future. Others will be carried out soon. The theoretical interpretation is based on chiral dynamics. The analysis of the Σ-term has been discussed in [22].

References

1.   G. Höhler, in Landolt-Börnstein, Vol. 9 b2, ed. H. Schopper (Springer, Berlin, 1983).
2.   J. Gasser, H. Leutwyler: Phys. Reports **87** 77 (1982)
3.   A. Badertscher et al.: these proceedings
4.   E.C. Aschenauer et al.: these proceedings
5.   M. Pavan et al.: these proceedings
6.   C. Schütz: these proceedings
7.   Ulf-G. Meißner: these proceedings
8.   R. Koch: Z. Phys. C **29** 597 (1985); Nucl. Phys. A **448** 707 (1986)
9.   W. Beer et al.: Phys. Lett. B **261** 16 (1991)
10.   J. Spuller et al.: Phys. Lett. B **67** 479 (1977)
11.   M. Fukugita et al.: KEK preprint 94-11 (1994)
12.   R. Koch, E. Pietarinen: Nucl. Phys. A **336** 331 (1980)
13.   G. Höhler: πN Newsletter **3** 66 (1991); **6** 154 (1992)
14.   V. Stoks, R. Timmermans, J.J. de Swart: Phys. Rev. C **47** 512 (1993)
15.   F.G. Markopoulou-Kalamara, D.V. Bugg: Phys. Lett. B **318** 565 (1993)
16.   R.A. Arndt, R.L. Workman, M.M. Pavan: Phys. Rev. C **49** 2729 (1994)
17.   M.L. Goldberger, H. Miyazawa, R. Oehme: Phys. Rev. **99** 986 (1955)
18.   J.R. Carter et al.: Nucl. Phys. B **58** 378 (1973)
19.   E. Pedroni et al.: Nucl. Phys. A **300** 321 (1978)
20.   B. Tromborg et al.: Phys. Rev. D **15** 725 (1977); Helv. Phys. Acta **51** 584 (1978)
21.   M.E. Sadler: private communication to G.H.
22.   M.E. Sainio: these proceedings

MEASUREMENT OF THE STRONG INTERACTION SHIFT AND WIDTH OF THE 1S LEVELS IN PIONIC HYDROGEN AND DEUTERIUM

A. Badertscher[1], E.C. Aschenauer[3], M. Bogdan[1], D. Chatellard[2], J.-P. Egger[2], K. Gabathuler[3], P.F.A. Goudsmit[1], E. Jeannet[2], H.J. Leisi[1], E. Matsinos[1], A.J. Rusi El Hassani[4], H.Ch. Schröder[1], D. Sigg[1], L.M. Simons[3] and Z.G. Zhao[1]

[1] Institute for Particle Physics, ETH Zurich, CH-5232 Villigen PSI
[2] Institut de Physique de l'Université, Rue Breguet 1, CH-2000 Neuchâtel
[3] Paul Scherrer Institut, CH-5232 Villigen PSI
[4] Ecole Mohammadia des Ingénieurs, Rabat, Morocco

1 INTRODUCTION

The pionic hydrogen atom offers the unique possibility to study the πN interaction at zero energy. With a precise measurement of the energy and line shape of a pionic K X-ray line one can determine the strong interaction shift ε_{1s} and width Γ_{1s} of the 1s level. The shift is defined as the difference between the measured X-ray energy and the calculated electromagnetic transition energy, and Γ_{1s} is the total absorption width, including the partial width from radiative capture. These two quantities are directly related to the elastic and charge exchange scattering lengths by Deser's formula given below [1].

$$\varepsilon_{1S} = -\frac{4E_{1S}}{r_B}\, a_{\pi^- p - \pi^- p} \tag{1}$$

$$\Gamma_{1S} = \frac{8E_{1S}}{r_B} q (1 + 1/P)\, (a_{\pi^- p - \pi^\circ n})^2 \tag{2}$$

where E_{1s} is the electromagnetic binding energy of the $\pi^- p$ atom in the 1s state, r_B the Bohr radius, P the Panofsky ratio and q the c.m. momentum of the π°. The scattering lengths in equations (1) and (2) have still to be corrected for electromagnetic and mass splitting effects to obtain the purely hadronic quantities. In the case of deuterium the shift and width can still be related with the Deser formula to the πd scattering lengths; the extraction of the πN scattering lengths involves a theoretical treatment to account for multiple scattering and absorption effects [2].

The 3p-1s X-ray energies to be measured were 2.89 keV for pionic hydrogen and 3.07 keV for deuterium and the shifts measured with earlier experiments are -7.12 ± 0.32 eV (attractive) [3] in hydrogen and +5.5 ± 1.3 eV [4] in deuterium. The widths were expected to be about 0.9 eV and 1 eV [5]; they have been measured for the first time with this experiment. The goal of the experiment was to reach a precision of ±1% for the shift and about ±10% for the width in pionic hydrogen and similar (absolute) precisions for deuterium.

2 EXPERIMENTAL METHOD

An energy measurement of 3 keV X-rays with a precision of $\leq$25 ppm, which is necessary for a 1% measurement of the hydrogen shift, is only possible with a high resolution crystal spectrometer. The principle of the crystal spectrometer experiment was to measure the Bragg angle *difference* of the pionic K_β line to a close-by calibration line. As a calibration line we used the (electronic) argon K_α doublet, which was recently remeasured at NIST with a precision for the $K_{\alpha 1}$ energy of 6 ppm [6].

The experimental setup consisted of three main parts: The cryogenic gas target mounted in the center of the cyclotron trap (a device to increase the pion stop rate in gaseous hydrogen or deuterium), the crystal assembly mounted on a turn table and consisting of six cylindrically bent quartz(110) crystals [7] and the CCD X-ray detector [8] mounted on a X-Y table.

For the determination of the width the measured line shape had to be deconvoluted with the instrumental resolution function. The resolution function was measured with the pionic Be(4f-3d) transition because it has a negligible natural line width and its energy is close to the pionic hydrogen (deuterium) transition.

3 RESULTS

Figure 1 shows the measured X-ray lines obtained from the on-line analysis.

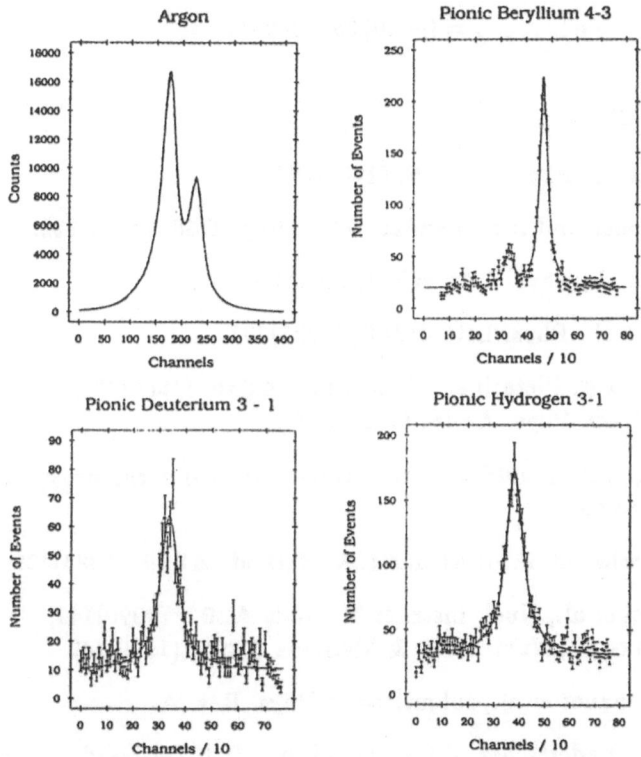

Figure 1. Measured X-ray lines. The beryllium and the Ar K_α are calibration lines.

The instrumental resolution was obtained from a fit of the Be(4f-3d) line and is 0.66 ± 0.03 eV (FWHM). The small peak in the beryllium line is from the parallel transition $\pi^- \text{Be}(4d\text{-}3p)$.

Preliminary values for the shift and width of the 1s states deduced from the pionic hydrogen and deuterium data can be given: $\epsilon_{1s}^{\pi^- p} = -7.09 \pm 0.06$ eV (attractive) and $\Gamma_{1s}^{\pi^- p} = 0.97 \pm 0.15$ eV for hydrogen and $\epsilon_{1s}^{\pi^- d} = +2.46 \pm 0.10$ eV and $\Gamma_{1s}^{\pi^- d} = 0.96 \pm 0.20$ eV for deuterium. For the width a small correction for a potential Doppler broadening due to Coulomb deexcitation [9] has been applied and a systematic error added to the experimental error. Inserting the hydrogen values into Deser's formula and correcting the scattering lengths for electromagnetic and mass splitting effects [10], one gets the following preliminary values for the hadronic scattering lengths:

$$a_{\pi^- p - \pi^- p}^h = b_0 - b_1 = \frac{1}{3}\left(2a_1 + a_3\right) = (0.0873 \pm 0.0007)m_\pi^{-1} \tag{3}$$

$$a_{\pi^- p - \pi^0 n}^h = \sqrt{2}b_1 = \frac{\sqrt{2}}{3}\left(a_3 - a_1\right) = (-0.136 \pm 0.011)m_\pi^{-1} \tag{4}$$

From the shift in deuterium a preliminary value for the deuteron elastic scattering length is obtained:

$$a_{\pi^- d - \pi^- d}^h = (-0.0265 \pm 0.0011)m_\pi^{-1}. \tag{5}$$

References

[1] S. Deser et. al. Phys. Rev. **96**, (1954)774.

[2] A.W. Thomas and R.H. Landau, Phys. Rep. **C58**, (1980)122.

[3] W. Beer et al., Phys. Lett. **B261**, (1991)16.

[4] E. Bovet et al., Phys. Lett. **B153**, (1985)231.

[5] R. Koch and E. Pietarinen, Nucl. Phys. **A336**, (1980)331;
R. Koch, Nucl. Phys. **A448**, (1986)707.

[6] J. Schweppe, National Institute of Standards and Technology, USA, private communication.

[7] A. Badertscher et al., Nucl. Instr. & Methods **A335**, (1993)470.

[8] G. Fiorucci et al., Nucl. Instr. & Methods **A292**, (1990)141;
D. Varidel et al., Nucl. Instr. & Methods **A292**, (1990)147.

[9] E.C. Aschenauer et al., submitted to Phys. Rev. A.

[10] D. Sigg, A. Badertscher, P.F.A. Goudsmit, H.J. Leisi and G.C. Oades, in preparation.

IS ISOSPIN SYMMETRY VIOLATED IN THE PION-NUCLEON SECTOR AT THRESHOLD?

E.C. Aschenauer[3], A. Badertscher[1], M. Bogdan[1], D. Chatellard[2], J.-P. Egger[2], K. Gabathuler[3], P.F.A. Goudsmit[1], E. Jeannet[2], H.J. Leisi[1], E. Matsinos[1], A.J. Rusi El Hassani[4], H.Ch. Schröder[1], D. Sigg[1], L.M. Simons[3] and Z.G. Zhao[1]

[1] Institute for Particle Physics, ETH Zurich, CH-5232 Villigen PSI
[2] Institut de Physique de l'Université, Rue Breguet 1, CH-2000 Neuchâtel
[3] Paul Scherrer Institut, CH-5232 Villigen PSI
[4] Ecole Mohammadia des Ingénieurs, Rabat, Morocco

1 INTRODUCTION

The recently performed crystal-spectrometer X-ray experiments [1] allow (for the first time) to address the problem of isospin-symmetry violation in the $\pi - N$ sector at at threshold *directly*, i.e. without extrapolation to threshold from data obtained at higher energies where s- and p-wave interactions are mixed [2]. The way to detect violation of the isospin symmetry in general would be as follows. One assumes isospin symmetry to hold, and then determines constraints to the two s-wave scattering lengths (e.g. $b_0 = \frac{1}{3}(a_1 + 2a_3)$ and $b_1 = \frac{1}{3}(a_3 - a_1)$) from independent measurements in at least three different reactions. Inconsistencies among different constraints should indicate violation of the isospin symmetry.

2 PIONIC HYDROGEN

Two constraints are obtained from the strong-interaction shift ε_{1S} and the width Γ_{1S} of the $1S$ level [1]. These quantities are related to the (total) elastic and single-charge-exchange scattering lengths by Deser's formula (eqs. (1) and (2) of ref. [1]). The purely hadronic scattering lengths are obtained by applying to the total scattering lengths the electromagnetic and mass-splitting corrections [2]. The resulting constraints are displayed in Fig. 1 (area marked as (a) from the shift and area (b) from the width).

3 PIONIC DEUTERIUM

A third constraint to the $\pi - N$ scattering lengths is obtained from the strong-interaction shift of the $1S$-level of pionic deuterium. Deser's formula relates this shift to the (real) π^--deuteron elastic scattering length $a_{\pi-d}$ [2]. Following Thomas and Landau [4], we write the (hadronic) $\pi^- - d$ elastic scattering length as the sum of a single-scattering part

$$a_{\pi^- - d}^{(0)} = 4\frac{m_p + m_\pi}{m_d + m_\pi} b_0 \quad , \tag{1}$$

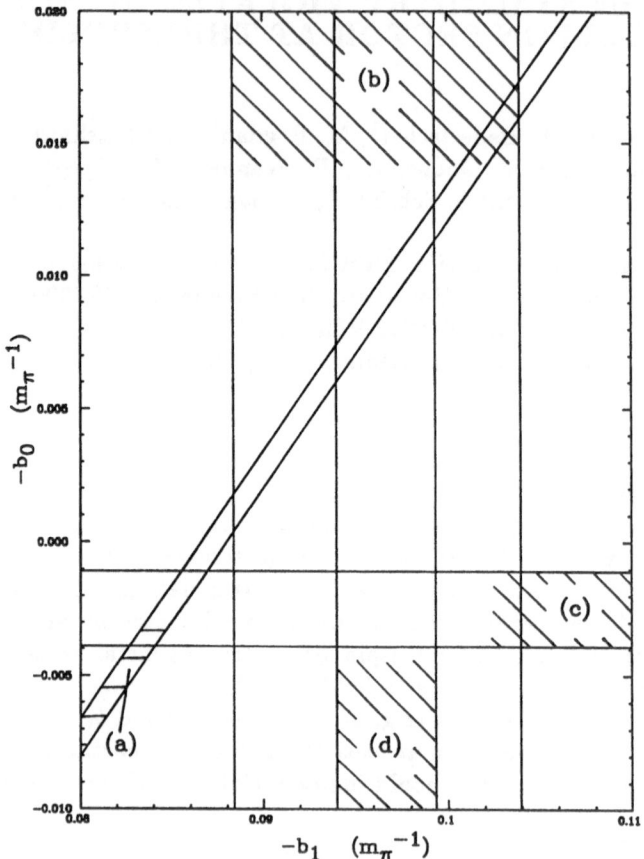

Figure 1. Constraints to the scattering lengths b_0 and b_1 derived from pionic hydrogen (areas (a) and (b)), from pionic deuterium (area (c)) and single-charge-exchange (area (d); ref. [3]).

where $m_{p,\pi,d}$ are the proton, charged pion and deuteron masses, and a higher-order term which is (according to ref. [4]) given by

$$a_{\pi-d}^{(\text{high.ord.})} = -0.0318 \pm 0.0028 \ m_\pi^{-1} \ . \tag{2}$$

Using the value for $a_{\pi-d}$ (eq. (5) of ref. [1]) and (2), we find from (1) a constraint to b_0 (area (c) in Fig. 1). It is to be noted that the uncertainty in b_0 is dominated by the theoretical uncertainty in eq. (2), rather than by the experimental error in the measured transition energy.

4 DISCUSSION

The following conclusions are based on the assumption that the calculations of Thomas and Landau [4] are complete.

The constraints on b_0 and b_1 from our pionic hydrogen and deuterium preliminary results are marginally consistent with each other ((a), (b) and (c) in Fig. 1). From this we conclude that *our present measurements are not inconsistent with isospin symmetry*. Recently, we have gathered more data on pionic hydrogen which will enable us to obtain a considerably improved determination of b_1 from the width of the $1S$ level (by about a factor of two).

Other determinations of s-wave scattering lengths - in contrast to ours - all rely on extrapolations to threshold based on data taken at higher energies. If we include as a constraint the precise b_1 determination of Siegel and Gibbs [3], which was deduced from single-charge-exchange data (area (d) in Fig. 1), then there is a clear indication for violation of isospin symmetry.

References

[1] A. Badertscher et al., *"Measurement of the strong-interaction shift and width of the 1S levels in pionic hydrogen and deuterium"*, preceeding contribution.

[2] A. Badertscher et al., ETHZ-IPP preprint PR-94-10 (1994).

[3] P.B. Siegel and W.R. Gibbs, Phys. Rev. **C33** (1986) 1407.

[4] A.W. Thomas and R.H. Landau, Phys. Rep. **C58**, (1980)122.

Is $f^2_{\pi NN}=0.0755$? A look at VPI FA93 and new Experimental Results

Marcello M. Pavan

Department of Physics, University of British Columbia,
Vancouver, B.C., Canada, V6T-1Z1

1 Introduction

The controversy surrounding the values of the pion-nucleon coupling constant
(f^2) and sigma term ($\Sigma_{\pi N}$) has helped re-ignite experimental and theoretical
interest in low energy πN physics. (To whit: after the first πN Newsletter in
1983, the next eight have appeared since May 1990). This article will direct its
attention to f^2 and on some very recent developments which support a value
~5% smaller than the standard result [1]: f^2 ~0.0755 vs. 0.0790.

2 The VPI FA93 Partial-Wave Analysis

The πN partial-wave analyses (PWA) of Arndt et al. at the Virginia Polytechnic Institute (VPI) have been subject to considerable scrutiny over the years
(e.g. [2]), and in particular the SM90 solution [3], which formed the basis for
their original claim of a small f^2, consistent with the (also controversial) Nijmegen results from NN PWAs [4]. In the SM90 analysis, f^2 and its uncertainty
were inferred [5] from various dispersion relations (DRs), but the absence of DR
constraints on the solution prompted a litany of criticisms.

In response to these criticisms, we (the VPI group and this author) have
devised a method to determine f^2 from VPI partial-wave analyses constrained
by fixed-t and forward DRs, in a manner which establishes a natural criterion
for determining its value and uncertainty from πN scattering data. The details
have been published elsewhere [6], so here the method will be clarified briefly by
responding to the most persistent criticisms of SM90 relevant to extracting f^2.

1. Fixed-t and forward DR constraints were not used: Constraints
on the partial-wave solution are imposed at many kinematical points (T_{lab} <700
MeV, t >-0.3 GeV^2/c^2) by the fixed-t DRs for the invariant $B_\pm(\nu, t)$ amplitudes
(via the Hüper plot), and at $t=0$ by the DRs for $C^\pm(\nu)$. After each fit cycle, the
change in the real part of the amplitude required to satisfy the DR is determined
and then input as 'quasi-data' into the next iteration. The procedure is repeated
until the fit stabilizes. **2. Determinations of f^2 via the Hüper plot were
t-dependent:** As f^2 is fixed before each set of runs (see point 6.), the coupling

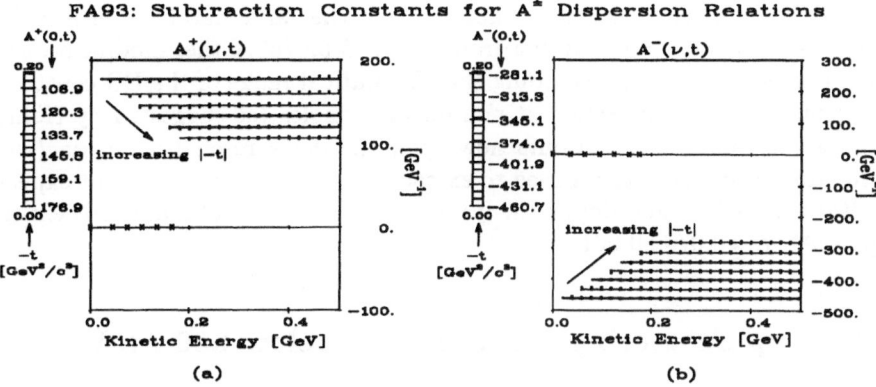

Fig. 1. These dispersion relations were *not* used as constraints in VPI FA93, nonetheless, the consistency shown is superior to that of the Karlsruhe solutions (e.g. [1]).

in the final solution is independent of t by construction. **3. The VPI representation of the partial-waves does not contain the pion pole, so f^2 cannot be reliably extracted:** DRs which explicitly contain the Born term are used to extract f^2. **4. The form of the Coulomb barrier correction in the VPI analysis could contribute to a systematic shift in f^2:** As a brute-force check, the Coulomb barrier *and* rotation corrections were reintroduced into the amplitudes, but the coupling extracted from the Hüper plot was not significantly different. This is not surprising, since each 'side' of the Hüper plot is sensitive mainly to π^+ or π^- data, which feel the Coulomb effects with opposite sign. One would expect then that the intercept, which gives f^2, will not be affected. **5. Effects of the Δ mass splitting are absent:** Provisions are made for a separate P33 for π^+ and π^-. However, the effect makes little difference to the result. **6. Uncertainties assigned to f^2 from the Hüper plots are too small:** f^2 is *fixed* prior to each analysis, but Höhler has claimed [7] that the data can *force* a value of f^2 different from the input. On the contrary, we find that stable, consistent solutions are obtained for a *range* of 'reasonable' f^2 inputs, and the value extracted is *always* the same as the input. Moreover, the χ^2s characterizing these solutions (i.e. constraint, π^+, π^-, scx, total) have a smooth parabolic dependence on f^2 with a pronounced minimum. *This suggests a natural criterion for determining f^2 and its uncertainty:* the 'best' value gives the lowest overall χ^2, and the uncertainty is estimated from the depth and location of the minima in each charge channel and the constraints.

Following these procedures, an 'optimal' solution (FA93) was determined [6] with $f^2 = 0.076 \pm 0.001$. FA93 exhibits excellent consistency with the constraint DRs, as well as with the DRs for the $A^\pm(\nu, t)$ amplitudes, despite their *not* being used as constraints (see fig 1). In fact, the agreement is superior to that shown by the Karlsruhe solutions [1]. The technique can be extended to allow 'mapping out' χ^2 as a function of f^2 and other DR parameters, e.g. the scattering length

$a(\pi^- p)$. Many such exploratory efforts (using the same database as FA93) recently have shown f^2 to be stable around 0.0755-0.0760. Other modifications are being pursued, including extending the kinematical range of the DR constraints and using $C^{\pm}$ DR constraints at non-forward angles (as suggested by Höhler). However, as the DR constraints now span the delta resonance region (which dominates in the amplitudes used to extract f^2), these changes are not expected to alter the basic conclusion, namely: from the VPI πN database as it exists in Sept.1994, $f^2 = 0.076 \pm 0.001$.

3 New πp Scattering Experiments below 300 MeV

New πp scattering data at energies around the first resonance are beginning to emerge which could have consequences for f^2. Refer to figure 2, which shows preliminary data from two TRIUMF groups (Brack IV [8] solid squares; Pavan [9] solid circles, stars, diamonds) plotted as ratios to the KH80 solution. The open symbols (Brack I [10] squares; Bussey [11] circles; Sadler [12] stars; Frank [13] diamonds) are published data which are included in the FA93 database. The FA93 solution is plotted as the solid double line.

The Brack IV data, taken using an active target to detect the recoil protons in coincidence with the scattered pions, overlap consistently with Brack I, which used solid (CH_2) targets and scintillator telescopes to coincidently detect the scattered pions and recoil protons. Both sets have $\delta \sim 2.5\%$ (1σ) normalization uncertainties. At 141 MeV, these is also agreement with all three types of Pavan, et al. measurements ($\delta \sim 1.5\%$) (which also internally consistent): coincidence measurements with scintillator telescopes using both solid and liquid targets, and a single arm measurement with a liquid target detecting only the scattered pion. This agreement between Brack and Pavan (3 experiments, 4 different techniques) strengthens the credibility of all the Brack data. Furthermore, the agreement of Pavan, et al. with recent Sadler, et al. data near 267 MeV, and with the unitarity bound for 193 MeV π^+ to within $\sim 1\%$ (not shown), suggests that these data are also well under control. In general, FA93 describes the Pavan, et al. results at all energies much better than KH80.

One notices in figure 2 striking differences between the new π^- data below the resonance and KH80, (up to 20% at 117 MeV!), and large deviations as well for π^+ (10% at 87 MeV down to 4% at 141 MeV). This signifies S11 and P33 partial-wave strength significantly smaller than in the KH80 predictions. This decrease is confirmed in the FA93 solution up to 300 MeV. But below the resonance, the FA93 solution compromises between the Brack I and Bussey data sets (which dominate here). Following inclusion of the new results, a new VPI analysis should favour the TRIUMF data, and S11 and P33 will decrease. Consequently, f^2 somewhat lower than 0.076 might follow.

Other important new experiments should be presenting results soon which will have something to say about f^2: from LAMPF, charge exchange differential cross sections from 140 to 263 MeV [14], plus $\pi^+ p$ partial total cross sections from

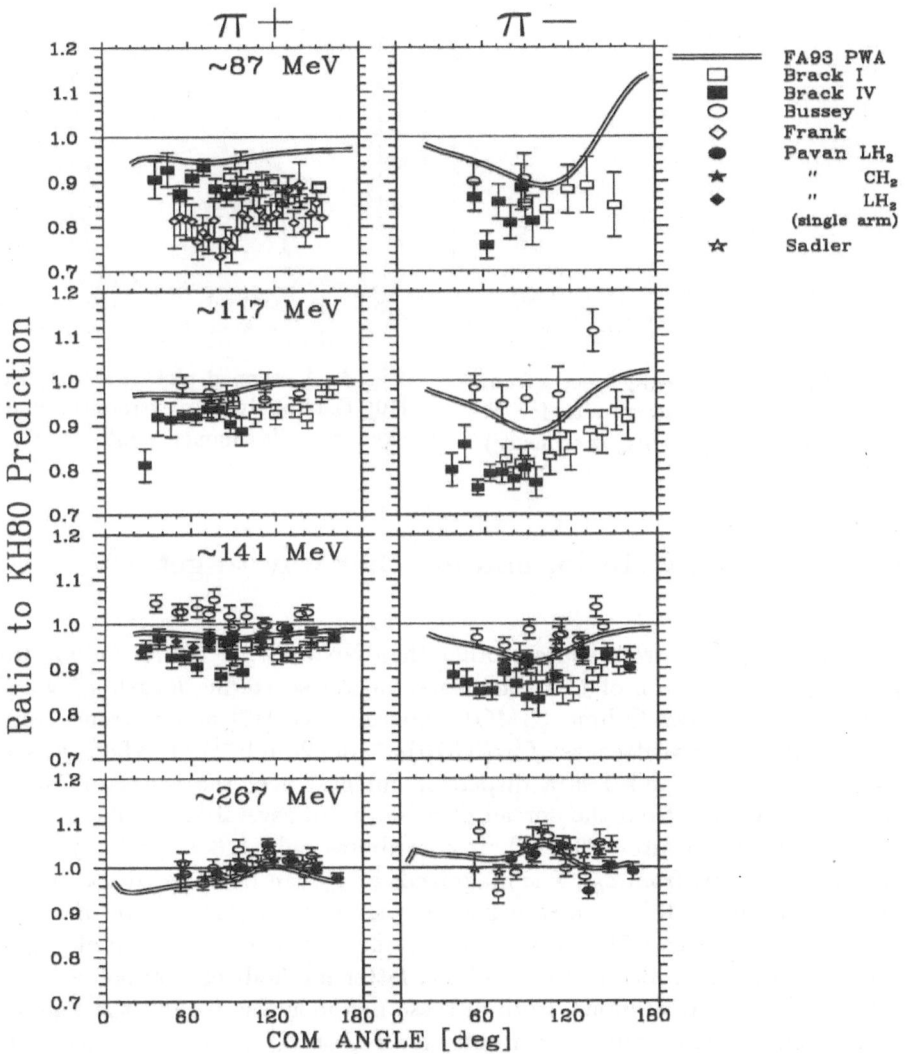

Fig. 2. Differential cross sections for $\pi^{\pm}p$ scattering as a ratio to the KH80 prediction. The filled in points are preliminary data which do *not* appear in the VPI database.

30 to 600 MeV [15]; and from TRIUMF, $\pi^{\pm}p$ polarization data from 30 to 240 MeV [16]. Preliminary results from [15] tend to support the TRIUMF differential cross sections. With the new data, there will soon be almost a complete set (differential, partial total, polarization) of quality data for all charge channels at energies from 30 to 500 MeV, which should provide a sound foundation for future f^2 extractions.

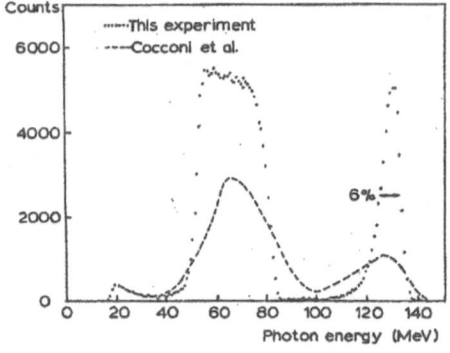

Fig. 3. Stopped $\pi^- p$ γ-ray spectra from the best two published results for P. (reprinted from [20]='This experiment')

Fig. 4. A stopped pion spectrum from [10]. Data analysis to extract the Panofsky ratio will commence fall '94 [21].

4 The Panofsky Ratio, and another way to get f^2

A method to get f^2 from sources other than NN and πN PWAs follows from the intimate connection of f^2 to the isovector πN scattering length a_0^- via the Goldberger-Miyazawa-Oehme (GMO) sum rule (see [17] and references cited therein). It can be written as: $f^2=(0.57\mu)\cdot a^- - (0.025 \text{ mb}^{-1})\cdot$ J; where μ is the charged-pion mass, and J is a dispersion integral over the isovector forward amplitude. Judging from the spread of values from several Karlsruhe and VPI solutions, J=-1.065 mb $\sim$2% (1σ). It contributes only 1/3 to f^2, so f^2 is determined primarily from a_0^-. The magnitude of a_0^- can be accessed directly, via the Deser formula [18], by measuring the pionic-hydrogen 1s state line width Γ_{1s}, or indirectly (see [17]), by determining the threshold electric dipole E_0^+ for π^- photoproduction on the neutron. The latter method requires some extrapolation and theoretical input. Both of these methods also require the Panofsky ratio P. There are no published results for Γ_{1s} or $E_0^+(\pi^- p)$ at threshold, but new preliminary results have been reported at this workshop by Badertscher (pionic hydrogen) [18] and Kovash (electric dipole) [19]. After accumulation of more statistics in [18], and the resolution of a problem in [19] (the threshold E_0^+ amplitude extracted via the angular distributions and the total cross sections are inconsistent), an uncertainty of about 3-4% for $|a_0^-|^2$ is ultimately expected from each, or 2-3% from the weighted average.

Given the potential for a useful determination of f^2, it seemed worth reviewing the reliability of the 20 year old measurement yielding the canonical value of the Panofsky ratio, and to see whether a new measurement is desirable. Refer to figure 3, which shows the stopped $\pi^- p$ γ-ray energy spectrum as measured by Spuller, *et al.* (see [20] for details and references), from which P = 1.546 ± 0.009 was extracted, compatible with earlier results. The largest corrections required to get P (roughly 0.5*N('π^0 box')/N(γn peak)) came from: a) the low energy tail ($\sim$1.5-3%), arising from $e^\pm$ production followed by bremstrahlung in the collimators used to obtain decent γ-ray energy resolution; b) in-flight reactions ($\sim$0.6%), due mainly from the broad energy distribution in the large LH_2 target from heavily energy-degraded beams ($T_\pi \sim$ 30 MeV at the front surface); and c) empty target subtraction, where residual H_2 gas gave rise to corrections of 1-2% to the radiative capture peak. Corrections arising from the spectral tails were relatively small ($\sim$ 0.2%), but the response function was measured only for 129 MeV gammas. In total, corrections of about 2% (1.5%) and 4% (3%) for the π^0 (γ) yields in their two runs, respectively, resulted in an estimated 0.4% systematic uncertainty, for an overall (systematic + statistical) uncertainty of 0.6%. Despite less than ideal conditions, the systematics in the experiment seem manageable, even if the uncertainty assigned is a little optimistic (as is often the case). There appears little cause to place the result in great doubt.

Serendipitously, a new measurement using modern facilities might already exist. The Kovash, *et al.* experiment [19], while designed to measure in-flight radiative pion capture, also employed simultaneously a separate stopped-pion trigger for calibration. It ran on the TRIUMF M13 channel with narrow beams of good momentum resolution at three energies below 25 MeV, using a thin, low-mass LH_2 target, and a NaI crystal with excellent energy resolution. A sample spectrum appears in figure 4, showing much improved energy resolution and a suppressed low energy tail compared to figure 3. With these (and other) improvements, a very good determination of the Panofsky ratio could be possible. Data analysis will commence in fall 1994 [21].

With new results for a_0^- and the Panofsky ratio imminent, it could be possible in the near future to determine the πNN coupling f^2 to about ± 0.002, which would be a good check of the πN and NN PWA determinations.

5 Concluding Remarks

The claim of a pion-nucleon coupling constant 4-5% below the long-assumed value of 0.079 is now on firm footing following the VPI FA93 partial-wave analysis [6], but this is not as controversial as it once was. Calculations of deuteron properties from Bonn-like NN potentials using new tensor-force generating mechanisms [22] *can* accommodate 0.075. Bugg [23] has found 0.0771$\pm$0.0014 from a PWA using the Hamilton DR, and (with Machleidt [24]) 0.0765$\pm$0.002 from an analysis of high NN partial-waves; so one of the original authors of 0.079 and a relentless scrutinizer of VPI analyses appears now to favour a lower coupling. The Nijmegen NN PWA result [4] (0.0745$\pm$0.0005) is still lower than the above determinations, but consistent with the current status of f^2 from the GMO sum rule [17]. In the future, the inclusion of new TRIUMF and LAMPF πN scattering data into the database may conceivably lower the VPI value as well. With all this in mind, this author recommends using f^2=0.0755$\pm$0.0015, and one hopes that the uncertainty can be substantially reduced in the near future.

References

1. R. Koch, E. Pietarinen: Nucl. Phys. **A336** 331 (1980)
2. G. Höhler, Proc. Few Body XII, Vancouver (July 1989), Nucl. Phys. **A508** 525 (1990)
3. R.A. Arndt, *et al.*, Phys. Rev. **D43** 2131 (1991)
4. V. Stoks, R. Timmermans, J.J. de Swart: Phys. Rev. **C47** 512 (1993)
5. R.A. Arndt, *et al.* Phys. Rev. Lett. **65** 157 (1990); R.A. Arndt, *et al.* Phys. Rev. **D44** 289 (1991)
6. R.A. Arndt, R.L. Workman, M.M. Pavan, Phys. Rev. **C49**, 2729 (1994)
7. G. Höhler, private communication ; see also G. Höhler, *et al.*, Handbook of πN Scattering (Fachinformationszentrum, Karlsruhe, Germany) p.431
8. J.T. Brack, *et al.*, submitted to Phys. Rev. C, Sept.1994
9. M.M. Pavan, *et al.*, πN Newsletter No.8, Proc. 5th Intl. Symposium on Meson-Nucleon Physics, Boulder, Colorado, (Aug 1993) eds. G. Höhler, B. Nefkens, W. Kluge (1993); n.b. final results due fall 1994
10. J.T. Brack, *et al.*, Phys. Rev. **C34**, 1771 (1986)
11. P. J. Bussey, *et al.*, Nucl. Phys. **B58** 363 (1973)
12. M. E. Sadler, *et al.*, Phys. Rev **D35** 2718 (1987)
13. J.S. Frank, *et al.*, Phys. Rev **D28** 1569 (1983); n.b. the π^- data not shown due to its very large normalization uncertainty
14. M.E. Sadler, *et al.*, πN Newsletter No.9, Proc. 5th Intl. Symposium on Meson-Nucleon Physics, Boulder (Aug 1993) in eds. B. Nefkens, M. Clajus (1993)
15. B. Kriss, *et al.*, πN Newsletter No.8
16. Measurements using the CHAOS spectrometer commence Sept 1994 (G. Hoffman, private communication)
17. R.L. Workman, R.A. Arndt, M.M. Pavan: Phys. Rev. Lett. **68** 1653 (1992)
18. A. Badertscher, *el.*, these proceedings.
19. M. Kovash, these proceedings; see also Kailin Liu, Ph.D. Thesis University of Kentucky 1994 (unpublished)
20. J. Spuller, *et al.*: Phys. Lett. B **67** 479 (1977); also J. Spuller: Ph.D. Thesis U.B.C. 1977 (unpublished)
21. M. Kovash, private communication
22. R. Machleidt, G.Q. Li, πN Newsletter No.9; G. Janssen, K. Holinde, J. Speth, Phys. Rev. Lett. **73** 1332 (1994)
23. F.G. Markopoulou-Kalamara, D.V. Bugg, Phys. Lett. **B318** 565 (1993)
24. D.V. Bugg, R. Machleidt, Los Alamos preprint nucl-phys:9404017

Part VI: Electromagnetic Meson Production

Threshold Pion Photo- and Electroproduction in Chiral Perturbation Theory

Véronique Bernard

Centre de Recherches Nucléaires et Université Louis Pasteur de Strasbourg,
Physique Théorique, Bat. 40, BP28, 67037 Strasbourg Cedex, France

1 Introduction

Threshold pion photo- and electroproduction off nucleons allow us, by the use of a well-understood probe, to test our understanding of the strong interaction at low energies, i.e. in the non–perturbative regime. Over the last few years, renewed interest in these reactions has emerged. This was first triggered through precise new data on neutral pion photoproduction [1] which lead to a controversy about their theoretical interpretations. Furthermore, new data on π^0 electroproduction [2] being obtained with an unprecedented accuracy have given further constraints on the understanding of these fundamental processes in the non–perturbative regime of QCD. While the study of these reactions, also with charged pions in the final state, continues on the experimental as well as on the theoretical side, complementary information can be gained from the two-pion production process $\gamma N \rightarrow \pi\pi N$, where N denotes the nucleon and γ the real or virtual photon. Presently available data on this reaction focus on the resonance region (excitation energies close to the first strong resonance excitation of the nucleon, the $\Delta(1232)$) and above, however, an attempt to measure the $\pi^0\pi^0 p$ channel very close to threshold is actually being made in Mainz [3].

I will report here on a systematic analysis of the processes $\gamma^{(*)} N \rightarrow \pi N$ and $\gamma N \rightarrow \pi\pi N$ in the threshold region making use of baryon chiral perturbation theory (CHPT). Other aspects of nucleon chiral perturbation theory are discussed in these proceedings by Ulf-G. Meißner. In general, CHPT allows one to systematically investigate the strictures of the spontaneously broken chiral symmetry of QCD. It is based on the observation that in the three flavor sector of QCD, the quark masses are small and that the theory in the limit of vanishing quark masses admits an exact chiral symmetry. The latter is dynamically broken which leads to the appearance of massless pseudoscalar excitations, the Goldstone bosons. In the real word, the quark masses are not exactly zero and thus the Goldstone bosons acquire a small mass. The interaction of these particles with each other and matter fields like e.g. the nucleons are weak at low energies as mandated by Goldstone's theorem. This fact is at the heart of CHPT which amounts to a systematic and simultaneous expansion of the QCD Green functions in small momenta and quark masses. To perturbatively restore unitarity it is mandatory to consider pion loop diagrams. Here we will work in the one–loop approximation which has been shown to be of sufficient accuracy for many threshold phenomena. For a review see H. Leutwyler [4] and Ulf-G. Meißner [5].

Talk given at the workshop on Chiral Dynamics: Theory and Experiments, MIT, Cambridge, USA, July 25 - July 29, 1994.

2 Effective Lagrangian

In this section I will be rather short. For more details see A. Manohar and Ulf-G. Meißner, these proceedings. To systematically work out the consequences of the spontaneously broken chiral symmetry at low energies, one makes use of an effective Lagrangian of the asymptotically observed fields, in our case the Goldstone bosons (pions) and the nucleons (*i.e.* proton and neutron). We work in flavor SU(2) and mostly in the isospin limit $m_u = m_d = \hat{m}$ and consider terms up to and including $\mathcal{O}(q^4)$ where q denotes a generic small momentum, in the heavy mass formulation. This formulation has the advantage that there is a one-to-one correspondence between loops and powers of q. Indeed, this is not the case in the relativistic calculation where things are complicated by the fact that the nucleon mass does not vanish in the chiral limit. The effective Lagrangian can be written as:

$$\mathcal{L}_{eff} = \mathcal{L}_{\pi N}^{(1)} + \mathcal{L}_{\pi N}^{(2)} + \mathcal{L}_{\pi N}^{(3)} + \mathcal{L}_{\pi N}^{(4)} + \mathcal{L}_{\pi\pi}^{(2,4)} \tag{2.1}$$

where $\mathcal{L}_{\pi N}^{(1)}$ represents the non–linear σ model coupled to nucleons. Four parameters enter the theory to lowest order, namely the pion decay constant F, the axial coupling constant $\mathring{g}_A$, the nucleon mass $\mathring{m}$ ($\mathring{m}$ only enters implicitly through the non-relativistic energy–momentum relation $p = \mathring{m}v$. It does not appear in the Lagrangian to lowest order) and the leading term in the quark mass expansion of the pion mass M_π. F, $\mathring{g}_A$ and $\mathring{m}$ denotes quantities in the chiral limit, thus differing from the physical ones through terms proportional to the current quark mass. The vertices from $\mathcal{L}_{\pi N}^{(1)}$ needed to perform a one–loop calculation of photo- and electroproduction are the pseudovector πNN vertex and the well-known Weinberg and Kroll Ruderman terms. Notice that working in the Coulomb gauge $\epsilon_0 = 0$ (as we will do in the following) there is no direct coupling of the photon to the nucleon (proportional to $\epsilon \cdot v$). This of course simplifies the calculation since many diagrams vanish. $\mathcal{L}_{\pi N}^{(2,3,4)}$ contain $1/m$ corrections and counterterms. The a priori unknown coefficients of these counterterms are the so-called low energy constants. For a detailed discussion of their estimates see G. Ecker, these proceedings. I will come back to the ones entering the particular reactions discussed here in section 3.1. Note that $\mathring{\kappa}_{S,V}$, the isoscalar and isovector anomalous magnetic moments of the nucleon in the chiral limit, enter $\mathcal{L}_{\pi N}^{(2)}$ as such low energy constants. Let me finally point out that another simplification arises in the heavy mass formulation from the fact that the algebra only involves the velocity operator v and the covariant spin operator S. For more details, see for example ref.[6].

3 Threshold Pion Photoproduction

Let us consider the reaction $\gamma(k) + N_1(p_1) \rightarrow \pi^a(q) + N_2(p_2)$ with $N_{1,2}$ denoting protons and/or neutrons and 'a' refers to the charge of the produced pion. In the case of real photons ($k^2 = 0$) one talks of photoproduction whereas for virtual photons (radiated off an electron beam) the process is called electroproduction. Of particular interest is the threshold region where the photon has just enough energy to produce the pion at rest or with a very small three-momentum. In this kinematical regime, it is advantageous to perform a multipole decomposition since at threshold only the S-waves survive. These multipoles are labelled E, M, $L_{l\pm}$, where E, M, L stands for electric, magnetic, longitudinal respectively (L of course does not appear in photo-production where the photons are transverse), $l = 0, 1, 2, \ldots$ the pion orbital angular momentum and the $\pm$ refers to the total angular momentum of the pion-nucleon system, $j = l \pm 1/2$. They parametrize the structure of the nucleon as probed with low energy photons.

3.1 CHPT calculations

Let us first concentrate on the photoproduction case. We will consider here only the neutral channel. A detailed discussion of the charged ones in the relativistic formalism is given in Ref.[7] and the experimental status is reported by J. Bergstrom and M. Kovash (these proceedings). The T-matrix depends on four amplitudes and takes the following form close to threshold where it is legitimate to keep only the S- and P-waves:

$$T \cdot \vec{\epsilon} = i\vec{\sigma} \cdot \vec{\epsilon}(E_{0+} + \hat{k} \cdot \hat{q}P_1) + i\vec{\sigma} \cdot \hat{k}\, \vec{\epsilon} \cdot \hat{q}\, P_2 + (\hat{q} \times \hat{k}) \cdot \vec{\epsilon}\, P_3 \qquad (3.1)$$

The quantities $P_{1,2,3}$ represent the following combinations of the three P-waves, E_{1+}, M_{1+} and M_{1-},

$$\begin{aligned}
P_1 &= 3E_{1+} + M_{1+} - M_{1-} \\
P_2 &= 3E_{1+} - M_{1+} + M_{1-} \\
P_3 &= 2M_{1+} + M_{1-}
\end{aligned} \qquad (3.2)$$

These four amplitudes are easily calculable within CHPT. Let us first discuss the $\mathcal{O}(q^3)$ results. I should point out that since ϵ, the polarization vector of the photon, is a quantity of order q in the chiral counting (it enters the covariant derivative at the same level as ∂_μ) E_{0+} and the P_i are therefore quantities of order q^2.

- At tree level one has contributions from $\mathcal{L}_{\pi N}^{(2,3)}$ which bring corrections of order E/m^2 and E^2/m^3, respectively, where E is a generic small momentum which in the present case can either be ω, the π^0 center of mass energy, $|\vec{q}|$ its three momentum (for the P-waves) and M_π.
- Loops start at this order. The well-known rescattering graph and the very important, as we will see later, triangle graph contribute to E_{0+} while, as shown in Fig.1 two other graphs contribute to the P-waves. It turns out that these loop contributions are finite, of the type E^2/F_π^3 and cancel for P_3.
- Three counterterms appear at $\mathcal{O}(q^3)$. One, κ_p, the anomalous magnetic moment of the proton in the chiral limit, enters $\mathcal{L}_{\pi N}^{(2)}$ as discussed in the previous section. Note that in contrast to the relativistic result [8] κ_p is almost entirely given by the counterterm. It contributes to all the multipoles except P_3 at variance with the two other counterterms which only contribute to P_3. These, b_{22} and b_{14} which come from the lagrangian $\mathcal{L}_{\pi N}^{(3)}$, have been first written down by Ecker [9]. They are infinite but they enter through the combination $b = 4b_{14} + g_A b_{22}$ which just cancels these infinities. Effectively there is thus only one counterterm b entering P_3. There are two ways of determining these counterterms (both will be

investigated here). The first one which is in the spirit of CHPT, is to fit them to the total and differential photocrossection. One then *predicts* the multipoles and all the electroproduction observables. An alternative way is to use the principle of resonance saturation demonstrated by Ecker *et al.* [10] in the meson sector. The low energy constants are evaluated through the exchange of resonances (in our case the Δ, ρ and ω). This is often called the QCD version of Vector Meson Dominance. This introduces of course some scale dependence since the couterterms are obtained at a certain scale which is given by the resonance masses. For more detail, see G. Ecker, these proceedings.

Fig.1 One-loop diagrams contributing to E_{0+} (two first) and P_i's (two last).

I will not enter into the discussion of the $\mathcal{O}(q^4)$ terms in detail here. Let me just say that the loops are calculated with one vertex from $\mathcal{L}_{\pi N}^{(2)}$ and one from $\mathcal{L}_{\pi N}^{(1)}$, or with vertices from $\mathcal{L}_{\pi N}^{(1)}$ but a propagator from $\mathcal{L}_{\pi N}^{(2)}$ or are $1/m$ corrections to the loops appearing at $\mathcal{O}(q^3)$. It turns out that there are two more counterterms from $\mathcal{L}_{\pi N}^{(4)}$ contributing to E_{0+} at this next order. Thus, the general expressions for E_{0+} and the P-waves to $\mathcal{O}(q^4)$ and $\mathcal{O}(q^3)$, respectively, are:

$$E_{0+} = \frac{\alpha}{m^2}E + \left(\frac{\beta}{m^3} + \frac{\gamma}{F^3} + \kappa_p \delta\right)E^2 + \left(\frac{\epsilon}{m^4} + \frac{\mu}{mF^3}\right)E^3 + d_5 M_\pi^2\omega + d_6\omega^3$$
$$P_{1,2} = \frac{\alpha'}{m^2}|\vec{q}| + \left(\frac{\beta'}{m^3} + \frac{\gamma'}{F^3} + \kappa_p \delta'\right)|\vec{q}|E \tag{3.3}$$
$$P_3 = \text{small Born} + d_7|\vec{q}|\omega$$

where d_i are quantities proportional to the low energy constants, $\alpha, \beta, \delta, \epsilon$ coming from tree diagrams are real functions of ω and γ and μ which are loop contributions are complex function of ω for $\omega > M_{\pi^+}$ (same holds for the ' quantities), thus restoring unitarity in a perturbative manner. An important fact is that to lowest order E_{0+} and the P's have the same chiral power. Indeed, $P_{1,2}$ are proportional to $|\vec{q}|$ and not $|\vec{q}||\vec{k}|$ as it is usually assumed. However since $|\vec{k}|$ is a very slowly varying function close to threshold it turns out that this approximation is rather good. Note that apart from a small Born term P_3 is essentially given by a counterterm.

3.2 Low energy theorems

A low energy theorem was derived in the seventies [11], giving E_{0+} for neutral pion photoproduction at threshold as a series in $\mu = M_\pi/m = 1/7$ up to and including $\mathcal{O}(\mu^2)$ in terms of physical quantities only. This "LET" was based on chiral symmetry and on some so-called harmless assumptions. It turned out that it was in contradiction with the latest experimental informations [1]. The theorem has been therefore reconsidered by several authors [12] (and confirmed by some) and the interpretation of the experimental data has been critically examined [13,14]. Before entering the discussion of this LET, let me first emphasize that CHPT is *not* a model but is a method for solving QCD at low energy. As it is well-known and most clearly spelled out by Weinberg [15] CHPT embodies the very general principle of gauge invariance, analyticity, crossing and PCAC (pion pole dominance). Therefore to lowest order, one recovers the venerable current algebra LETs, which are based on these principles. The effective chiral lagrangian is simply a tool to calculate these but also all next-to-leading order corrections in a systematic and controlled fashion. Thus any theory or model which claims to embody PCAC should lead to the same result as CHPT if one goes to the same level of sophistication. Performing a one-loop CHPT calculation, E_{0+} at threshold is given by:

$$E_{0+} = -\frac{eg_{\pi N}}{8\pi m}\mu\left\{1 - [\frac{1}{2}(3+\kappa_p) + (\frac{m}{4F_\pi})^2]\mu + \mathcal{O}(\mu^2)\right\} \quad , \qquad (3.4)$$

with $\kappa_p = 3.71$ the anomalous magnetic moment of the proton. The second term in the square brackets was not appearing in the original "LET". It is the new one found in the CHPT calculation. It stems from the so-called triangle diagram and its crossed partner. If one expresses E_{0+} to lowest order in terms of the conventional Lorentz invariant function A_1, $E_{0+} = \mu A_1$, one finds that these diagrams give to leading order, $\delta A_1 = (eg_{\pi N}/32F_\pi^2)\,\mu$ due to the presence of a logarithmic singularity and consequently *contributes at next-to-leading order* in the quark mass expansion of E_{0+}. Thus this novel term originates from the presence of Goldstone bosons which leads to the existence of non analytical terms ($\mu \sim \sqrt{\hat{m}}$) invalidating the assumptions ($\delta A_1 = \mathcal{O}(\mu^2)$) used to derive what one should *not* call a "LET" but a "LEG" (low energy guess as defined by G. Ecker. this workshop). Clearly, the expansion in μ is slowly converging, the coefficient of the term of order μ^2 is so large that it compensates the leading term proportional to μ. Therefore, for a meaningful prediction one has to go further in the expansion. To order μ^3 one gets -1.14 using the conventional units of $10^{-3}/M_{\pi^+}$ to be compared with -3.35 at leading order and 0.89 at next-to-leading order. These results show that the LET for E_{0+} is practically useless and that it turns out to be very hard to determine this quantity theoretically, making it not the best test of chiral dynamics.

However, it is very easy to derive LETs for the P-waves within CHPT. These have never been looked at before. They take the following forms:

$$\frac{1}{|\vec{q}|}P_1 = \frac{eg_{\pi N}}{8\pi m^2}\left\{1 + \kappa_p + [-1 - \frac{\kappa_p}{2} + \frac{g_{\pi N}^2}{48\pi}(10 - 3\pi)]\mu + \mathcal{O}(\mu^2)\right\} \quad ,$$

$$\frac{1}{|\vec{q}|}P_2 = \frac{eg_{\pi N}}{8\pi m^2}\left\{-1 - \kappa_p + [3 + \kappa_p - \frac{g_{\pi N}^2}{12\pi}]\frac{\mu}{2} + \mathcal{O}(\mu^2)\right\} \quad .$$

$$(3.5)$$

They are very fastly converging functions of μ. Indeed the terms of $\mathcal{O}(\mu)$ contributes for 6% (less than 0.1%) to P_1 (P_2) respectively. They thus constitute a very good test of chiral dynamics. Writing them in the conventional units $|\vec{q}||\vec{k}| \, 10^{-3}/M_{\pi^+}$, one gets 10.3 and -10.9 for P_1 and P_2, respectively. Notice that already at next to leading order P_1 and P_2 differ. Consequently, the value $E_{1+} \equiv 0$ is excluded. I will end the discussion on the P-waves here just emphasizing that they depend weakly on energy, they show no cusp effect and they have a small imaginary part.

3.3 Results and discussion

All the CHPT results presented in this section are preliminary. Let me first concentrate on the real part of E_{0+}. The reanalyses of the data agree on its value at threshold, $E_{0+} \sim -2. \, 10^{-3}/M_{\pi^+}$ (*accidentally* the experimental result and the LEG are in agreement) but disagree on its energy dependence, the difference between the

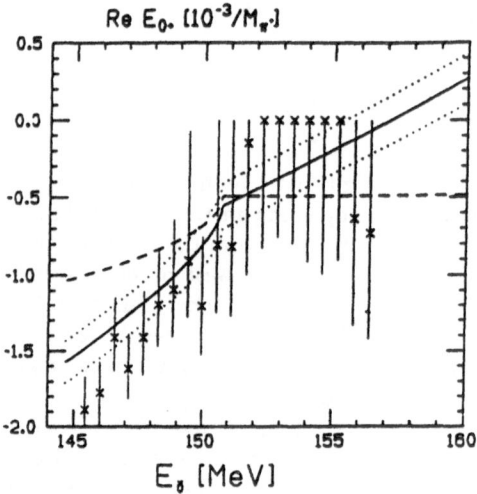

Fig.2 Real part of E_{0+}.

reanalyses coming essentially from different assumptions on the P-waves. We just saw that P_1 and P_2 are constrained since they obey LETs, these were however not known at the time of the reanalyses. Bergstrom [13] and Tiator [13] show a rather strong energy dependence while Bernstein [13] has a weak one. At present this weak energy dependence seems to be ruled out by the LETs which are violated in Bernstein's reanalysis. In fig.2 is shown the CHPT result. There, some isospin breaking effects have been taken into account by differentiating in the loops the charge pion mass from the neutral one. The two curves represent the two different ways of estimating the low energy constants discussed previously. The full curve corresponds to the best fit to the differential and total cross-sections, the band corresponding to a ± 1 standard deviation. The dotted curve corresponds to the determination via resonance exchange where the values of the off-shell parameters of the $\Delta\gamma N$ vertex and the Δ propagator are again obtained by a best fit to the data constrained so that these values lie within the error bars of the theoretical estimates. One notes that the value of d_7 is somewhat independent of the method used. The value of E_{0+} at threshold lies between -1 to -1.5 somewhat smaller in magnitude than the experimental one. It is very much constrained by the bell shape type of the angular distribution of the differential cross-sections for a photon energy larger than 150 MeV as seen in Fig.3 which shows how good the fits are. The energy dependence is weaker than in the reanalysis of Bergstrom and Tiator, their value of $\mathrm{Re}E_{0+}$ at the π^+n threshold being close to zero to be compared with the CHPT one of -0.5. Thus the difference of -2. between the two threshold cannot be obtained within a $\mathcal{O}(q^4)$ calculation. We had already seen that the μ expansion of E_{0+} is very slowly converging, this of course demands a two-loop calculation to test the validity of the results presented here. Of course more precise measurements are needed too.

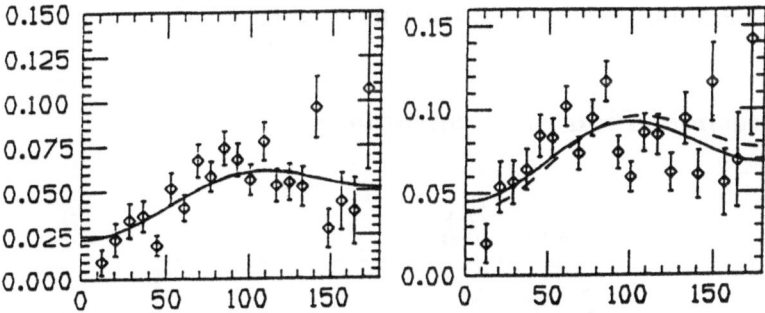

Fig.3 Angular distribution of $d\sigma/d\Omega$: $E_\gamma = 151.4$ (left), $= 153.7$ MeV (right).

The imaginary part of E_{0+} is a very important quantity to look at and this for the following reasons. First, its result is genuine since no counterterm is involved in its calculation. Second, it is related to the change of its real part through a dispersion relation. This can also be seen very easily by using a simple but however realistic model where $E_{0+} = -a-b\sqrt{1 - E_\gamma/E_\gamma^{(+)}}$ with $E_\gamma^{(+)}$ the photon energy at the π^+n threshold (a good fit is obtained with $a = 0.3$ and $b = 5.7$ in the units $10^{-3}/M_{\pi^+}$). Another point has been stressed by Bernstein [16]. The imaginary part of E_{0+} is related to the πN scattering lengths through the Fermi Watson theorem. Thus a measure of the imaginary part would lead to another way of determining these important quantities. Within CHPT to $\mathcal{O}(q^4)$, $\mathrm{Im}E_{0+}$ stays below $10^{-3}/M_{\pi^+}$ for $E_\gamma < 160$ MeV. It is too small, as we have already seen, to account for the strong energy dependence of $\mathrm{Re}E_{0+}$. For all these reasons it would be very interesting to have a measure of $\mathrm{Im}E_{0+}$. This can be done through a measurement of polarized observables. Two of them turn out to be particularly sensitive to this quantity, these are the polarized target asymmetry T and the recoil polarization P as can easily be seen by looking at their definitions:

$$T \propto \mathrm{Im}((E_{0+} + \cos\theta P_1)(P_2 - P_3)^\star)$$
$$P \propto \mathrm{Im}((E_{0+} + \cos\theta P_1)(P_2 + P_3)^\star)$$

(3.6)

The CHPT results are shown in Fig. 4. One sees that though P stays very small, T reaches 15 to 20% over a wide range of angles. This makes this observable a good candidate for a measurement of $\mathrm{Im}E_{0+}$.

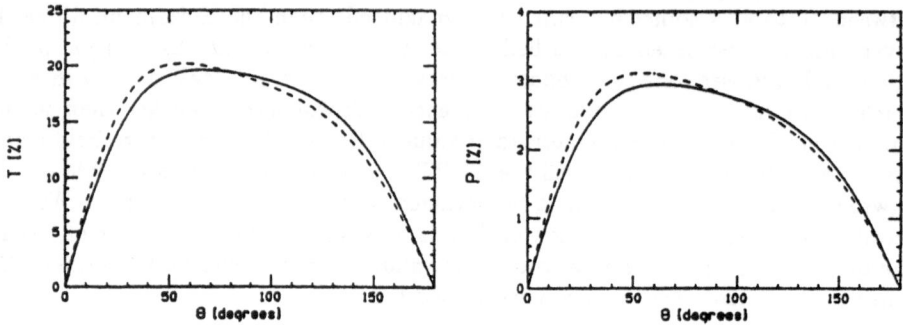

Fig.4 Polarized observables P and T for $E_\gamma = 151.4$ MeV.

4. Threshold Pion Electroproduction

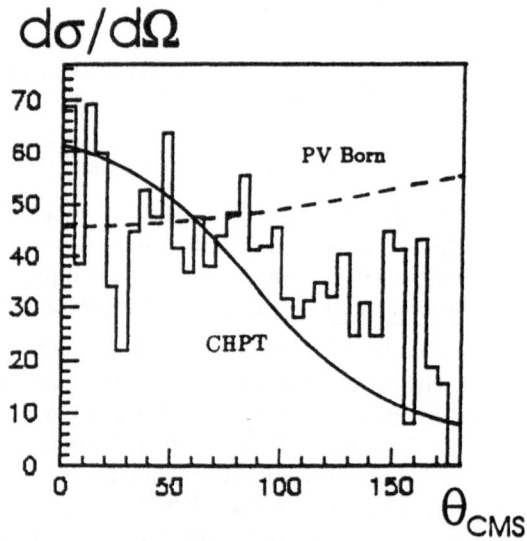

Fig.5 Angular distribution of $d\sigma_T/d\Omega$ for $k^2 = -0.1$ GeV^{-2}.

I will now discuss some rather new data on electroproduction. Welch et al.[2] have published the S-wave cross section for the reaction $\gamma^* p \rightarrow \pi^0 p$ very close to the photon point. This measurement is a quantum step compared to previous determinations which mostly date back to the seventies when pion electroproduction was still a hot topic in particle physics. In this experiment, k^2 varied between -0.04 and -0.10 GeV^2 and the S-wave cross section could be extracted with an unprecedented accuracy (see fig.2 in [2]). This is also the kinematical regime where a CHPT calculation might offer some insight. Indeed, in ref.[17] it was shown that the k^2-dependence of this cross section seems to indicate the necessity of loop effects. With conventional models including e.g. form factors and the anomalous magnetic moment coupling the trend of the data can not be described. However, the corrections from the one loop diagrams to the tree level prediction are substantial. This gives further credit to the previously made statement that a calculation beyond next-to-leading order should be performed. Very recently, Distler *et al.* [3] have measured the angular distribution of the differential transverse cross section at $k^2 = -0.1$ GeV^2. Even though this value is somewhat too high for a good test of CHPT (there, the one-loop order corrections are of the order of 50%) it is interesting to see that the preliminary experimental results show the same trends as the CHPT ones and are at variance with the pseudo-vector Born approximation (see Fig.5). It would be of particular interest to have experimental informations at $k^2 = -0.05$ GeV^2.

The last topic I want to address in this section concerns the determination of the nucleon axial radius from charged pion electroproduction. Let me briefly explain how the axial form factor comes into play. The basic matrix element to be considered is the time-ordered product of the electromagnetic (vector) current with the interpolating pion field sandwiched between nucleon states. Now one can use the PCAC relation and express the pion field in terms of the divergence of the axial current. Thus, a commutator of the form [V,A] arises. Current algebra tells us that this gives an axial current between the incoming and outgoing nucleon fields and, alas, the axial form factor. The isospin factors combine in a way that they form a totally antisymmetric combination which can not be probed in neutral pion production. These ideas were formalized in the venerable low-energy theorem (LET) due to Nambu, Lurié and Shrauner [18] for the isospin–odd electric dipole amplitude $E_{0+}^{(-)}$ in the chiral limit,

$$E_{0+}^{(-)}(M_\pi = 0, k^2) = \frac{eg_A}{8\pi F_\pi}\left\{1 + \frac{k^2}{6}r_A^2 + \frac{k^2}{4m^2}(\kappa_V + \frac{1}{2}) + \mathcal{O}(k^3)\right\} \tag{3.7}$$

Therefore, measuring the reactions $\gamma^*p \rightarrow \pi^+n$ and $\gamma^*n \rightarrow \pi^-p$ allows to extract $E_{0+}^{(-)}$ and one can determine the axial radius of the nucleon, r_A. This quantity measures the distribution of spin and isospin in the nucleon, i.e. probes the Gamov–Teller operator $\sigma \cdot \tau$. A priori, the axial radius is expected to be different from the typical electromagnetic size, $r_{em} \simeq 0.85$ fm. It is customary to parametrize the axial form factor $G_A(k^2)$ by a dipole form, $G_A(k^2)/g_A = (1 - k^2/M_A^2)^{-2}$ which leads to the relation $r_A = \sqrt{12}/M_A$. The axial radius determined from electroproduction data is typically $r_A = 0.59 \pm 0.04$ fm whereas (anti)neutrino-nucleon reactions lead to somewhat larger values, $r_A = 0.65 \pm 0.03$ fm. This discrepancy is usually not taken seriously since the values overlap within the error bars. However, it was shown in ref.[19] that pion loops modify the LET (3.7) at order k^2 for finite pion mass. In the heavy mass formalism, the coefficient of the k^2 term reads

$$\frac{1}{6}r_A^2 + \frac{1}{4m^2}(\kappa_V + \frac{1}{2}) + \frac{1}{128F_\pi^2}(1 - \frac{12}{\pi^2}) \tag{3.8}$$

where the last term in (3.8) is the new one. This means that previously one had extracted a modified radius, the correction being $3(1 - 12/\pi^2)/64F_\pi^2 \simeq -0.046$ fm^2. This closes the gap between the values of r_A extracted from electroproduction and neutrino data. It remains to be seen how the $1/m$ suppressed terms will modify the result (3.8). Such investigations are underway.

5. Threshold Two Pion Photoproduction:

At threshold in the center-of-mass frame (*i.e.* $\vec{q}_1 = \vec{q}_2 = 0$), the two-pion photoproduction current matrix element can be decomposed into amplitudes as follows if we work to first order in the electromagnetic coupling e,

$$T \cdot \epsilon = \chi_f^\dagger \{i\vec{\sigma} \cdot (\vec{\epsilon} \times \vec{k})[M_1 \delta^{ab} + M_2 \delta^{ab}\tau^3 + M_3(\delta^{a3}\tau^b + \delta^{b3}\tau^a)]\}\chi_i \qquad (4.1)$$

with $\chi_{i,f}$ two-component Pauli-spinors and isospinors and we used the gauge $\epsilon_0 = 0$. To leading order in the chiral expansion, only the amplitudes M_2 and M_3 are non vanishing, with $M_2 = -2M_3$ [20]. Therefore the production of two neutral pions is strictly suppressed. One can derive LETs for $M_{2,3}$. They read:

$$M_2 = -2M_3 = \frac{e}{4mF_\pi^2}(2g_A^2 - 1 - \kappa_V) \qquad (4.2)$$

There exists in the literature some incomplete determination of these LETs [21]. The authors did not use the most general effective Lagrangian and thus did not get the term proportional to κ_V. At next order in the chiral expansion, one has to consider kinematical $(1/m)$ corrections, one-loop contributions and tree graphs with insertion of the $\Delta(1232)$ (chirally expanded). Note that, to this order, the Δ contributions are absent in the $\pi^0\pi^0$ channel. Here we use the principle of resonance saturation to estimate the counterterms. At this order one has a LET for M_1:

$$M_1 = \frac{eg_A^2 M_\pi}{4mF_\pi^2} \qquad (4.2)$$

and the $M_{2,3}$ have a somewhat more complicated expressions (see ref.[20]). Their loop contributions have a non-zero imaginary part even at threshold which comes from the rescattering graph. Due to unitarity the pertinent loop functions have a right hand cut starting at $s = (m + M_\pi)^2$ (the single pion production threshold) and these functions are here to be evaluated at $s = (m + 2M_\pi)^2$ (the two- pion production threshold).In Fig.6 are shown the total cross sections for $\gamma p \rightarrow \pi^+\pi^- p$, $\pi^+\pi^0 n$ and $\pi^0\pi^0 p$. Here the calculation was done with the correct phase-space and approximating the amplitudes in the threshold region through their threshold values in the case of the two last reactions. For the first one the first correction above threshold is also shown [20]. At $E_\gamma = 320$ MeV, the total cross section for $\pi^0\pi^0$ production is 0.5 nb whereas the competing $\pi^+\pi^0 n$ final state has $\sigma_{tot} = 0.07$ nb. Double neutral pion production reaches $\sigma_{tot} = 1.0$ nb at $E_\gamma = 324.3$ MeV in comparison to $\sigma_{tot}(\gamma p \rightarrow \pi^0\pi^+ n) = 0.26$ nb and $\sigma_{tot}(\gamma p \rightarrow \pi^+\pi^- p) < 0.1$ nb. This means that for the first $10 \ldots 12$ MeV above $\pi^0\pi^0$ threshold (chiral window), one has a fairly clean signal and much more neutrals than expected. Of course, the above threshold correction for all the channels should be calculated systematically. However the first correction, which vanishes proportional to $|\vec{q}_i|$ (i=1,2) at threshold, has been calculated and found to be small. It is therefore conceivable that the qualitative features described above will not change if even higher corrections are taken into account. We hope to report on these in the not too distant future. From the experimental side TAPS seems to have seen some π^0's close to threshold. For more detail, see T. H. Walcher, these proceedings.

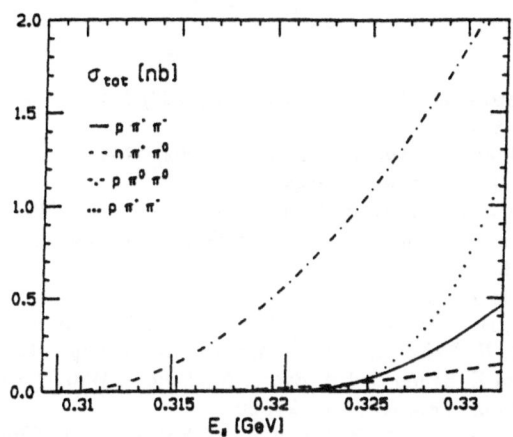

Fig.6 Total cross sections (in nb) for the γp initial state

6 Summary and Outlook

Pion photo- and electroproduction in the threshold region is a good testing ground of the chiral nucleon structure. A one-loop CHPT calculation gives a satisfactory description of most of the existing observables. The pertinent results of our investigation can be summarized as follows:

- The LET for E_{0+} is practically useless. At variance with what is usually believed the P-waves are the quantities providing a good test of chiral dynamics.
- We have stressed the importance of the imaginary part of E_{0+}.
- We have obtained a better understanding of the axial radius of the nucleon as measured in charged electroproduction.
- We have seen that there is a window of about 10 MeV above $\pi^0\pi^0$ threshold in which one should detect much more neutrals than expected in two pion photoproduction.

At this stage some more efforts are needed on the theoretical side to pin down the isospin breaking effects. Also a two loop calculation is mandatory to determine the validity of the results presented here, especially in the case of the S-wave observables. On the experimental side, it is clear that in order to test CHPT, one has to measure close to threshold and to get to very accurate data. I want to stress again that it would be very important to have a measure of the imaginary part of E_{0+} as well as a good determination of the energy dependence of its real part.

7 Acknowledgements

I would like to thank the organizers for their invitation and kind hospitality. The work reported here has been done in collaboration with J. Gasser, N. Kaiser, T.-S.H. Lee and U.-G. Meißner to whom I express my gratitude.

References

[1] E. Mazzucato et al., *Phys. Rev. Lett.* **57** (1986) 3144; R. Beck et al., *Phys. Rev. Lett.* **65** (1990) 1841

[2] T. P. Welch et al., *Phys. Rev. Lett.* **69** (1992) 2761

[3] T. H. Walcher, these proceedings

[4] H. Leutwyler, see for exemple these proceedings and "Principles of Chiral Perturbation Theory", lectures given at the Workshop "Hadrons 1994", Gramado, Brasil, BUTP-94/13

[5] Ulf-G. Meißner, *Rep. Prog. Phys.* **56** (1993) 903

[6] V. Bernard, N. Kaiser and Ulf-G. Meißner, *Z. Phys.* **A348** (1994) 317; in preparation

[7] V. Bernard, N. Kaiser and Ulf-G. Meißner, *Nucl. Phys.* **B383** (1992) 442

[8] J. Gasser, M.E. Sainio and A. Švarc, *Nucl. Phys.* **B307** (1988) 779

[9] G. Ecker, *Phys. Lett.* **B336** (1994) 508

[10] G. Ecker, J. Gasser, A. Pich and E. de Rafael, *Nucl. Phys.* **B321** (1989) 311; G. Ecker, J. Gasser, H. Leutwyler, A. Pich and E. de Rafael, *Phys. Lett.* **B223** (1989) 425

[11] A. I. Vainshtein and V. I. Zakharov, *Nucl. Phys.* **B36** (1972) 589; P. de Baenst, *Nucl. Phys.* **B24** (1970) 633

[12] V. Bernard, N. Kaiser, J. Gasser and Ulf-G. Meißner, *Phys. Lett.* **B268** (1991) 291 and references therein

[13] A. M. Bernstein and B. R. Holstein, *Comments Nucl. Par. Phys.* **20** (1991) 197; J. Bergstrom, *Phys. Rev.* **C44** (1991) 1768; L. Tiator, "Meson Photo- and Electroproduction", lecture given at the II TAPS Workshop, Alicante, 1993

[14] R. Davidson and N. C. Mukhopadhyay, *Phys. Rev. Lett.* **60** (1988) 748; E. Mazzucato et al., *Phys. Rev. Lett.* **60** (1988) 749; S. Nozawa, T.-S.H. Lee and B. Blankleider, *Phys. Rev.* **C41** (1990) 213; A. N. Kamal, *Phys. Rev. Lett.* **63** (1989) 213

[15] S. Weinberg, *Physica* **A96** (1979) 327

[16] A. M. Bernstein, πN *Newsletter* **9** (1993) 55

[17] V. Bernard, N. Kaiser, T.-S.H. Lee and Ulf-G. Meißner, *Phys. Rev. Lett.* **70** (1993) 387; *Phys. Reports* **246** (1994) 315

[18] Y. Nambu and D. Lurié, *Phys. Rev.* **125** (1962) 1429; Y. Nambu and E. Shrauner, *Phys. Rev.* **128** (1962) 862

[19] V. Bernard, N. Kaiser and Ulf-G. Meißner, *Phys. Rev. Lett.* **69** (1992) 1877

[20] V. Bernard, N. Kaiser and Ulf-G. Meißner, *Nucl. Phys.* **A580** (1994) 475

[21] R. Dahm and D. Drechsel, in Proc. Seventh Amsterdam Mini-Conference, eds. H. P. Blok, J. H. Koch and H. De Vries, Amsterdam, 1991; M. Benmerrouche and E. Tomusiak, *Phys. Rev. Lett.* **73** (1994) 400

Pion-production experiments at NIKHEF

H. P. Blok

Department of Physics and Astronomy, Vrije Universiteit, de Boelelaan
1081, 1081 HV Amsterdam, The Netherlands
and
NIKHEF-K, P.O. Box 41882, 1009 DB Amsterdam, The Netherlands

Abstract: The results are described of two experiments on the ^{1}H(e,e$'\pi^0$)p reaction
that have been performed at NIKHEF. In the first one the total cross section was measured close to threshold. In the second one the pion angular distribution was measured
up to 12 MeV above threshold. Preliminary results are given. Experiments planned for
the future are also shortly described.

1 Introduction

Two experiments on the ^{1}H(e,e$'\pi^0$)p reaction near threshold have been performed at NIKHEF-K in Amsterdam. The motivation for these studies is discussed in other presentations in this workshop [1]. In short they are the following.
Data for the ^{1}H(γ, π^0) p reaction near threshold taken at Saclay and Mainz [2]
seemed to indicate a violation of the prediction from the Low Energy Theorem
(LET). A major issue was how to take into account charge-exchange via π^+ production, for which the threshold lies 6.8 MeV above the one for π^o production.
After quite some discussions there is now agreement that the LET has to be
applied at the π^o threshold [3] and that extrapolation of the data to this point
yields a value for the E_{0+} amplitude consistent with the LET prediction (for a
discussion of this "old" LET prediction see the contribution by Bernard). Analysis of the data also indicates that the E_{0+} amplitude varies rather strongly as
a function of W as one approaches the π^+ threshold [4,5] (the analysis of [6],
which yields a much weaker dependence of E_{0+} on W, is probably not realistic
because it gives quite small values for the p-multipoles).

In electroproduction one can investigate in addition the L_{0+} multipole, for
which there is also a LET prediction [7]. Further the Q^2 dependence can be
studied. This was the motivation for the first experiment (see section 2), which
was performed close to threshold and the results of which have been published
[8]. Since it may be expected that the presence of the π^+ channel will also

influence the value of L_{0+}, the second experiment was designed to study the W-dependence of the cross section up to about 12 MeV above the π^0 threshold, thus covering the region around the π^+ threshold. This time angular distributions were measured, which allows one to determine the L_{0+} and E_{0+} multipoles (including sign) separately. This experiment is described in section 3 and some preliminary results of the analysis are presented. Finally in section 4 we describe shortly some planned experiments on pion production.

2 Measurement of the total cross section for the ^{1}H(e,e'π^0)p reaction close to threshold

The first ^{1}H(e,e'π^0)p experiment [8] was performed with the 1% duty-factor MEA accelerator [9]. Data were taken for values of Q^2 of 0.05 and 0.10 (GeV/c)^2 at $E = 500$ MeV, while data at $Q^2 = 0.05$ (GeV/c)^2 were also taken at $E = 350$ MeV, thus enabling an investigation of the L/T ratio. The target was a H_2 gas cell, operated at 0.3 MPa and 30 K. With an average current of $7\,\mu$A this yielded a luminosity of 2×10^{36} e^- at/sec. The measurements were performed by detecting the proton (instead of the π^0) and the scattered electron in the QDQ and QDD high-resolution spectrometers.

Near threshold the protons are emitted in a narrow cone around the momentum transfer q. For values of the invariant mass W up to about 2 MeV above threshold this cone fits within the acceptance solid angle of the QDQ spectrometer, which has a half opening angle of 4^0. By measuring the e-p coincidences the total cross section for π^0 production near threshold can thus be determined. This cross section can be written as

$$\frac{d^3\sigma}{dE_{e'}d\Omega_{e'}} = 4\pi\Gamma\frac{p_\pi^*}{q_L}\frac{W}{M}(a_0 + bp_\pi^{*2}) \,, \tag{1}$$

where Γ is the virtual-photon flux factor

$$\Gamma = \frac{\alpha}{2\pi^2}\frac{E_f}{E_i}\frac{q_L}{Q^2}\frac{1}{1-\epsilon} \tag{2}$$

with $q_L = (W^2 - M^2)/2M$ the equivalent photon energy. Further p_π^* is the pion momentum in the πN centre of mass frame and

$$a_0 = |E_{0+}|^2 + \epsilon_L|L_{0+}|^2 \,, \tag{3}$$

where $\epsilon_L = \epsilon Q^2/q_0^{*2}$, while bp_π^{*2} gives the contribution to the cross section due to p-multipoles.

The measured cross section is an average over the considered region in W taking into account the acceptances of the spectrometers. The corresponding volume element $d\Omega_{e'}dE_{e'}$ in the determination of our final values for a_0 was determined by Monte-Carlo calculations.

Fig. 1 displays the world's data including our results. The marked increase in precision, which is due to the fact that we were able to measure near threshold with high resolution in the invariant energy, is clearly seen. The theoretical predictions of Devenish and Lyth [10], which are based on a parameterization of non-threshold data, as well as those of Dombey and Read [11] and Benfatto et al. [12], which are based upon extensions of LETs, are globally consistent with our results.

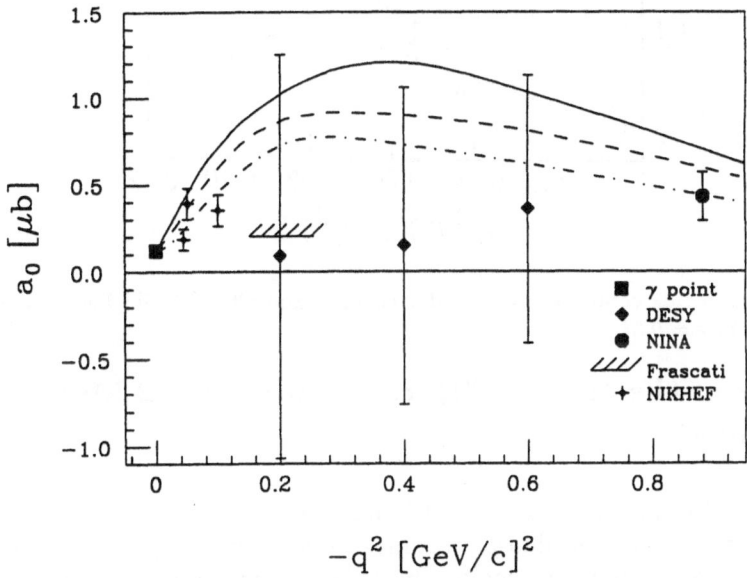

Fig. 1. Plot of the world's data for a_0 and some theoretical predictions (calculated for $\epsilon = 1$). The previous data are for $\epsilon \geq 0.9$, while our data are for $\epsilon = 0.6$ and 0.8.

According to these predictions the value of E_{0+} is rather small. This makes it possible to determine L_{0+} by subtracting from a_0 the small contribution from E_{0+} and dividing by ϵ_L. Extrapolation of the values thus obtained to the photon point yields $|L_{0+}(Q^2 = 0)|^2 = 0.13 \pm 0.05 \,\mu b$, to be compared with the LET prediction of [7] of $0.16 \,\mu b$.

The data have also been compared to predictions from Chiral Perturbation Theory [13] (see fig. 2). Compared to the pseudo-vector (PV) prediction ChPT yields a decrease in the value of L_{0+} and a (smaller) increase in the value of E_{0+} at larger values of Q^2. For this reason a determination of E_{0+} and L_{0+} separately, either by an explicit Rosenbluth separation (see [14]), or by determining the pion angular distribution and the σ_{LT} interference cross section (see the next section), is a necessary next step.

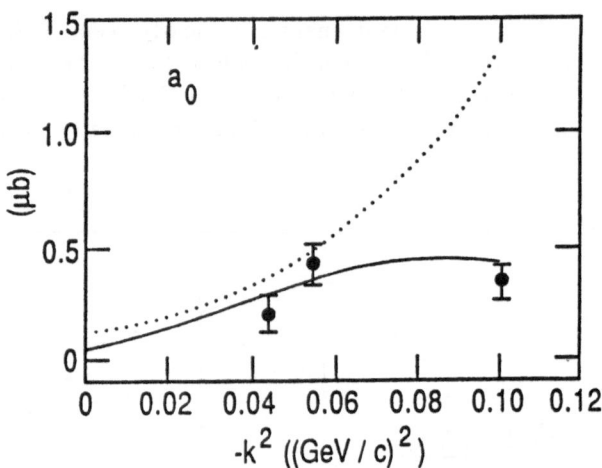

Fig. 2. Data for a_0 compared to PV (dotted line) and ChPT (solid line) predictions, calculated for $\epsilon = 0.58$.

3 Study of the ^{1}H(e,e$'\pi^0$)p reaction up to 12 MeV above threshold

The ^{1}H(e,e$'\pi^0$)p reaction was studied in more detail in a second experiment, in which the π^0 angular distribution was measured at Q^2=0.1 (GeV/c)^2 for values of $W \approx$ 2-12 MeV above threshold. The cross section can be written as:

$$\frac{d^5\sigma}{dE_{e'}d\Omega_{e'}d\Omega_\pi^*} = \Gamma\frac{d\sigma_v}{d\Omega_\pi^*} \tag{4}$$

with

$$\frac{d\sigma_v}{d\Omega_\pi^*} = \epsilon_L\sigma_L + \sigma_T + \epsilon\sigma_{TT}\cos 2\phi + \sqrt{2\epsilon_L(\epsilon+1)}\sigma_{LT}\cos\phi \tag{5}$$

where $\phi=\phi_\pi^*$ is the angle between the (e,e) and (q,p$_\pi$) planes. The σ 's are functions of the pion angle θ_π^*. If one restricts oneself to the s-multipoles together with the large p-multipoles M_{1+} and M_{1-}, the σ 's can be written as

$$\sigma_x = \frac{p_\pi^*}{q_L}\frac{W}{M}R_x \tag{6}$$

with

$$
\begin{aligned}
R_L =&\ |L_{0+}|^2\\
R_T =&\ |E_{0+}|^2 + \tfrac{1}{2}|2M_{1+} + M_{1-}|^2 + \tfrac{1}{2}|M_{1+} - M_{1-}|^2\\
&+ 2\cos\theta_\pi^* \operatorname{Re}\{E_{0+}^*(M_{1+} - M_{1-})\}\\
&+ \cos^2\theta_\pi^*(\tfrac{1}{2}|M_{1+} - M_{1-}|^2 - \tfrac{1}{2}|2M_{1+} + M_{1-}|^2)\\
R_{TL} =&\ \sin\theta_\pi^* \operatorname{Re}[L_{0+}^*(M_{1+} - M_{1-})]\\
R_{TT} =&\ {-}3\sin^2\theta_\pi^* [\operatorname{Re}(M_{1+}^*M_{1-}) + |M_{1+}|^2]\ .
\end{aligned}
\tag{7}
$$

Since the M_{1+} and M_{1-} multipoles are rather well constrained by LETs and extrapolations from phase-shift analyses at higher energies, the value of L_{0+} can be extracted from a measurement of σ_{LT} and the value of E_{0+} from the $\cos\theta_\pi^*$ dependence of $\sigma_0 = \epsilon_L\sigma_L + \sigma_T + \epsilon\sigma_{TT}\cos 2\phi$. The remaining part of σ_0 can then be used for a consistency check.

Data were obtained in the beginning of June 1994 by using the extracted beam from the stretcher ring AmPS [15]. The electron energy was 525 MeV, the current 10 μA, and the duty factor about 30%. The target was a cryogenic H_2 target [16] operated at 35 K and 1.5 MPa, just above the critical point. There the density varies strongly with both the pressure and temperature. The chosen density of 22 mg/cm^3 was a compromise between count rate and the amount of energy loss and angle straggling in the target gas. With an effective target length of 2.5 cm, as viewed by the QDD spectrometer, which was the limiting one, the luminosity was $2\times10^{37}e^-$ at/sec.

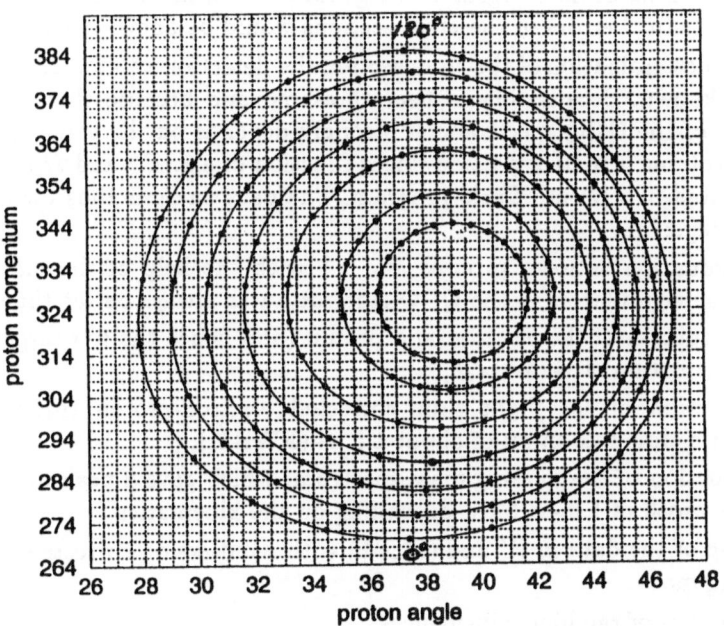

Fig. 3. Relation between the momentum and in-plane (lab) angle of the proton. The contours are for constant values of W, while the dots give the angle θ_π^* in steps of 15°. The right half of the figure is for $\phi = 0°$ and the left half for $\phi = 180°$.

The scattered electrons, which determine the values of Q^2 and W, were detected in the QDD spectrometer. The resolution in W is estimated to be better than 0.3 MeV. The protons, which are emitted in a cone around q, were detected in the QDQ spectrometer. For every value of W (and a given Q^2) there is a relation between the direction of the proton (which is directly related to the direction $(\theta_\pi^*, \phi_\pi^*)$ of the emitted pion) and its momentum (see fig. 3). Data were taken in four kinematics with central values at $W = 6$ MeV above threshold for θ_π^* of $0°$, $90°$ (for both $\phi = 0°$ and $180°$) and $180°$. Since the momentum bite of the QDQ spectrometer is $\pm 5\%$ and its opening angle $\pm 4°$, globally the region W=2-12 MeV above threshold and $\theta_\pi^* = 0° - 180°$ was covered, with data around $\theta_\pi^* = 90°$ for both $\phi = 0°$ and $180°$.

In order to reconstruct θ_π^* (and ϕ_π^*) from the data the transformation from the wire chamber information in the focal plane of the spectrometer to the target vector [17] has to be known accurately, and for an extended target. This transformation has been carefully calibrated in the past with so-called sieve-slit measurements [18]. As a check on the calibration some sieve-slit data were taken during the π^0 measurements. Also the kinematically overcomplete ^{1}H(e,e'p) reaction was used as a check on the reconstruction of the target vectors and on the corrections for energy loss in the target. At the same time the coincidence efficiency was determined with this reaction to be $99 \pm 1\%$. Data were collected for different values of the integrated charge for the four kinematics such that the absolute statistical errors in the cross sections would be about the same.

For the determination of the cross section the detection volume $dE_{e'} d\Omega_{e'} d\Omega_\pi^*$ has to be known. Because one is near threshold and because the effective θ_π^* acceptance of the QDQ spectrometer strongly depends on the value of W, this detection volume is a strongly non-linear function of e.g. $E_{e'}$ and $\Omega_{e'}$. Hence it was calculated in a Monte Carlo program. The data analysis thus included the following steps:

- discrimination between protons and π^+ with the help of the information from the scintillators in the QDQ (the ratio was always better than 20:1),
- reconstruction of the electron and proton (and hence the π^0) target vector and calculation of W and Q^2,
- construction of the flight time corrected electron-proton timing, which yielded a time resolution of about 1.5 ns (FWHM),
- construction of the missing mass spectrum, which gave a peak at 135 MeV with a width of 1.5 - 3 MeV depending on the kinematics,
- binning of the data in W and $(\theta_\pi^*, \phi_\pi^*)$ with windows on corrected time and missing mass,
- subtraction of random coincidences,
- division by integrated charge, target thickness, dead time and efficiency factors and by the calculated detection volume in order to get cross sections.

Some results of a first analysis are given below. It is stressed that these results are preliminary since various checks on, e.g., the W-calibration, the target thickness, etc. still have to be performed and no radiative corrections have been applied yet. The measured cross sections as a function of W integrated over all Ω_π^* of that kinematics and over the acceptance in Q^2, which is globally 0.10 ± 0.01 $(\text{GeV}/c)^2$, are shown in fig. 4 for the four kinematics. It is immediately clear that there is a large difference in cross section for $\theta_\pi^* = 0°$ and $180°$ and for $\phi_\pi = 0°$ and $180°$ at $\theta_\pi^* = 90°$. The latter implies a large value of σ_{LT}.

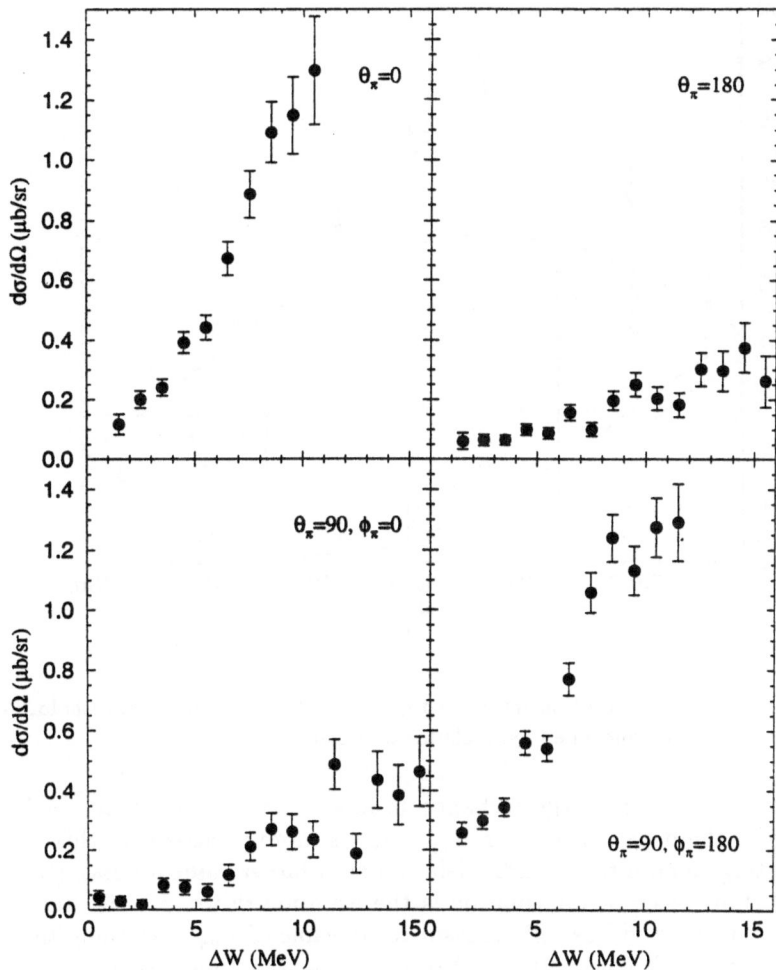

Fig. 4. Global cross sections as a function of W above threshold for the four kinematics.

The cross section (integrated over Q^2 and $W = 4 - 8$ MeV above threshold) as a function of θ_π^* for $|\phi_\pi^*| < 30°$ and $|\phi_\pi^* - 180°| < 30°$ is shown in fig. 5 . Since the cross section near $\theta_\pi^* = 180°$ is smaller than near $\theta_\pi^* = 0°$, the coefficient of the $\cos\theta_\pi^*$ term must be positive. Given that $M_{1+} > 0$, this tells that E_{0+} is positive at $Q^2 = 0.1$ $(\text{GeV/c})^2$, whereas it is negative at the photon point.

The dotted curve is the result of a calculation by Blaazer [19] using the Dressler operator [20], which is similar to the Blomquist-Laget operator. In this calculation E_{0+} is just positive at $Q^2 = 0.1$ $(\text{GeV/c})^2$, but still rather small, which results in the cross section being about the same at $\theta_\pi^* = 0°$ and $\theta_\pi^* = 180°$. The full curve gives the results of a calculation by Nozawa and Lee [21]. Strictly spoken their model is not valid very close to threshold, but it includes

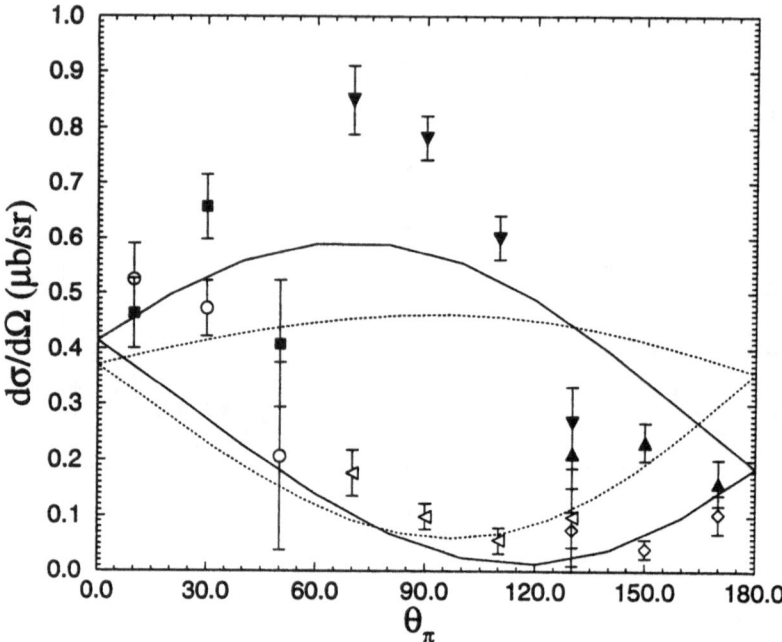

Fig. 5. Pion angular distribution integrated over $W = 4$-8 MeV above threshold. Open symbols are for $\phi = 0°$ and filled symbols for $\phi = 180°$.

the ^{1}H(e,e'π^+)(π^+, π^0) charge-exchange process, and is seen to give a better description of the data (since their E_{0+} value is more positive). In this respect it is interesting to note that ChPT yields an even larger value for E_{0+} [13]. The other important piece of information is the asymmetry for $\phi = 0°$ and 180°. This is due to σ_{LT} and basically measures the value of L_{0+}, assuming that the p-multipoles are well constrained. Also here it is seen that larger values than predicted by the calculations are needed.

4 Future plans

The future plans at NIKHEF involve the following experiments:
- investigation of π^0 production near threshold on the neutron with the ^{2}H(e,e'π^0)d reaction (coherent π^0 production),
- study of the ΔN interaction and $NN \leftrightarrow \Delta N$ coupling with the ^{2}H(e,e'π^0)d reaction in the Δ region,
- study of Δ-excitation and propagation with coherent π^0 production on ^{4}He (^{4}He(e,e'π^0)^{4}He),
- investigation of quasi-free pion production on bound nucleons with the ^{4}He(e,e'π^0p)^{3}H, ^{4}He(e,e'π^+n)^{3}H and ^{4}He(e,e'π^-p)^{3}He reactions.

Reactions a) and b) can be performed with the external beam and the two-spectrometer setup. The studies c) and d) will be performed using the internal beam of the AmPS ring and an open ^{4}He gas cell. The detection of π^0 particles (or the neutron) will be avoided by detecting instead the recoiling nucleus. In the last reaction mentioned under d) with three charged particles in the final state detection of the ^{3}He nucleus instead of either the proton or π^- is advantageous, since it greatly enhances the phase-space that can be covered.

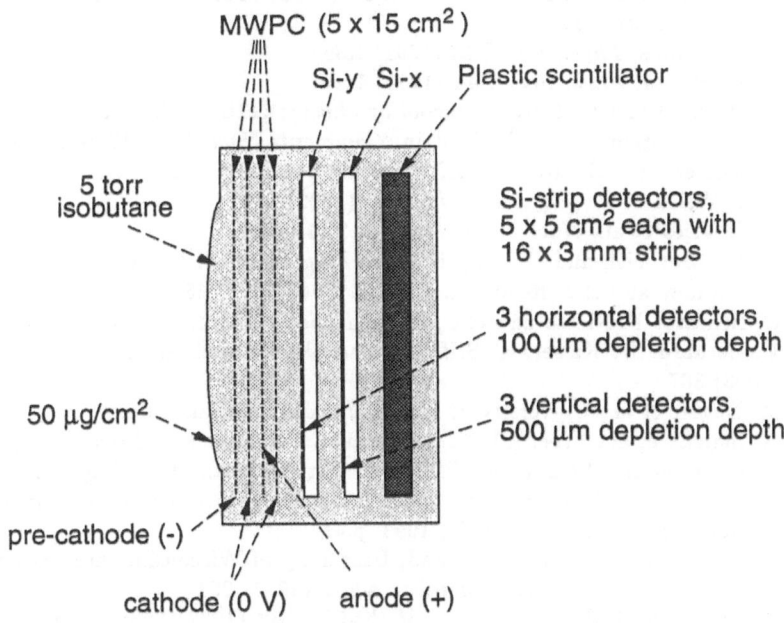

MWPC (5 x 15 cm^2)

Si-y Si-x   Plastic scintillator

5 torr isobutane

Si-strip detectors,
5 x 5 cm^2 each with
16 x 3 mm strips

3 horizontal detectors,
100 µm depletion depth

50 µg/cm^2

3 vertical detectors,
500 µm depletion depth

pre-cathode (-)

cathode (0 V)   anode (+)

Fig. 6. Global layout of the recoil detector.

A special detector is being developed to detect the recoiling ^{3}H and ^{3,4}He nuclei (but it can also be used, e.g., to study pn correlations with the ^{4}He(e,e'pn)^{2}H reaction), which have typical energies of 1-10 MeV. A global lay-out of the detector is shown in fig. 6. It consists of a low-pressure MWPC followed by two layers of Si strip detectors and a plastic scintillator. For particles of short range the MWPC and first Si layer act as a ΔE or time of flight - E system. For more penetrating particles that do not give enough pulse height in the MWPC the two Si layers provide particle identification and energy determination via the $\Delta E - E$ method. The scintillator is used as an E counter for higher energy deuterons and protons. Position information is provided by the MWPC (y), the first Si layer (x) and the second Si layer (y again). The expected energy resolution is about 200 keV and the position resolution is 3 mm. The detector with accompanying analog and digital electronics will be tested in the fall of 1994 and the first experiment with it is planned for 1995.

Acknowledgements
I would like to thank the organizers for their invitation and hospitality. The results presented in section 3 are part of the thesis work of drs. B van den Brink.

References

1. V. Bernard, these proceedings; Ulf-G. Meißner and B. Schoch, these proceedings
2. E. Mazzucato et al., *Phys. Rev. Lett.* **57** (1986) 3144; R. Beck et al., *Phys. Rev. Lett.* **65** (1990) 1841
3. H.W.L. Naus, *Phys. Rev.* **C43** (1991) R365
4. J. Bergstrom, *Phys. Rev.* **C44** (1991) 1768
5. D. Drechsel and L. Tiator, *Journal of Physics* **G18** (1992) 449
6. A. M. Bernstein and B. R. Holstein, *Comments Nucl. Part. Phys.* **20** (1991) 197
7. S. Scherer and J.H. Koch, *Nucl. Phys.* **A534** (1991) 461
8. T. P. Welch et al., *Phys. Rev. Lett.* **69** (1992) 2761
9. C. de Vries et al., *Nucl. Instr. Methods* **223** (1984) 1
10. R.C.E. Devenish and D.H. Lyth, *Nucl. Phys.* **B93** (1975) 109
11. N. Dombey and B.J. Read, *Nucl. Phys.* **B60** (1972) 65
12. G. Benfatto F. Nicolo and G.C. Rossi, *Nuovo Cimento* **14A** (1973) 425
13. V. Bernard, N. Kaiser, T.-S.H. Lee and Ulf-G. Meißner, *Phys. Rev. Lett.* **70** (1993) 387
14. M. Distler et al., University of Mainz, private communication
15. G. Luijckx et al., *Proc. XVth International Conference on High-Energy Accelerators*, Hamburg, 1992, p.464; P.K.A. de Witt Huberts, *Nucl. Phys.* **A553** (1993) 845c; P.K.A. de Witt Huberts, *Proc. International Conference on Particles and Nuclei (PANIC) XIII*, Perugia, 1993, p.417
16. O. Unal and J.F.J. van den Brand, University of Wisconsin, Internal Report
17. H. Blok et al., *Nucl. Instr. Methods* **A262** (1987) 291
18. E. Offerman et al., *Nucl. Instr. Methods* **A262** (1987) 298; L. de Vries et al., *Nucl. Instr. Methods* **A292** (1990) 629; B. van den Brink, NIKHEF, Internal Report EMIN 91-06
19. F. Blaazer, Vrije Universiteit, Amsterdam, private communication
20. E.T. Dressler, *Can. J. Phys.* **66** (1988) 279
21. S. Nozawa, B. Blankleider and T.-S.H. Lee, *Nucl. Phys.* **A513** (1990) 459; S. Nozawa and T.-S.H. Lee, *Nucl. Phys.* **A513** (1990) 511; T.-S.H. Lee and S. Nozawa, private communication
22. H.P. Blok, *Proc. of the Workshop on Future of Nuclear Physics in Europe with Polarized Electrons and Photons*, Orsay, 1990, p.22

Pion Photoproduction and Compton Scattering at Saskatoon (SAL)

J.C. Bergstrom

Saskatchewan Accelerator Laboratory, University of Saskatchewan,
Saskatoon, Saskatchewan Canada S7N 0W0

1. Introduction

It appears that experimental verification of the Low Energy Theorems for neutral pion photoproduction from the nucleon is currently a non-issue. As everyone knows, the historical development of Low Energy Theorems for pion photoproduction culminated in the pivotal work of de Baenst [1] in the early 1970's, who utilized the rather general principles of gauge invariance and PCAC. These results, which I refer to as the classical LETs, expressed the S-wave photopion amplitude called E_{0+} as a power series in the pion mass and were strictly model-independent up to certain specified orders of the expansion parameter. Of course, these LETs do not preclude the possible existence of model-dependent and model-independent terms of higher order, and perhaps of large magnitude, in the physical amplitude. Nevertheless, there was some concern when Saclay in 1986 [2] and Mainz in 1990 [3] reported measurements of the $p(\gamma, \pi^0)$ reaction near threshold which appeared in marked contradiction to the LET prediction. A subsequent re-examination of these claims by others [4, 5] clarified the situation for awhile by demonstrating that the data in fact were not at odds with the LET prediction. Since then the waters have been muddied by proponents of chiral perturbation theory [6] who have argued that the classical LETs are incomplete, which would seem to imply the apparent agreement between the classical LET for $p(\gamma, \pi^0)$ and the experimental data is a coincidence. On the other hand, other workers disagree with this argument [7], and I do not wish to enter the debate here. Whatever the interpretation, everyone agrees that the present chiral calculations for $E_{0+}(\pi^0)$ deviate significantly from the classical LETs, and therefore attempts to experimentally verify an LET for π^0 production are at present difficult if not impossible. This situation is further aggravated by the apparent slow convergence of the chiral expansion.

Leaving aside consideration of the LETs, which are strictly valid only at threshold, let me focus on three photoproduction problems of immediate interest to us in Saskatoon, namely the behavior of E_{0+} just above the π^0 threshold in the proton and ^{3}He, and near-threshold π^+ production from the proton. Finally, I will briefly describe part of the Compton scattering program at SAL.

2. Pion Photoproduction

2.1 $p(\gamma, \pi^0)$ at Low Energies

As the Mainz results [3] presented in Fig. 1 clearly demonstrate, the $E_{0+}(\pi^0 p)$ multipole seems to display a rather remarkable energy dependence, being nearly completely suppressed in the region of the π^+ threshold, beyond which it seems to recover its strength as is evident from multipole analyses of older data. The origin of this behavior likely lies in the isospin splitting of the pion masses which produces a Wigner (or unitarity) cusp from the familiar constraints of unitarity. A theoretical description of the energy dependence of E_{0+} in this region, which is free of phenomenological assumptions (as were employed, for example, in the curve in Fig. 1 [4]), certainly lies within the domain of chiral perturbation theory, but may be uninteresting to theorists since diagrams of two and more loops are involved. However as the quality of the data continues to improve, it may become sufficiently interesting to motivate that very tedious effort.

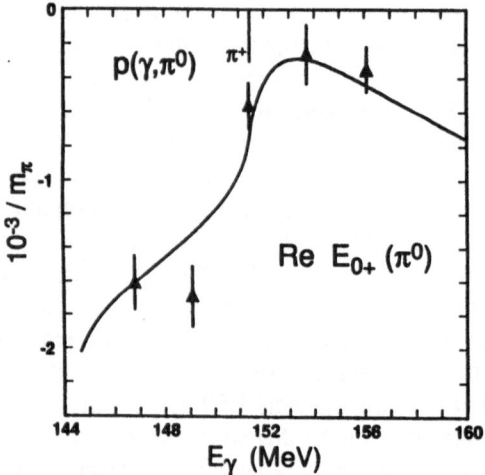

Fig. 1 The multipole $E_{0+}(\pi^0 p)$ depicted by the triangles represents Solution II of the Mainz analysis, described in Ref. [3]. The curve is a phenomenological description [4].

As part of our experimental photopion program at SAL, we are attempting to map out the energy dependence of E_{0+} with an energy resolution of about 0.2 MeV. The aim is to check if the suppression of the multipole is nearly complete near the π^+ threshold, as the Mainz results indicated, and to see if the slope of the amplitude is infinite at π^+ threshold, which is expected for a certain category of unitarity cusps such as the one illustrated in Fig. 1. Finally, we want to verify that E_{0+} achieves a minimum above π^+ threshold rather than at that threshold.

The measurement is being performed using the tagged photon facility at SAL, with a newly installed high-resolution focal plane detector system. The intrinsic resolution of each channel is about 0.17 MeV which increases to about 0.19 MeV when spectrometer optics are considered. The π^0 detector, which we call Igloo (Fig. 2), is a sort of box constructed from 68 lead glass blocks, each 19 x 19 x 30 cm in size. Each glass block has been calibrated using the tagged photon beam, and the corresponding response functions have been incorporated into the Monte Carlo program which simulates the complete π^0 detector. Photomultiplier gain stability is monitored continuously in Igloo by recording cosmic ray events which are effectively collimated in direction by a special trigger arrangement between the glass blocks. The aperture of Igloo is a square about 41 cm on each side, and is 90 cm long. The target is situated on the aperture axis but is shifted upstream slightly to improve the π^0 geometric acceptance. The geometric efficiency for π^0 detection is about 87% of 4π near threshold, and decreases to 77% at 30 MeV above threshold. In this configuration, Igloo is used to measure the total π^0 cross section with high statistical precision, but provides no useful information on the pion angular distribution.

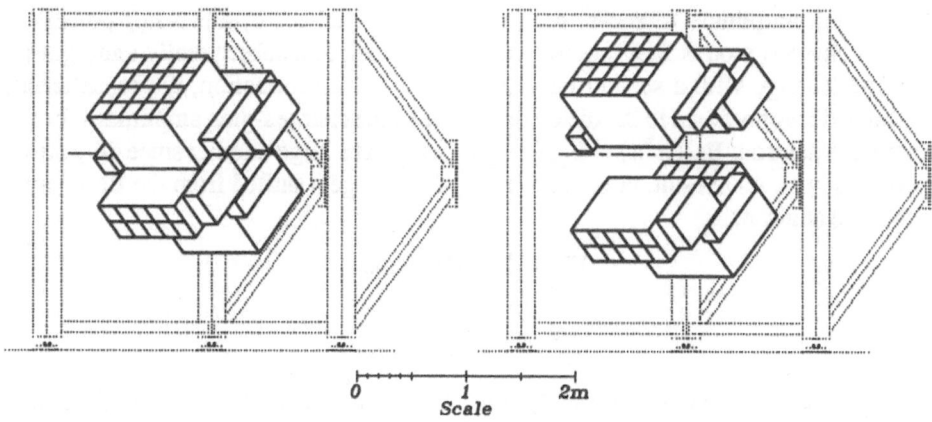

Fig. 2 The π^0 detector Igloo, shown in the closed configuration (left) used for measuring the total $p(\gamma, \pi^0)$ cross section, and open (right) for angular distribution measurements.

Angular distributions are measured by reconfiguring the detector. It has been designed so that it can be cleaved in half along a diagonal of the square aperture, and the two halves separated like the segments of a walnut shell. The angular resolution for reconstructed π^0 events is only about $\Gamma_{FWHM} = 30°$ so extraction of the P-wave contribution will rely on Monte Carlo simulations. The S-wave amplitude will be deduced from a joint analysis of the total cross section and the angular distributions, and will focus on the energy region within 25 MeV of threshold.

Let me make one final technical comment concerning Igloo which has to do with background rejection, often a problem with detectors of this type. Because the pions of interest have relatively low kinetic energies, the opening-angle between the two decay photons is large and therefore many potential pairs of photon "hits" on the lead-glass detectors are geometrically forbidden except perhaps to a background event. From

Monte Carlo simulations we have constructed a series of software masks, one for each tagged photon energy, which summarizes all possible detector pairs corresponding to an acceptable pion. These masks are quite effective in rejecting background events such as pions originating from photons beyond the highest tagged energies. One recognizes of course that this is equivalent to applying opening-angle cuts to the raw data without actually reconstructing the opening-angle itself.

2.2 ^{3}He(γ, π^0) at Low Energies

Our present effort is concentrated on the proton, the first phase of which is scheduled to conclude in mid August '94. Another part of our photopion program involving the Igloo detector will be a similar near-threshold measurement of π^0 production from ^{3}He. It is common knowledge that to the extent the proton spins are saturated, ^{3}He provides a convenient neutron target for a general class of spin-flip transitions, but this is definitely not true for π^0 production. The reason is rescattering [8]. Since the S-wave (γ, π^0) amplitude for the nucleon is an order of magnitude smaller than the charged pion amplitude, direct π^0 production is overwhelmed by a two-step process in which a charged pion is first produced on one nucleon and subsequently undergoes a charge-exchange with a second nucleon. For S-wave production, the rescattering amplitude is proportional to the difference between the charged-pion amplitudes for the proton and neutron, $E_{0+}(\pi^+ n) - E_{0+}(\pi^- p)$, which is a large number since they are of opposite signs. As a result, the effective E_{0+} for ^{3}He gets shifted from the Born value of the free neutron,

$$E_{0+}^B \approx 0.50 \times 10^{-3}/m_\pi$$

to

$$E_{0+}^{eff} \approx -2.8 \times 10^{-3}/m_\pi.$$

The latter estimate is based on an analysis of ^{3}He(γ, π^0) data from Saclay as described in Ref. [9]. Clearly one will have difficulty learning anything about the elusive E_{0+} of the neutron from a study of ^{3}He(γ, π^0).

For isospin $T = 1/2$ nuclei rescattering may also proceed through an intermediate analogue state and yields a second class of rescattering corrections. In the case of ^{3}He this corresponds to

$$\gamma + {}^3\text{He} \rightarrow \pi^+ + {}^3\text{H} \rightarrow \pi^0 + {}^3\text{He}.$$

Such processes are subject to the constraints of unitarity as expressed by Watson's theorem and were considered years ago in a classic paper by Fäldt [10]. In an isospin-symmetric situation such processes only contribute to the imaginary part of the S-wave π^0 amplitude, at least in the simple K-matrix formalism, unlike the rescattering previously discussed. However, when the isospin splitting of the pion masses is taken into consideration, the potential for some interesting physics arises since now the real part of the S-wave amplitude can be modified. This, roughly speaking, is the origin of the peculiar energy dependence of E_{0+} for the proton, and our main purpose in studying ^{3}He(γ, π^0) is to search for a similar modulation of the effective S-wave amplitude in the mass-3 system.

The cusp in the free proton amplitude may reflect itself in ^{3}He since pion production and rescattering can occur on the same nucleon in the second class of rescattering corrections. The question then arises as to whether one should expect to see a similar cusp-like behavior in the effective S-wave amplitude of the deuteron. On the surface, the answer appears to be no since the two-neutron intermediate state is required to have isospin T = 0. Anyway, the low binding energy of the deuteron certainly complicates a similar experimental search.

2.3 $p(\gamma, \pi^+)$ close to threshold

Work will commence at SAL in late summer of '94, in collaboration with the University of Alberta, on a precision measurement of the $\gamma p \rightarrow \pi^+ n$ cross section within 2 MeV of threshold, with the aim of obtaining the S-wave multiple $E_{0+}(\pi^+ n)$ at threshold. The standard value which has been used for years is, in units of $10^{-3}/m_\pi$,

$$E_{0+}(\pi^+ n) = 28.6 \pm 0.2$$

and derives from pre-1976 data. The quoted error should be viewed with some suspicion since those data extend only to within about 10 MeV of threshold. This multipole is of fundamental significance for reasons discussed later, and it is clearly time to update the experimental value.

The measurement will employ tagged photons in the energy range $E_\gamma = 150 - 160$ MeV and a 5 cm thick liquid hydrogen cell. Close to threshold the resulting neutrons are confined to a narrow forward cone and maintain energies well above zero. They are detected in a position-sensitive liquid-scintillator array situated in the forward cone at a distance of 3 meters from the target. This array, which I call the Flys' Eye because it reminds me of one, consists of 84 cells covering neutron angles $\theta_n = 1° - 8°$ with about 1° resolution, with a hole in the center to allow free passage of the photon beam. The geometrical layout is illustrated in Fig. 3. Thin scintillators in front of the array will veto charged particles, while pulse-shape discrimination (PSD) techniques serve to reject photon-induced triggers. Time-of-flight measurements will provide an average neutron kinetic energy resolution of about 0.3 MeV. Determination of θ_n plus the neutron kinetic energy yields the photon energy E_γ, and independent measurements of E_γ in the photon tagger then serves to suppress the background. The detector permits the simultaneous collection of total cross section data within the first 2 MeV above threshold in 0.5 MeV steps.

I stated that $E_{0+}(\pi^+ n)$ was of fundamental significance. Perhaps it might even be interesting. It is fundamental because it describes the inelastic response of the proton to a photon leading to the π^+ channel, just as the polarizability describes the elastic response to the photon. Furthermore, it can be calculated to a relatively good degree of accuracy and, unlike the π^0 channel, the LET derived from chiral perturbation theory is identical to the classical LET of de Baenst and others. This is because 1-loop diagrams apparently start contributing at the order $(m_\pi/M)^2$, while the LET expansion terminates at one power less. Finally, $E_{0+}(\pi^+ n)$ is much less sensitive than $E_{0+}(\pi^0 p)$, relatively speaking, to the effects of isospin splitting of the pion and nucleon masses.

My challenge to the χ_{PT} theorists is this: do they have sufficient confidence in their calculation of $E_{0+}(\pi^+ n)$ that one could use it to determine the πNN coupling constant by comparing with the experimental E_{0+}? As you know, there has been a heated debate in recent years concerning $f^2_{\pi NN}$. The standard value of 0.0790, which seems to be employed in the χ_{PT} calculations, has shrunk to 0.0745 according to the Nijmegen and VPI analyses [11]. I remind you that this would have a 6% effect on the $p(\gamma, \pi^+)$ cross section near threshold, easily detectable in the forthcoming SAL experiment. If this idea has any merit, then the theorists must provide a realistic estimate of the "error bar" on their calculation, aside from $f^2_{\pi NN}$ itself.

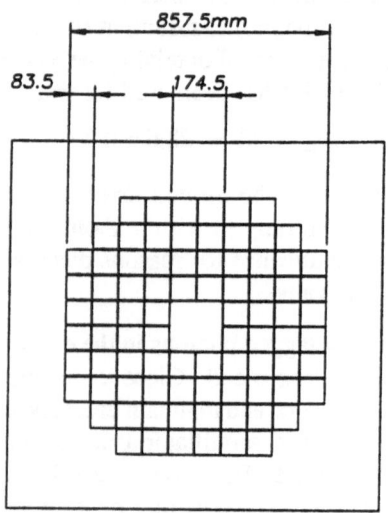

Fig. 3 Geometrical layout of the Fly's Eye neutron detector which will be employed in the measurement of $p(\gamma, \pi^+)$ near threshold.

2.4 Observations concerning $\gamma D \rightarrow np\pi^0$

Let me turn now to the reaction $\gamma D \rightarrow np\pi^0$ near threshold and make a few speculative remarks. At low energy the cross section receives contributions from both the sum and differences of the appropriate proton and neutron multipoles, also called the isovector-even and isoscalar combinations and denoted by A^+ and A^0, respectively. Their relative contributions however are very different.

The essential points can be demonstrated through a simple model of the deuteron using a square-well potential, ignoring the D-state. For the continuum final state we employ a phase-shifted S-wave with the phase shift given by the effective-range approximation. Thus, close to threshold we have two nucleon final states: $^3S_1$ (T = 0) and $^1S_0$ (T = 1). The pion is restricted to S and P waves. The predicted (γ, π^0) total cross sections for each of the final states are illustrated in Fig. 4. Let us now consider the salient features of these results. First, from the general isospin structure of the photopion operator it is easy to demonstrate that only A^+ contributes to the $^3S_1$

transition, while A^0 contributes solely to the $^1S_0$ transition. Now as far as P-wave pions are concerned, the corresponding amplitude A^0 is very small since the dominant elementary amplitudes M_{1+} for the nucleons are of the same sign and roughly equal in magnitude, unlike the S-wave situation. Consequently, P-wave production to the $^1S_0$ state should be strongly suppressed relative to the S-wave production.

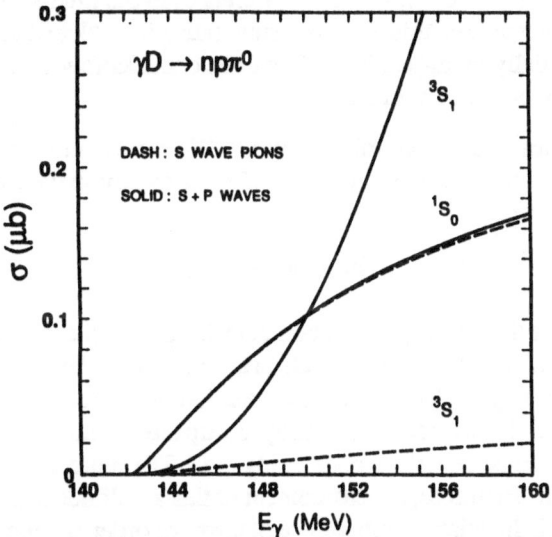

Fig. 4 Total cross section for $\gamma D \to np\pi^0$ near threshold as predicted by the simple model described in the text. Transitions to the $^3S_1$ and $^1S_0$ final states are displayed separately.

Now let us focus on S-wave pion production and employ the dynamics of our simple model. The strong final-state interaction in the $^1S_0$ continuum results in a considerable enhancement of this cross section relative to the $^3S_1$ transition. Specifically, we find the cross section leading to the $^1S_0$ state is roughly a factor of ten larger than the $^3S_1$ cross section within 10 MeV of threshold $(E_{th} = 142.20 \text{ MeV})$.

So, in summary, in the threshold region photoproduction to the $^1S_0$ $(T = 1)$ continuum is predominantly S-wave in nature, while production to the $^3S_1$ $(T = 0)$ continuum is mostly P-wave. Very close to threshold, say within a few MeV, the total cross section is strongly dominated by the former.

What do the two-step pion-rescattering processes do to this picture? They are known to have a profound effect on the cross section for elastic π^0 production from the deuteron, roughly doubling the value of the effective E_{0+} over the Born amplitude $E_{0+} (p\pi^0) + E_{0+} (n\pi^0)$. We can get a rough estimate of the effects by examining the isospin dependence of the rescattering amplitude as determined by the general isospin structures of the (γ, π) and (π, N) amplitudes. Ignoring intermediate π^0 exchange, one finds

$$\frac{< T = 1 \mid M_{\pi N}(2)\, M_{\gamma\pi}(1) \mid T = 0 >}{< T = 0 \mid M_{\pi N}(2)\, M_{\gamma\pi}(1) \mid T = 0 >} = \frac{3}{2}\, \frac{a_1}{a_3 - a_1}\, \cdot\, \frac{E_{0+}\,(n\pi^+) + E_{0+}(\,p\pi^-\,)}{E_{0+}\,(n\pi^+) - E_{0+}(\,p\pi^-\,)}$$

where a_1 and a_3 are the $t = 1/2$ and $t = 3/2$ πN scattering lengths.

Numerically, the above ratio is about 0.084, so taken at face value it would suggest the effects of rescattering are substantially reduced in the $^1S_0$ (T = 1) transition as compared to the $^3S_1$ (T = 0) transition. Differences in the final-state dynamics must also be considered, but at least this is a promising indication. Note that even if the $^3S_1$ transition is substantially modified, it would still be considerably weaker than the $^1S_0$ transition, from previous observations.

So how does all this relate to chiral dynamics? We have demonstrated that close to threshold, the $\gamma D \to np\pi^0$ cross section is almost totally driven by the isoscalar combination of S-wave multipoles,

$$A_S^0 = \frac{1}{2}\, [E_{0+}(p\pi^0) - E_{0+}(n\pi^0)]\, .$$

It is well known that the chiral perturbation calculations of the individual multipoles contain terms of order $(M/F_\pi)^2\mu^2$ which originate from the infamous "triangle" diagram, where $\mu = m_\pi/M$. These are the contentious terms in the corresponding Low Energy Theorems. However, since they completely cancel in the isoscalar combination, there can be no debate concerning the LET prediction for A_S^0, if one can think of such a thing. Therefore, a measurement of this amplitude near threshold could, in principle, establish the relative importance of terms of order μ^3 and higher, which of course lie outside the domain of the classical LETs, and at the same time sidesteps the triangle-diagram issue.

The previous discussion ignores the isospin splitting of the pion masses, which we believe induces a substantial energy dependence in the proton multipole $E_{0+}(p\pi^0)$, and probably in the neutron multipole as well. However since the signs of these induced changes are the same in both cases (positive) and are roughly the same magnitude (at least in a simple rescattering model), their influence on A_S^0 should be somewhat reduced due to their mutual cancellation.

The experimental measurement of the $\gamma D \to np\pi^0$ cross section a few MeV above threshold presents some formidable problems. The pion and one of the nucleons, it makes no difference which one, must be detected in coincidence to exclude the elastic channel $\gamma D \to \pi^0 D$. Although the nucleons are conveniently projected forward within roughly a 30° cone, their kinetic energies are very low. The energy distribution is gaussian-like, peaking at about 2.4 MeV. One might think of detecting the proton by employing an active target, but closer examination shows that electromagnetic shower buildup from the low energy bremsstrahlung photons masks the proton signal. Alternatively, one might consider detecting the neutrons, which easily emerge from a liquid deuterium target. Our Fly's Eye detector array seems ideally suited for this, except the detection efficiency for 2-4 MeV neutrons is very low and difficult to calibrate. So, although we have not overcome the technical obstacles, we continue to pursue the options.

3. Compton Scattering

Testing QCD at low energies is not a simple matter. Neutral pion photoproduction from the nucleons near threshold would serve as a convenient laboratory, were it not for the breaking of the isospin symmetry of the pion masses. As a result, rescattering effects such as $\gamma p \to n \pi^+ \to p \pi^0$ generate a unitarity cusp in the S-wave multipole, and as noted this certainly complicates matters for the χ_{PT} theorists. However, if these folks do rise to the challenge, then I would like to point out another physical process of similar computational challenge that might serve as a QCD testing ground, namely the unitarity cusp in the Compton scattering from the proton. The rescattering process is now $\gamma p \to n \pi^+ \to \gamma p$, and the cusp arises from a competition for flux from the competing π^+ channel: $\gamma p \to n \pi^+$ vs. $\gamma p \to \gamma p$, hence unitarity.

The cusp arises in the Compton multipole amplitude f_{EE}^{1-} which is the Compton analogue of the photopion multipole $E_{0+}(p \pi^0)$, roughly speaking. As can be seen in Fig. 5, it is quite spectacular and displays the classic shape of a unitarity cusp. The singularity, where the energy-derivative is infinite, is situated exactly at π^+ threshold. (Note that the model for $E_{0+}(p \pi^0)$ shown in Fig. 1 has a similar infinite derivative at π^+ threshold. However, the cusp shape is the second variety of the two possible shapes: if $E_{0+}(p \pi^0)$ displayed the same type of cusp as f_{EE}^{1-}, the minimum should occur at π^+ threshold, not above it as the data suggests.)

The first experimental evidence for the cusp in f_{EE}^{1-} was seen in a recent Compton scattering measurement at SAL [12], and reflects itself in the differential cross section as the abrupt change in slope evident in Fig. 6. Part of the abrupt change is due to the switching-on of the imaginary part of the amplitude at π^+ threshold, the remainder is due to the cusp in the real part. The curve in Fig. 6 is a prediction based in part on the f_{EE}^{1-} shown in Fig. 5, which in turn has its genesis in the same phenomenology which produced the curve describing $E_{0+}(p \pi^0)$ in Fig. 1. Clearly however it would be more meaningful to see the power of chiral perturbation theory attack this Compton amplitude, particularly the magnitude of f_{EE}^{1-} at the singularity.

The search for the Compton cusp was an after thought of an experimental program mainly devoted to higher energies, which explains the rather spare amount of data in Fig. 6. We plan to return to the threshold region in the near future at SAL and mount a more intensive effort of Compton scattering across the threshold region, and at smaller scattering angles where the cusp influence on the cross section is more enhanced.

I have not described the recent measurements at SAL on the polarizabilities of the proton, a favorite playground of chiral theorists. Let me conclude by saying that planning is underway with the Illinois group to try and learn something about the neutron polarizabilities through quasielastic Compton scattering on the deuteron, observing the forward - recoiling neutron with the Fly's Eye array. A simultaneous measurement of recoiling protons provides at least some mechanism for calibrating the apparatus, and perhaps also calibrating the nuclear model-dependence of this difficult proposal.

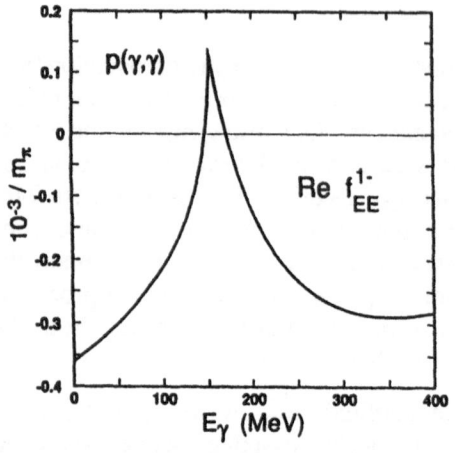

Fig. 5 Real part of the proton Compton multipole f_{EE}^{1-} illustrating the pronounced cusp at π^+ threshold. The curve is given by a semiphenomenological model as described in Ref. [12].

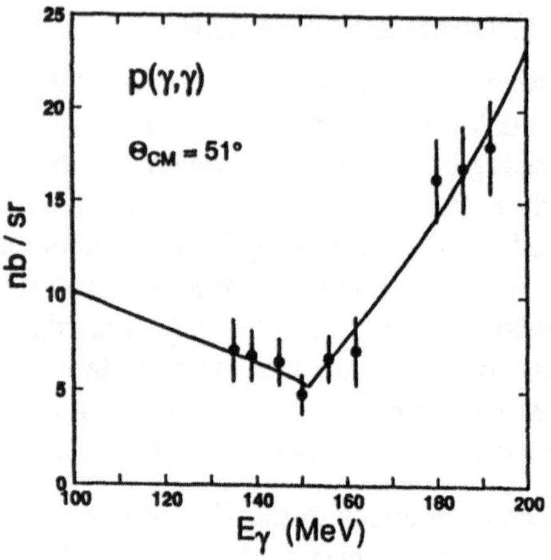

Fig. 6 Differential cross section for proton Compton scattering at $\theta_{cm} = 51°$ as measured at SAL [12]. The abrupt change in slope reflects both the cusp in $\mathrm{Re} f_{EE}^{1-}$ and the switching-on of $\mathrm{Im} f_{EE}^{1-}$. The line is the predicted cross section as described in Ref. [12].

References

1. P. de Baenst, Nucl. Phys. **B24**, 633 (1970).
2. E. Mazzucato *et al.*, Phys. Rev. Lett. **57**, 3144 (1986).
3. R. Beck *et al.*, Phys. Rev. Lett. **65**, 1841 (1990).
4. J.C. Bergstrom, Phys. Rev. **C44**, 1768 (1991).
5. A.M. Bernstein and B.R. Holstein, Comments Nucl. Part. Phys. **20**, 197 (1991).
6. V. Bernard, N. Kaiser, J. Gasser and Ulf-G. Mei β ner, Phys. Lett. **B268**, 291 (1991).
7. S. Scherer, J.H. Koch and J.L. Friar, Nucl. Phys. **A552**, 515 (1993).
8. J.H. Koch and R.M. Woloshyn, Phys. Lett. **60B**, 221 (1976).
9. J.C. Bergstrom, in Proceedings of the Seventh Mini-Conference at NIKHEF, Dec., 1991, Edited by H.P. Blok, J.H. Koch and H. de Vries.
10. G. Fäldt, Nucl. Phys. **A333**, 357 (1980).
11. V. Stoks, R. Timmermans and J.J. deSwart, Phys. Rev. **C47**, 512 (1993), and references cited therein.
12. E.L. Hallin *et al.*, Phys. Rev. **C48**, 1497 (1993); J.C. Bergstrom and E.L. Hallin, Phys. Rev. **C48**, 1508 (1993).

Threshold Photo/Electro Pion Production –. Working Group Summary

Ulf-G. Meißner[1], B. Schoch[2]

[1] Centre de Recherches Nucléaire, Physique Théorique, BP 28 Cr, F–67037 Strasbourg Cedex 2, France
[2] Universität Bonn, Institut für Physik, Nußallee 12, D–53115 Bonn, Germany

1 Introduction

Over the last few years, pion production off nucleons by real or virtual photons has become a central issue in the study of the non–perturbative structure of the nucleon, i.e. at low energies. Here, developments in detector and accelerator technology on the experimental side as well as better calculational tools on the theoretical one have allowed to gain more insight into detailed aspects of these processes and the physics behind them. One main trigger were the two papers by the Saclay and the Mainz groups [1,2], which seemed to indicate the violation of a so–called low energy theorem for the reaction $\gamma p \to \pi^0 p$. This lead to a flurry of further experimental and theoretical investigations. Another cornerstone was the rather precise electroproduction measurement $\gamma^* p \to \pi^0 p$ at NIKHEF [3]. Here, we wish to summarize the state of the art in calculating and measuring these processes in the threshold region. Furthermore, we outline what we believe have crystalized as the pertinent activities to be done in the near future.

2 Theoretical developments

The chiral perturbation theory (CHPT) machinery to calculate the reactions $\gamma N \to \pi^a N$, $\gamma^* N \to \pi^a N$ and $\gamma^{(*)} N \to \pi^a \pi^b N$, where N denotes the nucleon, π^a a pion of isospin $'a'$ and $\gamma (\gamma^*)$ the real (virtual) photon exists as described in some detail in V. Bernard's lecture [4]. It has become clear that to have precise predictions one has to calculate in the one loop approximation to order q^4 in the effective Lagrangian,

$$\mathcal{L}_{\text{eff}} = \mathcal{L}_{\pi N}^{(1)} + \mathcal{L}_{\pi N}^{(2)} + \mathcal{L}_{\pi N}^{(3)} + \mathcal{L}_{\pi N}^{(4)} \tag{2.1}$$

where the subsript $'\pi N'$ means that we restrict ourselves to the two flavor case (the pion–nucleon–photon system) and the superscript $'(i)'$ denotes the chiral dimension. While the first term in eq.(2.1) is given entirely in terms of well determined parameters, the string of terms of order q^2, q^3 and q^4 contains the so–called low–energy constants (LECs). At present, their determination induces the biggest uncertainty in

Summary talk presented at the Workshop on Chiral Dynamics: Theory and Experiments, Massachusetts Institute of Technology, Cambridge, USA, July 25 - 29, 1994, CRN 94-51 the chiral predictions since not enough sufficiently accurate low–energy data exist to uniquely pin them down. Therefore, in the case of threshold photo/electro pion production, we follow two approaches. The first one is "clean" CHPT, were the extensive photoproduction data (total and differential cross sections in the threshold region) are used to determine the appearing LECs (see the discussion in Bernard's lecture [4]). In that case, the predictive power resides in the photoproduction multipoles, polarization observables and the electroproduction processes. For the latter, some novel LECs appear, but they are all connected to single nucleon properties and thus can uniquely be fixed from the electromagnetic and axial radii of the nucleon. With that, the k^2 dependence (where $k^2 < 0$ denotes the photon four–momentum squared) of all electroproduction observables and multipoles is *parameter–free* fixed. For detailed calculations and predictions to order q^3 consult the *Physics Reports* [5]. Furthermore, one gains new insight due to the longitudinal coupling of the virtual photon to the nucleon spin. This rich field has only be glimpsed at and will serve as a good testing ground for the chiral predictions.

Now let us discuss what the limitations of such calculations are, i.e. to what values of $|k^2|$ and $\Delta W = W - W_{\text{thr}}$ (with W the center–of–mass energy of the πN system) these calcualtions to order q^4 can be trusted? The answer is, of course, dependent on the observable one looks at. Nevertheless, a good example is given by the one loop calculation of the S–wave cross section a_0 in $\gamma^* p \to \pi^0 p$ [6] in comparison to the data of ref.[3]. Although the trend of the data is well reproduced up to $k^2 = -0.1$ GeV^2, it is obvious that for such large four–momentum transfer ($|k| \simeq 2.3 M_\pi$) the one–loop corrections to the tree result are so large that they can not be trusted quantitatively, i.e. one would have to calculate further in the chiral expansion. From the extensive study of elastic pion–pion scattering in the threshold region[7], which is the purest process to test the chiral dynamics, we employ here the same rule of thumb advocated there, namely that as long as the one–loop corrections stay below 50% of the tree result, the predictions can be considered quantitative (accurate). For the S–wave cross section a_0 and many other observables, we conclude that for a rigorous test of the chiral dynamics, electroproduction experiments should be performed with

$$|k^2| \le 0.05 \,\text{GeV}^2 \qquad \Delta W \le 15 \,\text{MeV} \qquad (2.2)$$

where the number for ΔW is derived from photoproduction calculations and deserves more detailed studies. In any case, it is conceivable that one should stay as close to threshold (and the photon point) as experimentally possible.

The second approach to get a handle on the LECs involves the principle of resonance saturation (as discussed in some detail in these proceedings by Ecker [8] and by one of us [9]). Consider e.g. a contact term of order q^3 with an incoming photon and one outgoing pion. The resonance saturation hypothesis means that the numerical value of the LEC related to this operator is given by baryon excitations like the $\Delta(1232)$, the $N^*(1440)$ and others (as discussed here e.g. by Mukhopadhyay [10]) and also t–channel meson exchanges of scalar, vector or axial–vector type. Shrinking such types of diagrams to a point, the pertinent LEC is given in terms of masses and coupling constants related to the particles integrated out from the effective field theory. This, of course, induces some model–dependence but can on the other hand serve as a guideline to understand the numerical values of the LECs from some microscopic picture. However, we would like to stress again that it is preferable to have *enough* data to pin down all these low–energy constants, thereby testing the accurary of the resonance saturation idea (which works very well for strong and semi–leptonic interactions in the meson sector, but not for the non–leptonic weak interactions). This principle could then be used to estimate the contributions of higher order terms as it is frequently done in the meson sector. Also, we wish to point out that all this resonance physics seen in the threshold region is, of course, included in the phenomenological LECs. As long as one is not so ambitious to make statements throughout or above the region of nucleon excitations, *resonance contributions do not pose any problem to the consistent chiral expansion.*

We would also like to stress the importance of polarization observables. These can be calculated with ease and they serve as a sensitive filter in the multipole analysis which otherwise is difficult to perform and often not unique (in the sense that one has to determine more multipoles than the number of available observables). Another important topic is the neutron. Although no free neutron targets are at our disposal, measurements using the deuteron can serve as neutron probes. For that, it is mandatory to have a precise description of the weakly bound proton–neutron system. This can e.g. be achieved in the relativistic model of Gross and collaborators [11]. Clearly, some model–dependence is induced, but one hopes to be able to minimize this by a) measuring also the proton in the deuteron and comparing to its free space values and b) chosing particular kinematics where the corrections due to meson exchange currents and alike are minimized.

What are the theoretical improvements needed? First, in certain channels (where one has strong final–state interactions) it is mandatory to perform calculations including higher order effects. The most efficient machinery to do this is a combination of dispersion theory with CHPT as discussed in detail for pion form factors or $K_{\ell 4}$–decays in refs.[12]. In essence, consider an observable Q and write a dispersive representation,

$$Q = P(s)\,\Omega_\Lambda(s) \qquad (2.3)$$

where the reduced Omnès function $\Omega_\Lambda(s)$ accounts for the phase information (and sums up loops to very high orders), the polynomial $P(s)$ is constrained by chiral symmetry and s is a short–hand notation for the kinematical variables involved. Such a method has e.g. been applied succesfully in the determination of the πN σ–term and the scalar form factor of the nucleon as discussed by Sainio [13]. Also, it is an open question to what extent one should include the $\Delta(1232)$ as a dynamical degree of freedom in the effective field theory. We believe that for the threshold phenomena to be mapped out by the experiments within the constraints of eq.(2.2), it suffices to use the Δ to estimate LECs and thus treat it as a frozen (integrated out) d.o.f. The most important theoretical issue to be studied in more detail is the role of isospin breaking through virtual photons and the quark mass difference $m_u - m_d$. At present, one puts in the different masses for the neutral and charged pions by hand in a manner consistent with all the symmetries and Ward identities. It is well-known that the pion mass differnec is essentially of electromagnetic origin, leading to the believe that this approximation accounts for the most important aspects of isospin breaking. Nevertheless, this has to be clarified in a detailed and thorough study. In any case, threshold pion photo– and electroproduction can not be considered alone but it is intimately connected to other processes like Compton scattering, pion-nucleon scattering, $\pi N \rightarrow \pi\pi N$ and so on. This interplay of the various reactions is of particular importance for the determinations of the LECs.

Finally, a few words about the extension to three flavors are in order. Here, the theoretical situation is much less satisfactory mostly due to the large kaon loop contributions and huge cancellations with counterterms. Stated differently, the intrinsic small parameter is $(M_K/4\pi F_\pi)^2 \simeq (0.4)^2$ (modulo logarithms) and thus it is mandatory to perform higher order calculations than for the two flavor case. This becomes particularly evident in the scalar sector of the baryon masses and σ–terms as detailed in ref.[14]. As long as a consistent theoretical description of these 2– and 3–point functions is eluding us, we can not make statements about kaon and eta production off the nucleon. However, beautiful new data for K and η production in the threshold region are becoming available ($\gamma p \rightarrow \eta p$, $\gamma p \rightarrow \Sigma K$ or $\gamma p \rightarrow \Lambda K$) which should lead to more theoretical effort to understand the chiral SU(3) meson–baryon system.

3 Experimental developments

Driven by the earlier results [1], [2] of the π^0–production at threshold on the proton and the discussion about the validity of low energy theorems experimental activities started in several laboratories in order to study the photo–and electroproduction of charged and neutral pions. The experiment of the electroproduction of neutral pions on the proton at NIKHEF [3] demonstrated the possibility to perform these experiments with the precision to be able to discriminate between different theoretical approaches. H. Blok, Vrije Universiteit Amsterdam, showed preliminary results of the $^1H(e, e'p)\pi^0$–reaction at a momentum transfer of $|k|^2 = 0.1$ (GeV/c)^2, thereby, covering an invariant mass range of 2 to 15 MeV above the production threshold. M. Distler, Universität Mainz, presented the first data of the $^1H(e, e'p)\pi^0$–reaction with the new spectrometer set-up at MAMI. Measurements at two polarizations of the virtual photon ($\varepsilon_1 = 0.89$ and $\varepsilon_2 = 0.52$) allow a separation of the transverse and longitudinal contributions.

Due to the large acceptance in the threshold region the out–of–plane cross sections σ_{TL} and σ_{TT} can be determined. The momentum transfer of $|k|^2 = 0.1$ (GeV/C)^2 like in the NIKHEF–experiment exceeds, however, (see equation 2.2) the region of the validity of present calculations within the framework of CHPT. Nevertheless, for the first time these high quality data will allow the extraction of the S– and P–wave multipoles. In addition to these measurements results of the electroproduction of positive pions from an experiment carried out at MAMI have been announced. Besides the possibility to break down the amplitudes into their isospin pieces the extraction of the axial formfactor at low $|k|^2$ becomes possible. Again, the results of calculations based on CHPT can be checked and compared with results of neutrino induced reactions.

J. Bergstroem, Saskatoon, reported about an ongoing experiment $^1H(\gamma, \pi^0)$ at SAL (Saskatoon) and R. Beck (Mainz) described a planned experiment of the reaction $^1H(\vec\gamma, \pi^0)$ at threshold. This experiment will allow a clean separation of the S– and P–wave contribution to the cross section in the region below and above the threshold of positive pions.

The electric amplitudes of the charged channels are fixed to the first order by the Kroll–Rudermann term. So far, the published experimental data agree within a few percent with this prediction. However, in the π^+–case there exist no data of the differential cross section close to threshold. Either an extrapolation over more than 5 MeV was performed or the knowledge about the P–wave was used in order to extract the electric multipole from the measurement of the total cross section. J. Bergstroem reported about a recent experiment of the production of negative pions on the deuteron from SAL. F. Klein discussed results for this reaction from Mainz and addressed the problems which arise in order to get information for the elementary amplitude on the neutron.

These problems do not arise by investigating the inverse reaction $\pi^- p \rightarrow \gamma n$. M. Kovash presented new, again preliminary data from TRIUMF which have the accuracy to check the electric amplitude for negative pion production extracted, so far, from a measurement of the Panofski–ratio. As a by–product of this measurement the Panofski–ratio was measured, simultaneously, and, given the impressive quality of the raw data, an improved value for this ratio should be obtained as Pavan demonstrated in reviewing the previous measurements.

These new experimental results demonstrate the renewed interest for the physics questions and show the impressive progress in the experimental techniques.

On this basis even more difficult experiments can be envisaged. The measurement of the neutral pion production on the neutron constitutes such an experiment which would, combined with the older results, check the validity of the isospin decomposition of the production amplitudes. In this connection A. Bernstein proposed a measurement using a polarized beam and a polarized target in order to measure very close to threshold the phase shift of the $\pi^0 - p$ scattering process. Precise measurements on the proton and the neutron would yield a high sensitivity for isospin–breaking effects due to the difference of the current mass of the u– and d–quarks. High precision data from MAMI and ELSA (Bonn) at the η–threshold are now available for the proton as well as for the neutron. Also data for the kaon–production have been extracted from measurements with the SAPHIR-detector on ELSA. Besides total and differential cross sections the measurement of polarization observables allow in the near future a decomposition in all multipoles. The difficulties to describe these reactions within the framework of CHPT have been addressed in chapter 2.

4 Concluding remarks

Remarkable progress has been achieved in recent years concerning the understanding of the pion production at threshold. Due to the progress in the development of CHPT the foundations of Low Energy Theorems have been revisited and put into a new perspective (for a tutorial, see e.g. ref.[15]). CHPT provides a natural link for *all* low energy pion reactions. The development of a new generation of electron accelerators in combination with refined detector techniques allow precision experiments, a precondition to explore the implications of the spontaneously broken chiral symmetry of QCD in the baryon sector.

5 Acknowledgements

We would like to thank everybody in the working group for presenting their ideas and stimulating discussions.

References

[1] E. Mazzucato et al., *Phys. Rev. Lett.* **57** (1986) 3144

[2] R. Beck et al., *Phys. Rev. Lett.* **65** (1990) 1841

[3] T. P. Welch et al., *Phys. Rev. Lett.* **69** (1992) 2761

[4] V. Bernard, these proceedings

[5] V. Bernard, N. Kaiser, T.–S. H. Lee and Ulf-G. Meißner, *Phys. Reports* (1994), in print

[6] V. Bernard, N. Kaiser, T.-S.H. Lee and Ulf-G. Meißner, *Phys. Rev. Lett.* **70** (1993) 387

[7] J. Gasser and Ulf-G. Meißner, *Phys. Lett.* **B258** (1991) 219

[8] G. Ecker, these proceedings

[9] Ulf-G. Meißner, these proceedings

[10] R.M. Davidson and N.C. Mukhopadhyay, *Phys. Rev. Lett.* **60** (1988) 746; M. Benmerrouche and N.C. Mukhopadhyay, *ibid*, **67** (1991) 1070; M. Benmerrouche, N.C. Mukhopadhyay and J.F. Zhang, submitted to *Phys. Rev.* D

[11] F. Gross and D.O. Riska, *Phys. Rev.* **C36** (1987) 1928; F. Gross, *Czech. J. Phys.* **B39** (1989) 871; J.W. van Orden, F. Gross and N. Devine, in preparation

[12] J. Gasser and Ulf-G. Meißner, *Nucl. Phys.* **B357** (1991) 91; J. Bijnens et al., BUTP–94/4, to appear

[13] M.E. Sainio, these proceedings

[14] V. Bernard, N. Kaiser and Ulf-G. Meißner, *Z. Phys.* **C60** (1993) 111

[15] G. Ecker and Ulf-G. Meißner, "What is a Low–Enegy Theorem?", preprint CRN 94/51 and UWThPh-1994-33 (1994)

Part VII: Laboratory Summaries

Spin Physics with Polarized Electrons: MIT-Bates Program

Stanley Kowalski

Bates Linear Accelerator Center
Laboratory for Nuclear Science and Department of Physics
Massachusetts Institute of Technology
Cambridge, Massachusetts 02139 USA

I Introduction

The electromagnetic probe has long been an important tool for studying nucleon and nuclear structure and dynamics. High energy electron scattering can probe distances of less than 1 fm where quark degrees of freedom are expected to play an observable role. A new generation of medium energy accelerators, operating CW with energies of $0.5 - 6$ GeV, has recently come into operation. These enhanced capabilities, particularly the 100% duty factor and kinematic reach, will dramatically improve our ability to carry out coincidence studies. Storage ring facilities at MIT-Bates and NIKHEF make possible unique studies with thin polarized and unpolarized internal nuclear targets at relatively high luminosity.

Polarized electrons have until recently played only a relatively minor role in nuclear physics. At high energies, the SLAC parity violation [1] experiment provided a crucial early test of our understanding of electroweak processes. Later measurements at Bates and MAINZ extended these electroweak studies to lower energies.

Scattering of high energy polarized leptons from polarized targets has recently provided us with very detailed information on the spin structure of the nucleon. The resulting "spin crisis" has generated a flurry of theoretical and experimental activity. A large fraction of the nuclear physics programs at MIT-Bates, MAINZ and CEBAF involves the use of longitudinally polarized electrons. The spectrum of physics includes nucleon and nuclear structure, as well as electroweak studies. Recent theoretical efforts [2] have demonstrated that the measurement of spin observables provides a unique opportunity for addressing some long standing physics problems.

In this review, we summarize the status and results of some recent experiments at MIT-Bates which exploit the measurement of spin observables to study nucleon and nuclear structure. The use of polarized beams, polarized targets, recoil polarimetry and coincidence measurements of reaction products out-of-the scattering plane allows us to take full advantage of the capabilities of electron scattering. We also discuss some very promising experiments, utilizing unique new detector capabilities, which are planned for the near future.

II Neutron Form Factors

The electromagnetic structure of the nucleon is of fundamental importance to our understanding of models of the nucleon and nuclei. The neutron electric form factor, G_E^n, is the least well known of the four nucleon electromagnetic form factors.

The elastic scattering of unpolarized electrons from unpolarized nucleons ($I = \frac{1}{2}$), at a momentum transfer Q, involves a measurement of the cross section

$$\frac{d\sigma}{d\Omega} = \sigma_M \, f_{rec}^{-1} \, [(G_E^N)^2 \, + \, \tau(G_M^N)^2 \, \{1 \, + \, 2(1+\tau) \, \tan^2\frac{\theta}{2}\}]$$

where $\tau = -Q^2/4M_N^2$ and f_{rec}^{-1} is a kinematic recoil factor. A Rosenbluth separation of the electric and magnetic form factors, $G_E^N \, (Q^2)$ and $G_M^N \, (Q^2)$, allows for reasonable accuracy only when the two amplitudes are comparable. In the nucleon case the magnetic form factor dominates over the electric one at high momentum transfers. As a result, only the magnetic form factor is relatively well known over an extended range in momentum transfer. For the proton, reasonable knowledge of G_E^p exists only up to ~ 4 GeV^2. For the neutron, which is charge neutral, G_E^n is very small and as a result is very poorly known for all momentum transfers.

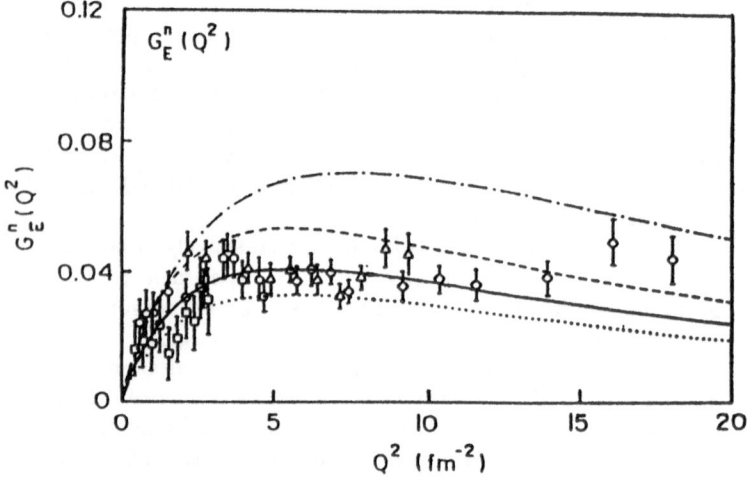

Fig. 1. The neutron electric form factor as deduced from $A(Q^2)$ by unfolding the Paris potential (solid line). Unfolding with other potentials yields form factors as shown by the different curves [3].

At present, much of our information on G_E^n comes from measurements of $A(Q^2)$ from electron-deuteron elastic and quasi-elastic scattering. The neutron electric form factor has recently been extracted from some precise new measurements [3] on $A(Q^2)$. The results for G_E^n are shown in Fig. 1 using the Paris potential wave function for the deuteron. Unfolding with other potentials yields different results for G_E^n. The resulting systematic dependences are as large as $\sim$50%. Scattering of thermal neutrons from atomic electrons measures the slope of G_E^n at $Q^2 = 0$ and the charge radius of the neutron. The result for the slope is

$$\left(\frac{dG_E^n}{dQ^2}\right)_{Q^2=0} = 0.0195 \pm .0004 \text{ fm}^2$$

Spin observables will allow us to make more accurate measurements of G_E^n. For the nucleon case experiments with both polarized beams and targets (or alternatively recoil polarimetry) are required. The differential cross section for inclusive scattering of longitudinally polarized electrons from a spin$\frac{1}{2}$ target may be written [2] as

$$\frac{d\sigma}{d\Omega d\omega} = \Sigma \pm \Delta(\theta^*, \phi^*),$$

where ω is the energy transfer and the angles θ^* and ϕ^* define the target spin direction relative to Q and the scattering plane as shown in Fig. 2.

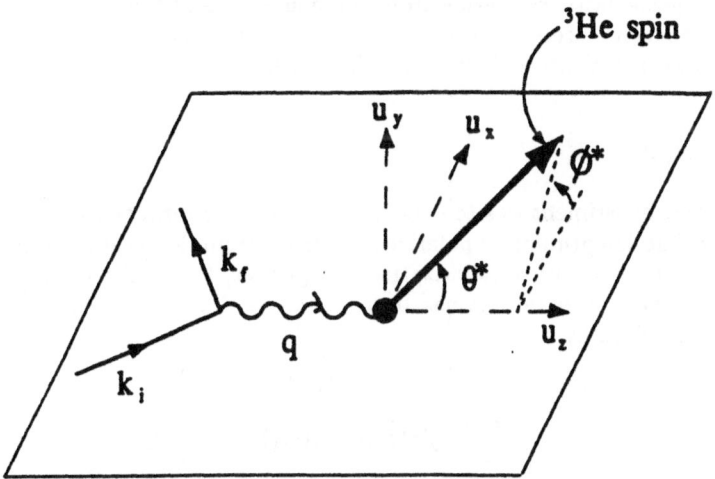

Fig. 2. Kinematic variables for electron scattering from polarized targets. The vector μ_z is along the direction of momentum transfer $\vec{q}$. The vector μ_y is normal to the electron scattering plane and $\mu_x = \mu_y \times \mu_z$ lies in the scattering plane. The target polarization direction is defined by the angles (θ^*, ϕ^*) as shown.

The plus (minus) sign corresponds to positive (negative) helicity of the incident electrons. The spin- independent cross section Σ has the usual Rosenbluth form

$$\Sigma = 4\pi\sigma_M[v_L R_L(q,\omega) + v_T R_T(q,\omega)],$$

where q is the momentum transfer, v_L and v_T are kinematic factors, R_L and R_T are the longitudinal and transverse response functions. These response functions are dependent on bilinear combinations of the nucleon electromagnetic form factors. The spin- dependent cross section Δ involves two other response functions:

$$\Delta = -4\pi\sigma_M\ [v_{T'} R_{T'}\ (q,\omega)\ \cos\theta^* + 2v_{TL'} R_{TL'}(q,\omega)\ \sin\theta^* \cos\phi^*],$$

where $v_{T'}$ and $v_{TL'}$ are kinematic factors independent of nuclear structure.

Experiments measure the asymmetry in the cross section under reversal of electron helicity: $A = \Delta/\Sigma$. The experimental asymmetry A_{exp} is related to A by $A_{exp} = P_e P_T A$, where P_e and P_T are the electron beam and target polarizations, respectively. The electron polarization $P_e = 0.40$ typically.

In the case of a polarized neutron target, the asymmetry may be expressed in terms of the electric and magnetic form factors as

$$A = \frac{-2\tau v_{T'}(G_M^n)^2 \cos\theta^* - 2\sqrt{2\tau(1+\tau)}\, v_{TL'} G_M^n G_E^n \sin\theta^* \cos\phi^*}{(1+\tau)v_L(G_E^n)^2 + 2\tau v_T(G_M^n)^2}$$

A measurement of the asymmetry for $\theta^* = \pi/2$ involves the interference term $G_M^n\, G_E^n$ which is directly sensitive to the small electric form factor and the relative sign.

Several experiments have been designed to exploit the measurement of this interference term and provide precise new information on G_E^n. The two most important ones include quasi-elastic scattering from the deuteron and from polarized ^{3}He. Initial measurements have recently been completed at MIT-Bates and at MAINZ. In the future, similar experiments at higher Q^2 will also be done at CEBAF.

i) ^{2}H$(\vec{e}, e'\vec{n})$p

The goal of this experiment was to determine the electric form factor of the neutron, G_E^n, by scattering longitudinally polarized electrons from deuterium quasi-elastically and measuring the transverse polarization component, P_x, of the recoiling neutron. In the impulse approximation, it has been shown [4] that the neutron polarization components are given by:

$$P_x = -2\sqrt{\tau(1+\tau)}\, G_M^n\, G_E^n/I_o,$$

$$P_z = \frac{E + E'}{M}\, \sqrt{\tau(1+\tau)}\, (G_M^n)^2\, \tan^2\frac{\theta}{2}/I_o,$$

where $\tau = Q^2/4M^2$ and the unpolarized cross section

$$I_o = (G_E^n)^2 + \tau(G_M^n)^2\, [1 + 2(1+\tau)\tan^2\theta/2].$$

The neutron polarization component, P_x, in the scattering plane normal to the neutron momentum is seen to be directly proportional to G_E^n.

The experiment was designed for quasifree kinematics. Calculations by Arenhövel [5] show that at these kinematics the recoil neutron polarization has almost no dependence on the deuteron model and is insensitive to the influences of final state interactions, meson- exchange currents, and isobar configurations.

In the Bates experiment [6], a longitudinally polarized electron beam, with an energy of 868 MeV, was incident on a liquid deuterium target (0.845 g/cm^3). The scattered electrons were momentum-analyzed in a magnetic spectrometer. The recoil neutrons were detected in coincidence with the scattered electrons, and their transverse polarization measured. Their energy was obtained from a measurement of the neutron time-of- flight from the target to the front scatterers in the polarimeter. The experiment operated at a luminosity of 3×10^{36} cm^{-2} s^{-1}, using a pulsed beam with a duty factor of 1%.

The neutron polarimeter [7] was designed and constructed at Kent State University especially for these measurements. It consists of four active scintillators acting as primary scatterers and sets of four plastic detectors as up- down analyzers. The design has been optimized for using n-p elastic scattering as a polarization analyzer. This polarimeter was tested and calibrated at the Indiana University Cyclotron Facility. The average analyzing power was measured to be $\bar{A}_y = 0.43$. The polarimeter is housed in a heavily shielded enclosure designed to reduce photon and neutron backgrounds.

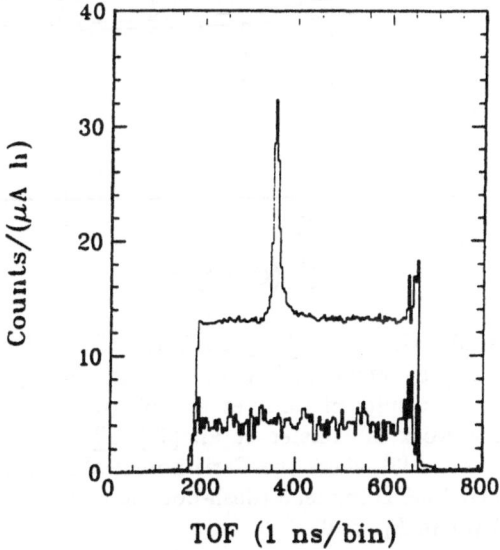

Fig. 3 Prompt neutron time-of-flight spectrum. The peak corresponds to a coincidence detection of a scattered electron and a neutron in the polarimeter.

Fig. 3 shows a time-of-flight spectrum of the neutrons measured with respect to the detection of an electron. The spectrum corresponds to triple coincidences between an electron in the magnetic spectrometer and a neutron in the first detector and in one of the eight rear detectors. Software corrections have been applied to eliminate events in the rear detectors of the polarimeter that are not correlated with electrons. The average beam current was $\sim 2~\mu A$ and the background due to accidental coincidences is clearly evident, making such a measurement extremely difficult. Fig. 4 displays the measured value of G_E^n. It is in agreement with the Galster parameterization.

The expressions for P_x and P_z can be used to calculate the ratio P_x/P_z:

$$\frac{P_x}{P_z} = -\left[\frac{2M}{E+E'}\right] \frac{G_E^n}{G_M^n} \tan^{-1}\theta/2$$

A measurement of this ratio would be largely insensitive to both the electron polarization and polarimeter analyzing power. The longitudinal polarization can be measured with the same polarimeter after precessing the neutron spin by $\pi/2$ with a dipole magnet.

Further experiments are planned at Bates at higher Q^2 using extracted CW beams from the South Hall Ring. Similar experiments have been carried out at MAINZ and are also planned for CEBAF at higher Q^2.

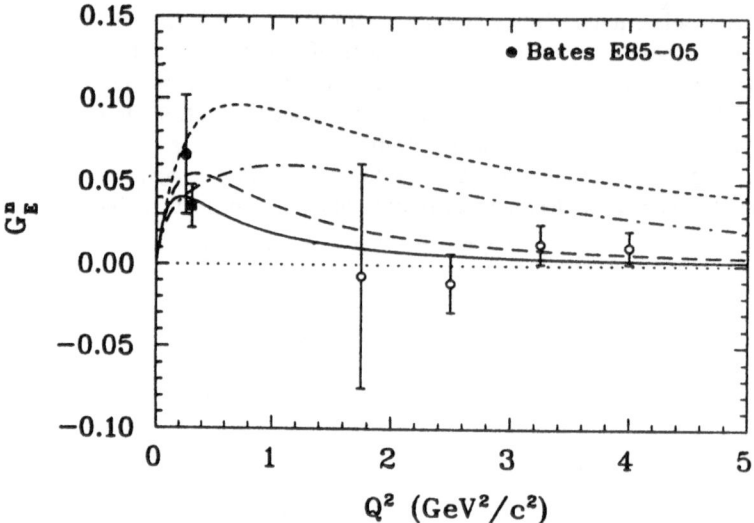

Fig. 4. Recently reported values of G_E^n: quasielastic $^2H(\vec{e}, e'\vec{n})$ reaction (shaded circle), exclusive $^3H\vec{e}(\vec{e}, e'n)$ by Meyerhoff et al. [35] (shaded square), and the inclusive quasielastic $^2H(e, e')$ results of Lung et al. [36] (open circles). The parameterizations are from the work of Galster et al. [37] $G_E^n = -\tau \mu_n (1+5.6\ \tau)^{-1}$ (long dashes), Platchkov et al [3] where the Paris potential fit is shown (solid), the Gari- Krümpelmann VMD-PQCD model 3 (dash-dot), and the parameterization $G_E^n = -\tau G_M^n$ (short dashes) where $F_{1n} = 0$.

ii) $^3H\vec{e}(\vec{e}, e')$

A program of inclusive scattering of longitudinally polarized electrons from high density polarized ^{3}He targets is underway at Bates. Its primary goal is to measure neutron form factors in the momentum transfer range of $Q^2 = 0.2$ to 0.75 GeV^2.

^{3}He is a very interesting nucleus for electromagnetic studies. It can be polarized and one expects that spin-dependent scattering from ^{3}He would be sensitive to neutron electromagnetic form factors. To a good approximation, the two protons in the ground-state are in a singlet configuration. Plane-wave impulse approximation calculations by Blankleider and Woloshyn [8] have demonstrated that in the vicinity of the quasi-elastic peak, spin- dependent effects are determined primarily by scattering from the neutron. More sophisticated calculations have also been performed by Friar et al. [9], Ciofi degli Atti and co-workers [10] and by Schulze and Sauer [11]. The latter results are based on a full spin-dependent spectral function to describe the ground state and compute the spin-dependent inclusive scattering in PWIA.

For the case of ^{3}He, $(I = \frac{1}{2})$, in analogy with the nucleon case, the asymmetry is given by

$$A = -\frac{v_{T'}R_{T'}\cos\theta^* + 2v_{TL'}\ R_{TL'}\sin\theta^*\ \cos\phi^*}{v_L R_L + v_T R_T},$$

where the v_k are kinematic factors. $R_L(Q^2,\omega)$ and $R_T(Q^2,\omega)$ are the longitudinal and transverse response functions contributing to the spin-independent cross section. $R_{T'}(Q^2,\omega)$ and $R_{TL'}(Q^2,\omega)$ are two new response functions that do not contribute to the unpolarized cross section. By choosing the target spin to be along $\vec{q}(\theta^* = 0°)$ or normal to $\vec{q}(\theta^* = 90°)$, one can select the asymmetry to be proportional to $R_{T'}$ or $R_{TL'}$, respectively.

All calculations show that the spin-dependent asymmetry for $\vec{q}$ parallel to the target spin is proportional to $(G_M^n)^2$. In comparison to a target of polarized neutrons, ^{3}He dilutes the asymmetry since proton scattering contributes to the unpolarized cross section. In addition, small amplitudes in the ^{3}He wave function which are due to polarized protons also contribute to the asymmetry.

Similarly, the spin-dependent asymmetry for $\vec{q}$ perpendicular to the target spin was expected to be sensitive to the neutron electric form factor. The calculations of Blankleider and Woloshyn [8] showed that the neutron contribution dominated allowing a direct measurement of G_E^n. The more recent calculations [10,11] for the response function $R_{TL'}$ show a substantial proton contribution which significantly reduces the sensitivity to G_E^n.

The first measurements of spin-dependent inclusive electron scattering from a polarized ^{3}He target were carried out at Bates in 1990 using two targets. In the first target, the polarization was achieved through metastability exchange optical pumping. The second target used a technique for polarizing ^{3}He involving spin exchange with optically pumped Rb vapor. The results for both experiments [12,13,14] are shown in Fig. 5 where the statistical and systematic errors have been combined in quadrature. The recent calculations by the Italian group are seen to be in very good agreement with our data.

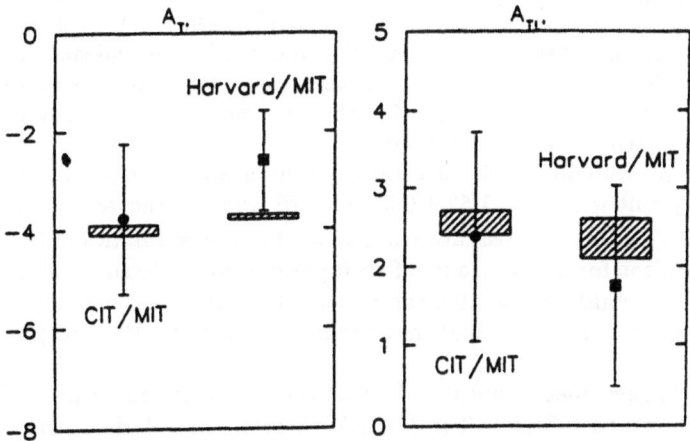

Fig. 5. Experimental results and theoretical calculations for: (a) the longitudinal asymmetry $A_{T'}$, and (b) the transverse asymmetry $A_{TL'}$. The calculations are from Ref. [10] and the experimental results are from the Caltech-MIT group [12,13] and the Harvard-MIT group [14].

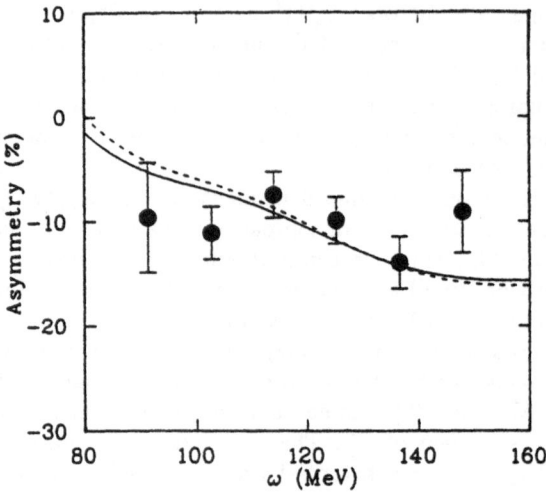

Fig. 6. The measured transverse asymmetry $A_{T'}$ as a function of energy loss ω. The dashed line is the calculation by Salmè et al. [10], and the solid line is the calculation by Schulze et al. [11].

A high-precision measurement [15] was performed at Bates last year using the Caltech target. The energy was 370 MeV, and 25 μA beams at 37% polarization were typical. The target polarization was ~38%. Fig. 6 shows the measured asymmetry, $A_{T'}$, over the quasi-elastic peak. The neutron magnetic form factor, G_M^n, has been extracted from this asymmetry based on recent PWIA calculations using spin-dependent spectral functions. The result at $Q^2 = 0.19$ GeV^2 is in agreement with the dipole parameterization, $\mu_n G_D$, and the ratio is $(G_M^n/\mu_n G_D)^2 = 0.998 \pm 0.117 \pm 0.059 \pm 0.030$, where the errors are statistical, systematic, and model dependence, respectively. The result for G_M^n is shown in Fig. 7, together with other experimental data. This experiment is the first measurement for the neutron magnetic form factor using spin-dependent inclusive electron scattering.

The same measurement [34] also yielded results for the $A_{TL'}$ asymmetry. The experimental result is $A_{TL'}^{exp} = 1.52 \pm 0.55 \pm 0.15\%$ and the theoretical prediction is, $A_{TL'}^{thy} = 2.1 - 2.9\%$. The theoretical variation is due to uncertainties in the wavefunction of ^{3}He, nucleon form factors and off-shell prescriptions. Given the sizeable proton contribution, it is unlikely that this can be used to provide accurate information on G_E^n. There is some expectation that measurements at higher Q^2 could provide useful data for G_E^n.

Further inclusive measurements of quasi-elastic asymmetries at high-Q^2 are planned at Bates and CEBAF. The MAINZ program on polarized ^{3}He emphasizes exclusive measurements.

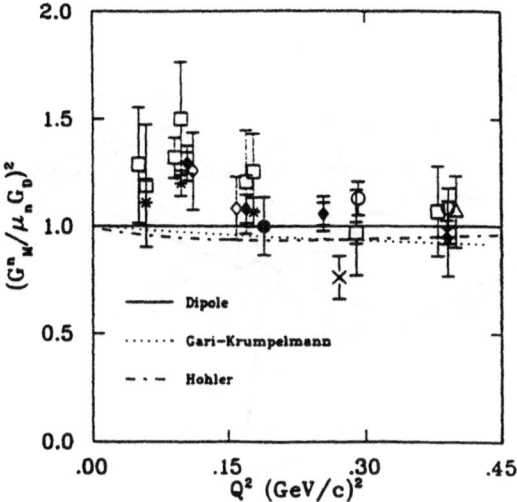

Fig. 7. The square of the neutron magnetic form factor $(G^n_M)^2$, in units of the standard dipole parameterization, $(\mu_n G_D)^2$, in the low $- Q^2$ region. The solid circle is from the data of Gao et al. [15]. The other data points are defined in Ref. [15].

III Out-of-Plane Spectrometry

Exclusive measurements involving electron scattering have been used for many years to study excitations in the nuclear continuum and to provide information on momentum distributions and medium effects in nuclei. An out-of-plane electron scattering capability is essential to fully exploit the potential of coincidence experiments. It has been shown that such a capability would give access to new observables and thus permit the isolation of important and otherwise inaccessible amplitudes.

The $A(\vec{e}, e'x)B$ cross section corresponding to the reaction depicted in Fig. 8 can be written in the one-photon exchange approximation as [16,17];

$$\frac{d\sigma}{d\Omega} = \Sigma(\theta^*, \phi^*) + h \, \Delta(\theta^*, \phi^*),$$

where

$$\Sigma = \sigma_M[v_L R_L + v_T R_T + v_{TL} R_{TL} \cos\phi^* + v_{TT} R_{TT} \cos2\phi^*],$$

and

$$\Delta = \sigma_M \, v_{TL'} \, R_{TL'} \, \sin\phi^*.$$

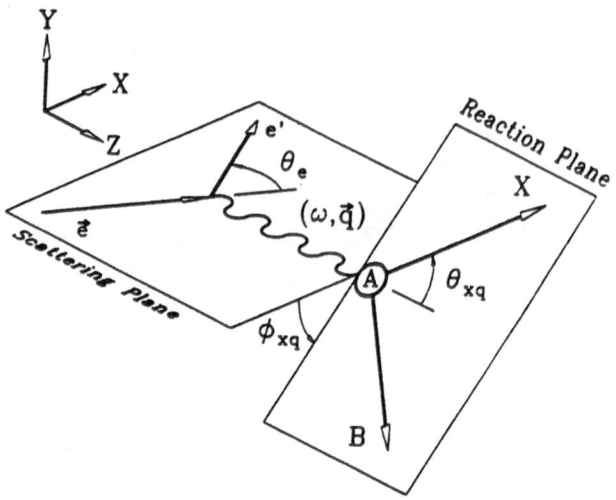

Fig. 8. Kinematic definitions for the $A(\vec{e}, e'x)$ B reaction.

The polar angles θ^* and ϕ^* define the direction of the decay product (e.g. a proton) relative to the direction of the momentum transfer and the scattering plane. The cross section is separated into helicity dependent and independent parts ($h = \pm 1$ is the electron helicity). In the usual case, where unpolarized targets are used and the polarization of the decay proton is not measured, the cross section depends on five nuclear response functions.

These five response functions are bilinear combinations of the transverse and longitudinal components of the nuclear current with respect to the direction of the momentum transfer $\vec{q}$. The goal of any experiment is to effect an accurate isolation of each response. R_L and R_T are readily determined using a Rosenbluth separation. R_{TL} and R_{TT} can be separated by measurements at values of ϕ^* on a cone centered on $\vec{q}$. The "fifth" structure function, $R_{TL'}$ – a new observable, breaks the symmetry for decay particles scattering above and below the plane and can be measured if the incident electrons are longitudinally polarized and decay particles are observed out-of-plane.

The longitudinal (R_L) and transverse longitudinal (R_{TL}) responses are sensitive to the wavefunctions and the basic interaction describing the structure of the system. The transverse (R_T) and transverse-transverse (R_{TT}) responses are very sensitive to exchange currents. The "fifth" structure function ($R_{TL'}$) is produced by the interference between two or more complex reaction amplitudes with different phases yielding an imaginary component for the new response.

Interference terms typically contribute at a level of (1-20)% to the total cross section. As a result, there are major experimental challenges in performing measurements with high statistical and systematic precision. The most straightforward approach involves sequential measurements with a spectrometer positioned at different θ^* and ϕ^*.

At MIT-Bates, we are implementing a unique system of four small magnetic spectrometers (OOPS) which will allow for simultaneous measurements [18]. The basic idea is depicted schematically in Fig. 9, where the spectrometers are shown symmetrically located about the momentum transfer $\vec{q}$. It is believed that the simultaneous measurement of several asymmetries will allow for the isolation of the interference structure functions with very high accuracy.

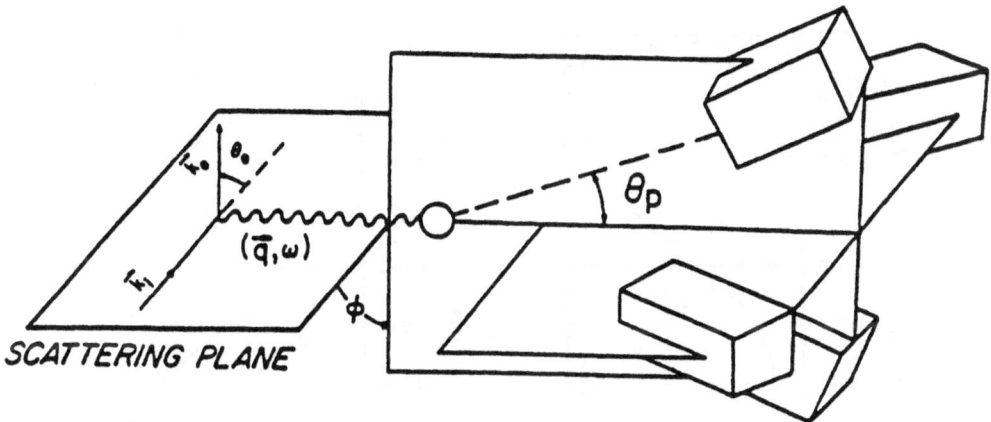

Fig. 9. Schematic layout of four out-of-plane spectrometers (OOPS) for $(\vec{e}, e'p)$ coincidence studies at MIT-Bates.

The required control of kinematic variables and other experimental parameters is expected to be greatly reduced in comparison with the usual technique of sequential measurements. Large out-of-plane angles are readily accommodated. This is essentially impractical with the normal large magnetic spectrometers. The Bates system will be fully operational in 1995.

i) N→ Δ Transition

The electroexcitation of the $\Delta(1232)$ resonance is predominantly M1. In a naive quark model, the nucleon and delta are each made up of three 1-S quarks and the transition is a pure spin-isospin flip excitation. If the quark wavefunction has a non-vanishing D-state component, C2 and E2 transition amplitudes would also be allowed. The isolation of quadrupole amplitudes in the N → Δ transition is of fundamental importance to our understanding of a possible delta deformation.

The first nucleon resonance has been studied with photoproduction and with inclusive and coincidence electron scattering. Rosenbluth separations have been used to extract longitudinal and transverse cross-sections. The data are not very accurate and the existence of a quadrupole excitation of the Δ at the level of a few percent is questionable. Recent results from Bonn [19] show a surprisingly large C2 amplitude [~13%] at low Q^2. It is very important to have confirmation of these results.

The out-of-plane spectrometer [OOPS] will be used to study the N → Δ excitation [18], using the reaction $^1H(\vec{e}, e'p)\pi^0$. Theoretical calculations (see Fig. 10) have shown that R_{TT} is sensitive to a resonant quadrupole contribution and that R_{TL} is sensitive to the interference of the Born terms and the resonant Δ excitation. This observable will be directly sensitive to background terms. It would vanish if only resonant amplitudes were present. It is expected that this new approach, using simultaneous measurements, will greatly improve our sensitivity to the expected small amplitudes.

E_l = 750 MeV θ_e = 30° Q^2 = −0.075 GeV²   h=0.4 M_Δ = 1230 MeV

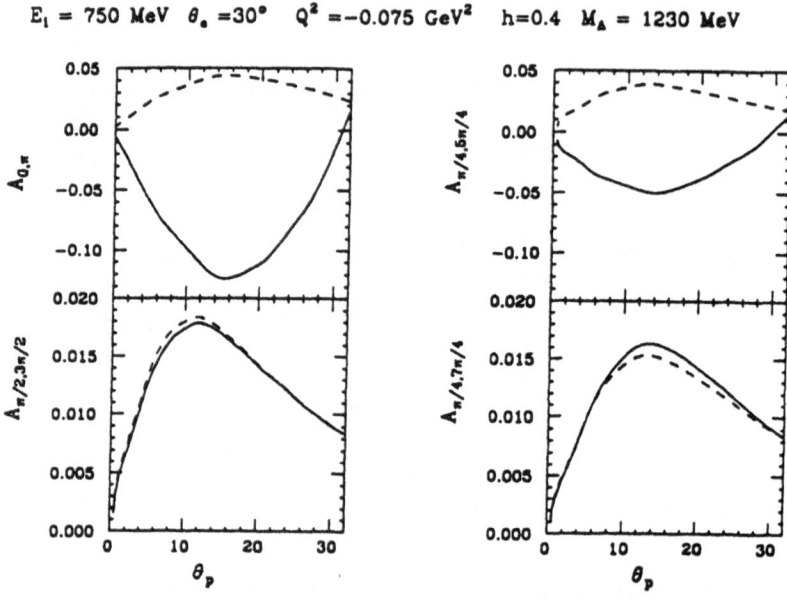

Fig. 10. Experimentally accessible asymmetries evaluated using the results of Laget [18]. The solid curves indicate results expected if the resonant quadrupole amplitude is identically zero, while the dashed curves include a resonant quadrupole excitation compatible with EMR = −0.04.

ii) Fifth Structure Functions: ^{2}H, ^{12}C

We have recently completed two pilot experiments involving the measurement of the "fifth" structure function in nuclei. They involved the out-of-plane coincident detection of protons, using longitudinally polarized electrons.

The reactions studied were ^{2}H$(\vec{e}, e'p)$ and ^{12}C$(\vec{e}, e'p)$. The objective was to test model descriptions of these nuclei and, in particular, to isolate the contributions of background terms from the resonant ones.

The initial experiments were carried out with a single OOPS module in the North Experimental Hall. Data were acquired at quasi-free kinematics and $Q^2 = 3.3$ fm^{-2}. The cross section, asymmetry, and structure function results [20] for ^{12}C are compared with theoretical calculations [21,22,23] in Fig. 11. Uncertainties are dominated by counting statistics. These first results are in good agreement with the model calculations. When the OOPS system is fully operational, we expect to continue these measurements with much improved accuracy.

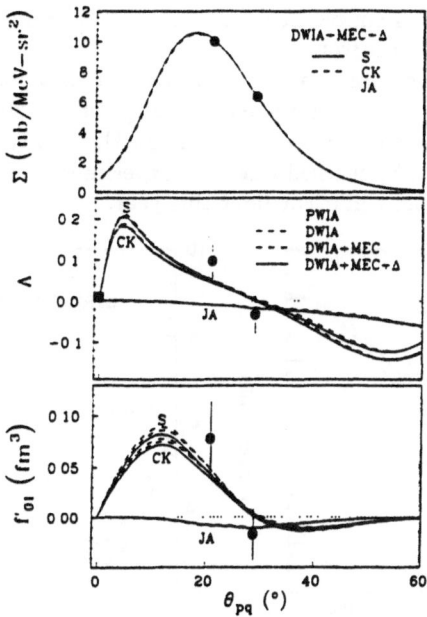

Fig. 11. The cross section, asymmetry, and fifth structure function are compared to theoretical predictions in the impulse approximation for three optical potentials. The square point at $\theta_{pq} = 0°$ in the asymmetry plot is a systematic uncertainty check based on measurements made with a deuterium target.

IV. Proton Recoil Polarimetry

A UVa - William and Mary - MIT collaboration [24] has constructed a proton polarimeter for the OHIPS magnetic spectrometer at Bates. It will be used to study $(e, e'\vec{p})$ and $(\vec{e}, e'\vec{p})$ reactions involving ^1H, ^2H, ^4He and other complex nuclei. Physics issues which will be addressed include the quadrupole amplitudes in the $N \to \Delta$ transition via the $^1H(\vec{e}, e'\vec{p})\pi^0$ reaction (issue of "deformation" in the nucleon and Δ wavefunction), the electromagnetic structure of the deuteron and ^4He and sensitive tests of nucleon-nucleus dynamics in the $^{16}O(e, e'\vec{p})$ reaction.

The use of a focal plane polarimeter represents an alternative approach for studying the $N \to \Delta$ transition. The goal will be to investigate the reaction $^1H(\vec{e}, e'\vec{p})\pi^0$ very accurately at the peak of the Δ resonance. Six of the eighteen response functions will be measured. Three of the response functions are very sensitive to the presence of a resonant quadrupole amplitude. The sensitivity is illustrated in Fig. 12 for the R_{LT}^n response. Polarized electrons will be used to separate the non- resonant multipoles providing additional leverage in isolating a possible small quadrupole component.

Another important experiment involves the deuteron: $^2H(\vec{e}, e'\vec{p})$. Such a measurement would determine G_E^p in the deuteron and test our detailed models of this nucleus. It would complement the ongoing measurements of G_E^n in the deuteron and give us added confidence in using this simple nucleus to obtain information about the electromagnetic properties of the neutron.

The proton polarimeter is based on the use of carbon as a polarization analyzer. It is designed to be used for protons in the energy range of 100-800 MeV. The device was calibrated using polarized protons at the Indiana University Cyclotron Facility. It has been installed in OHIPS, checked out, and experiments on the deuteron have started. $N \rightarrow \Delta$ measurements will occur later this year.

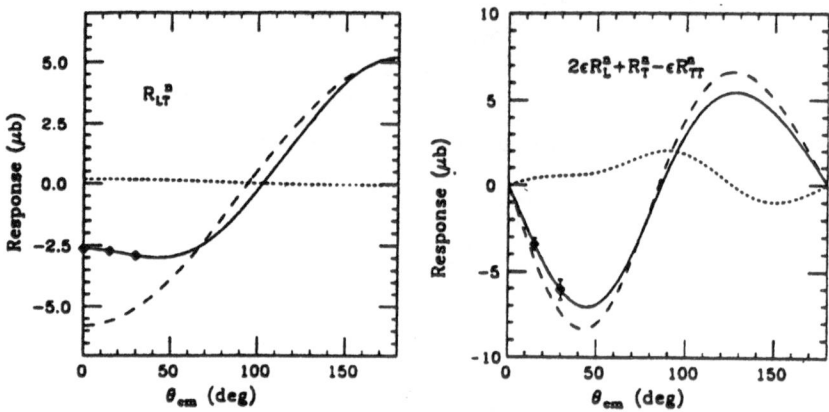

Fig. 12. Interference response functions for the $^1H(\vec{e}, e'\vec{p})\pi^0$ reaction at the Δ resonance. The proton polarization is normal to the reaction plane. The error bars are the projected result of the Bates experiment [24]. Solid curve: full calculation according to the Devenish-Lyth [25] parameterization. Long-dashed curve: S_{1+} and E_{1+} multipoles set to zero. Short-dashed curve: Born terms set to zero.

V Electroweak Physics

Experiments designed to probe the structure of the neutral currents continue to be of great interest to atomic, nuclear and particle physicists. All of the data are in impressive agreement with the Standard Model [26] of Glashow, Weinberg, and Salam. The best value for the weak mixing angle, a free parameter in the theory, is $\sin^2\theta_w = 0.233 \pm 0.002$ [27] as determined from measurements of the Z-boson mass.

The neutral currents are known to be parity violating and this aspect of their structure may be studied using polarized electrons. The parity violating asymmetry may be defined as

$$A = (\sigma_R - \sigma_L)/(\sigma_R + \sigma_L),$$

where $\sigma_R(\sigma_L)$ is the differential cross section for the scattering of electrons with right (left) helicity. Parity violating asymmetries were first observed in the scattering of polarized electrons from deuterons at SLAC [1] and in the spectra of heavy atoms [28] at low energies. These results provided some of the earliest confirmations for the predictions of the SM. They also motivated later efforts at MAINZ [29] and MIT-Bates to extend these measurements with improved accuracy to study the structure of the neutral currents.

The first Bates parity experiment involved elastic scattering from ^{12}C. In the case of a spinless and isoscalar nucleus the electromagnetic scattering is described by a single form factor and the asymmetry, A, may be written at the tree-level [30,31] as

$$A = \bar{\gamma}\frac{3}{2}G_F Q^2 (\sqrt{2}\,\pi\,\alpha)^{-1}$$

where G_F is the Fermi coupling constant, α is the fine structure constant, Q is the momentum transfer, and $\bar{\gamma}$ is the parity violating constant for an axial vector coupling to the hadronic matter. The asymmetry is completely independent of the electromagnetic form factor. It is a direct consequence of our assumption that the weak and electromagnetic currents couple to exactly the same operators. The cancellation is Q^2 independent and a precision test could be sensitive to possible extensions of the SM.

The ^{12}C experiment was carried out at Q = 150 MeV/c. A value of $\bar{\gamma} = 0.136\pm 0.032 \pm 0.009$ was measured [32] which is consistent with the prediction of the SM where, $\bar{\gamma}_{SM} = 0.155 \pm 0.002$.

Recently, there has been a lot of experimental and theoretical interest in strange quark contributions to hadronic matrix elements. New experiments using parity violating electron scattering as a neutral current probe of hadronic structure are underway at MIT-Bates and are also planned for MAINZ and CEBAF. They are designed to measure the neutral weak form factors G_E^z, G_M^z and G_A^z of the proton to provide an unambiguous signature for the presence of strange quarks in the nucleon.

The neutral weak magnetic coupling of the nucleon can be related to the electromagnetic couplings and a contribution, if any, from strange quarks

$$\mu_p^Z = \frac{1}{4}\,(\mu_p - \mu_n)\, - \mu_p\,\sin^2\theta_w - \frac{1}{4}\mu_s\,,$$

where μ_p and μ_n are the electromagnetic nucleon magnetic moments, θ_w is the weak mixing angle, and μ_s is the strange quark contribution.

The parity violating asymmetry for elastic scattering from a proton target can be written as a sum of three terms which reflect the interference between the electromagnetic and the neutral weak interactions:

$$A = \frac{G_F Q^2}{\pi\alpha\sqrt{2}} \frac{[\epsilon G_E^P\,G_E^Z\, + \tau G_M^P G_M^Z\, - \frac{1}{2}(1 - 4\sin^2\theta_w)\,\epsilon' G_M^P G_A^Z]}{\epsilon(G_E^P)^2\, + \tau(G_M^P)^2}$$

where $\tau = Q^2/4m_p^2$, $\epsilon = [1 + 2(1+\tau)\,\tan^2\theta/2]^{-1}$ and $\epsilon' = \sqrt{(1-\epsilon^2)\tau(1+\tau)}$.

The first two terms dominate at forward angles and the latter two terms contribute at backward angles. The term including the axial vector form factor, G_A^Z is suppressed by the factor $(1 - 4\sin^2\theta_w)$.

The form factors $G_{E,M}^Z$ contain contributions from strange quarks, $G_{E,M}^s$. The SAMPLE [33] experiment at MIT-Bates is designed to measure G_M^Z and extract the contribution of strange quarks to the static anomalous magnetic moment of the proton. Experiments are also planned at CEBAF and MAINZ to measure G_E^Z and G_M^Z.

SAMPLE will measure backward angle scattering ($\bar{\theta} \sim 150°$) at E=200 MeV and $Q^2 = 0.1$ GeV^2. A large solid angle (2 sr) air Čerenkov detector (10 mirrors) will be used to detect the scattered electrons from a 40 cm long liquid hydrogen target and $40\mu A$ of polarized beam ($P_e = 40\%$). Due to the high luminosity, individual event counting is not possible and the phototube currents are integrated over the $16\mu sec$ beam bursts. A schematic diagram of the experimental apparatus is shown in Fig. 13.

The predicted asymmetry is $\sim 8 \times 10^{-6}$. The contribution of the axial vector term to the asymmetry is $\sim 20\%$ and the weak radiative corrections to this term are about $20 - 30\%$ with large uncertainty. The goal is to measure the asymmetry to a statistical accuracy of 5%. We expect to achieve an overall error in G_M^s of $\Delta G_M^s = 0.22$. Theoretical predictions fall in the range $-1 < G_M^s(0) < 0$.

It would also be useful to have a direct measurement of the axial vector term. In quasi-elastic electron-deuteron scattering, at the same kinematics, the strange quark contributions to the proton and neutron add incoherently and approximately cancel in the asymmetry. However, the axial vector term and the uncertainty in the corresponding hadronic radiative correction contribute the same fraction to the asymmetry. Combining the two measurements puts constraints on the radiative corrections and improves the determination of the strange magnetic form factor. It is planned to do the deuterium measurements following those on the proton.

This next generation of parity violation experiments at low $- Q^2$ is very challenging. The physics is, however, very timely and important.

Fig. 13. Layout of the SAMPLE experiment. The polarized electron beam is incident from the right. The LH_2 target is in the hemispherical dome scattering chamber. The ten Čerenkov mirrors and phototubes are also shown.

VI Internal Target Physics

The Bates South Hall Ring was designed and constructed to provide high duty factor extracted beams and an internal target capability for nuclear research. In the internal target mode, beam energies of $(0.3 - 1.0)$ GeV and circulating currents of 80 mA are available with very long storage times. Windowless storage cell targets of polarized ^{1}H, ^{2}H and ^{3}He can be used with densities of $10^{14} - 10^{16}$ atoms/cm^2. Luminosities up to 10^{33}cm^{-2}s^{-1} will be possible.

The proposed physics program will emphasize spin- dependent electron scattering. A major component will be neutron electromagnetic form factor measurements on ^{2}H and ^{3}He in both inclusive and exclusive channels over the range $Q^2 = 0.1 - 0.8$ GeV^2. Other measurements are designed to measure small amplitudes in the structure of ^{3}He and to study the $N \rightarrow \Delta$ transition. These experiments would complement efforts with extracted beams.

The Bates Large Acceptance Spectrometer Toroid (BLAST) project is an initiative to develop and construct a large acceptance detector optimized for internal target physics with the SHR. BLAST is a non-focussing magnetic spectrometer (Fig. 14) with eight copper coils arranged in a toroidal configuration. Two opposing sectors would be instrumented with wire chambers, scintillation counters, and Čerenkov counters for tracking, time-of-flight, and particle identification.

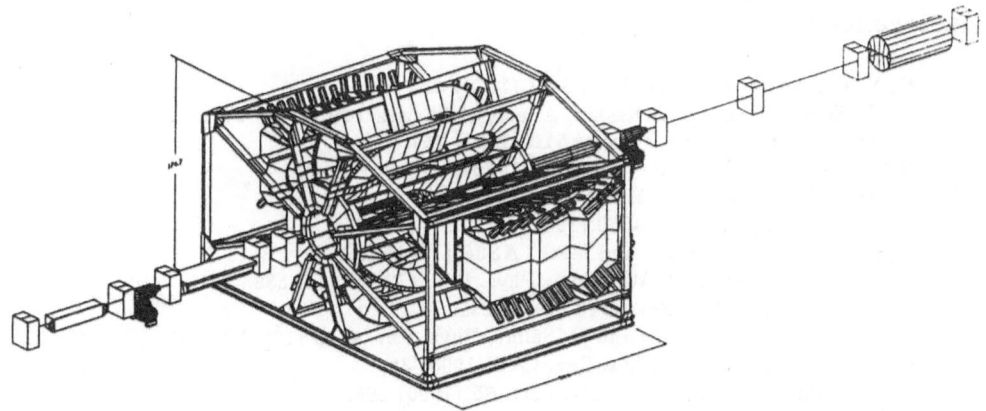

Fig. 14. Proposed Bates Large Acceptance Spectrometer Toroid (BLAST) detector. Detector would be located intersecting the South Hall Ring and used with polarized internal storage cell targets. Two opposing sectors are fully instrumental.

The cost of the detector is $\sim$6M\$ (US) and construction would be over a period of four years following approval. A large international collaboration has been involved in the design and many groups will participate in its construction and operation for physics.

VII Summary

The recent experiments at MIT-Bates, exploiting the use of polarized electrons and the measurement of spin observables, point to the importance of such studies in understanding nucleon and nuclear structure. In many instances we find both increased sensitivity and new observables which allow the accurate determination of small amplitudes. A major component of the proposed future program at MIT-Bates, NIKHEF, MAINZ and CEBAF, will involve the use of polarized beams, polarized targets and recoil polarimetry.

Modern electron accelerators, operating at 100% duty factor, will provide the ultimate capability for coincidence experiments. The stretcher/storage ring facilities at MIT-Bates and NIKHEF were designed to operate with internal targets. Internal target experiments with polarized beams and polarized targets at high luminosities represents an important new capability for nuclear research.

The next decade promises to be a very exciting one for electronuclear physics. For the first time the full power of the electron probe can be applied in the broadest sense to study nucleons and nuclei.

I would like to acknowledge the help of my colleagues at MIT-Bates and those at other laboratories who provided me with data and information about the research reported here. This work was supported in part by the U.S. Department of Energy under cooperative agreement NO. DE-FC02-94ER40818.

References

1. C. Y. Prescott *et al.*, Phys. Lett. **84B** (1979) 524.
2. T. W. Donnelly and A. S. Raskin, Annals of Physics **169** (1986) 247.
3. S. Platchkov *et al.*, Nucl. Phys. **A510** (1990) 740.
4. R. Arnold *et al.*, Phys. Rev. **C23** (1981) 363.
5. H. Arenhövel, Phys. Lett. **B199** (1987) 13.
6. T. Eden *et al.*, Phys. Rev. **C50** (1994) R1.
7. T. Eden *et al.*, Nucl. Instr. and Meth. **A338** (1994) 432.
8. B. Blankleider and R. M. Woloshyn, Phys. Rev. **C29** (1984) 538.
9. J. L. Friar *et al.*, Phys. Rev. **C43** (1990) 2310.
10. C. Ciofi degli Atti, E. Pace and G. Salmé, Phys. Rev. **C46** (1992) R1591.
11. R. Schulze and P. Sauer, Phys. Rev. **C48** (1993) 38.
12. C. E. Woodward *et al.*, Phys. Rev. Lett. **65** (1990) 698.
13. C. E. Jones *et al.*, Phys. Rev. **C47** (1993) 110.
14. A.K. Thompson *et al.*, Phys. Rev. Lett. **68** (1992) 2910.
15. H. Gao *et al.*, Phys. Rev. **C50** (1994) R546.
16. S. Boffi, C. Giusti, and F. D. Pacati, Phys. Rep. **226** (1993) 1.
17. S. Boffi, C. Giusti, and F. D. Pacati, Nucl. Phys. **A435** (1985) 697.
18. C. N. Papanicolas, Bates Proposal #87-09; S. Dolfini *et al.*, Bull. Am. Phys. Soc. **33** (1988) 1578.
19. B. Schoch, Bull. Am. Phys. Soc. **39** (1994) 1033.
20. J. Mandeville *et al.*, Phys. Rev. Lett. **72** (1994) 3325.
21. J. R. Comfort and B. C. Karp, Phys. Rev. **C21** (1980) 2162.
22. D. F. Jackson and I. Abdul-Jalil, J. Phys. **G6** (1980) 481.
23. M. M. Giannini and G. Ricco, Ann. Phys. (N.Y.) **102** (1976) 458.

24. R. W. Lourie and V. Burkert, Bates Proposal #89-03 (1989).
25. R. C. E. Devinish and D. H. Lyth, Nucl. Phys. **B43** (1972) 228.
26. S. L. Glashow, Nucl. Phys. 22 (1961) 579; S. Weinberg, Phys. Rev. Lett. **19** (1967) 1264; Phys. Rev. Lett. **27** (1971) 1688; A. Salam, Rev. Mod. Phys. **52** (1980) 525.
27. S. Fanchiotti and A. Sirlin, Phys. Rev. **D35** (1987) 785.
28. M. C. Noecker *et al.*, Phys. Rev. Lett. **61** (1988) 319.
29. W. Heil *et al.*, Nucl. Phys. **B327** (1989) 1.
30. G. Feinberg, Phys. Rev. **D12** (1975) 3575.
31. J. D. Walecka, Nucl. Phys. **A285** (1977) 345.
32. P. A. Souder *et al.*, Phys. Rev. Lett. **65** (1990) 694.
33. R. D. McKeown and D. Beck *et al.*, Bates Proposal #89-06.
34. J.-O. Hansen *et al.*, Preprint LNS-94-80, submitted for publication.
35. M. Meyerhoff *et al.*, Phys. Lett. **B327** (1994) 201.
36. A. Lung *et al.*, Phys. Rev. Lett **70** (1993) 718.
37. S. Galster *et al.*, Nucl. Phys. **B32** (1971) 221.

Recent achievements and future plans at Frascati

R. Baldini[1], E. Pasqualucci[2]

[1] INFN-Laboratori Nazionali di Frascati,
P.O.Box 13, 00044 Frascati, Italy
e-mail: Baldini@lnf.infn.it

[2] INFN-Sez. Roma 2,
Via della Ricerca Scientifica, 1, 00133 Roma, Italy
e-mail: Pasqualucci@roma2.infn.it

Abstract:
Recent achievements and future plans at Frascati are reported: the first measurement of the Neutron time-like Form Factors at ADONE, the new high luminosity Φ Factory DAΦNE , the KLOE detector, a real factory for Chiral Dynamics in K decay, the measurement of $\gamma\gamma \to \pi^0\pi^0$ and $\gamma\gamma \to \pi^+\pi^-$ cross sections at threshold and Quantum Mechanics paradoxes at DAΦNE .

1 Introduction

In the following recent achievements in strong interactions and future plans at the Frascati National Laboratories are shortly reported, namely :

- The last ADONE legacy: the first measurement of the Neutron time-like Form Factors.

- DAΦNE : the new high luminosity Φ Factory.

- KLOE: an "all purposes" detector, mainly for studying CP violation in K decay, but fully equipped to be at DAΦNE a real factory for Chiral Dynamics in K decay.

- Measurement of $\gamma\gamma \to \pi^0\pi^0$ and $\gamma\gamma \to \pi^+\pi^-$ cross sections at threshold.

Among the various interesting topics concerning DAΦNE there are also FIN-UDA, a detector for studying Hypernuclei and Kaon-Nucleon interactions at threshold [31] and Quantum Mechanics paradoxes on a macroscopic scale.

2 The last ADONE legacy

ADONE, the old Frascati e^+e^- storage ring, has been decommissioned in the summer '93, to leave space for the new Φ factory DAΦNE .

Historically ADONE was the first high energy storage ring and multihadronic production in e^+e^- annihilation was discovered with this machine, achieving a clearcut evidence for coloured quarks. ADONE operation as a collider was stopped in '78, but in '89 it was restored to measure for the first time the Neutron time-like Form Factors by the FENICE experiment [7].

The Nucleon Form Factors are still to be explored. The Nucleon space-like Form Factors have been studied in the last forty years. In spite of that only recently high precision data on the Neutron became available up to $Q^2 \sim -4M^2$[1]. In the time-like region, several measurements have been performed concerning the Form Factor of the Proton. However only recently measurements at high Q^2 and high precision measurements at low Q^2 have been published, the latter showing a surprising behaviour near threshold[2]. Regarding the Neutron time-like Form Factor, no data were available at all before FENICE.

From a theoretical point of view different models, which agree quite well for what concern space-like Form Factors, give very different predictions on the Neutron Form Factors. A comparison between them may be done looking at the quantity

$$r = \frac{\sigma(e^+e^- \to n\bar{n})}{\sigma(e^+e^- \to p\bar{p})}.$$

In Tab. 1 several predictions for r are shown. While PQCD calculations[3] predict $r \simeq (q_d/q_u)^2 = 0.25$, EVMD inspired[4] fits give values of r ranging from 1 up to 100, depending on the details of the model. Other calculations[6] related to the Skyrme model predict $r \sim 1$. According to the Λ time-like form factor[5], which has been poorly measured, it should be $r \sim 1$.

PQCD	~ 0.25
Cabibbo-Gatto (1961)	~ 14
Korner-Kuroda (1977)	~ 2
Voci et al. (1982)	~ 100
Dubnicka (1988)	~ 25
Ellis et al. (1990)	~ 1
Biagini et al. (1991)	≥ 1

Table 1: Some predictions for r

Therefore, from both theoretical and experimental point of view, data on the Neutron time-like Form Factors are strongly demanded.

Actually what is measured in the time-like region is the Magnetic Form Factor. In fact at threshold there is only one Form Factor, being $G_E = G_M$, and at high energy the contribution from G_M is dominant in any case.

In Fig. 1 the world data set for Proton and Neutron time-like Form Factors is shown, including the FENICE results on Proton and Neutron.

By averaging over the three FENICE points [8] [9], the ratio r turns out to be 1.5 ± 0.4, in agreement with some predictions, but in disagreement with the

PQCD one. Concerning the Proton Form Factor, the FENICE results are in agreement with the existing measurements [10]. In principle the PQCD prediction on r is quoted at somewhat higher energy. Actually, for instance according to the DIS structure functions, no peculiar behaviour is expected down to $Q^2 \simeq 4$ GeV^2.

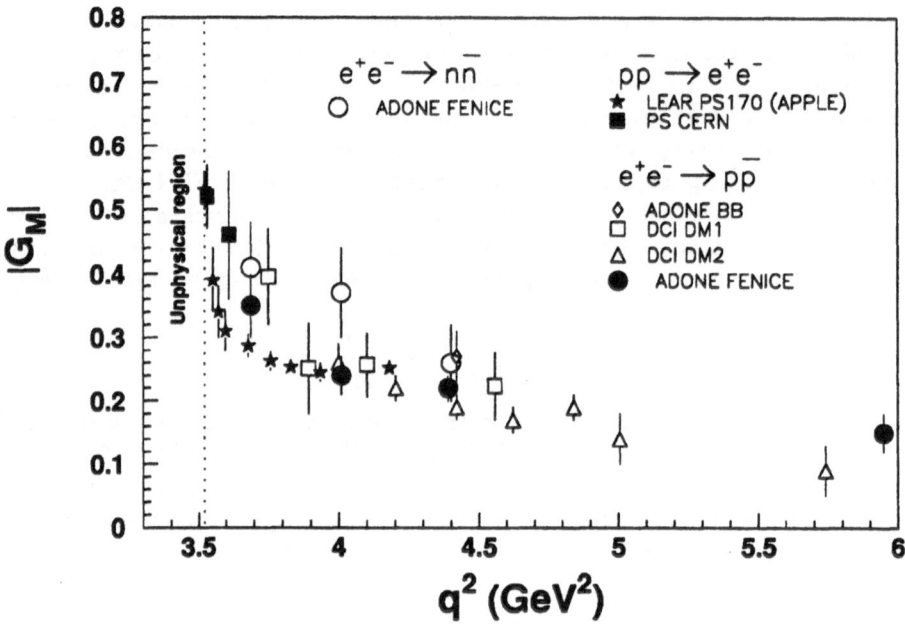

Figure 1: Proton and Neutron time-like Form Factors, world data set.

On the other hand concerning the Nucleon Form Factors further unexpected experimental behaviours are coming out. In particular space-like and time-like magnetic Form Factors, which are expected to be asymptotically the same, appear to have the same behaviour but they are different by a factor of $\simeq 3$, as it is shown in Fig. 2.

Also near threshold the Proton time-like Form Factor has a surprising behaviour, as mentioned before. The steep slope strongly suggests a narrow resonance below threshold, namely at ~1850 MeV. This is in agreement with the FENICE results [28] on the total e^+e^- annihilation cross section in this energy interval. In fact there is an interference pattern consistent with the interpretation of a narrow resonance, still at 1850 MeV, as shown in Fig. 3.

FENICE has measured also the J/Ψ decay into $n\bar{n}$ [11], which turns out to be the same as $p\bar{p}$ and is consistent with small electromagnetic corrections. On the other hand electromagnetic corrections, depending on the Form Factors near the J/Ψ, are expected rather large.

In conclusion something is missing in the interpretation of the Nucleon Form Factors and more theoretical efforts are needed. Of course much more data would be welcome in the time-like region, but they are not expected neither at present nor in the next future. However an upgraded version of DAΦNE could provide them.

WORLD NUCLEON E.M. FORM FACTOR

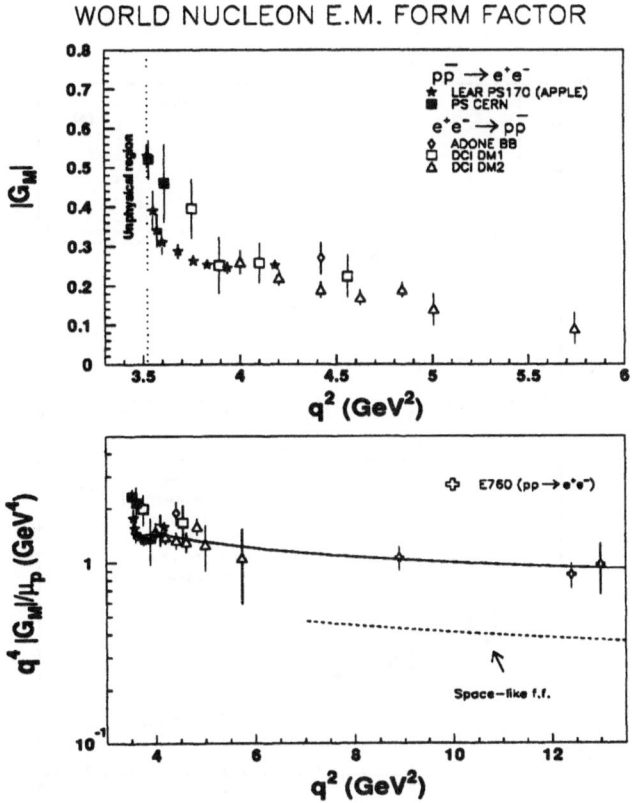

Figure 2: $Q^4 \times |G_M^p|$ in the time-like region.

3   DAΦNE and KLOE

The DAΦNE [12] [30] project (Double Annular ϕ-factory for Nice Experiments) has been approved in June 90 by the Italian Institute for Nuclear Physics (INFN). It is totally funded, the commissioning is planned for the end of 96 and it will be installed in the present ADONE building at the Frascati National Laboratories.

DAΦNE is an e^+e^- collider optimized to work at the Φ mass with a goal luminosity L $\sim 10^{33}$ cm^{-2}s^{-1}. The strategy to reach the target luminosity is: first of all to obtain the same single bunch luminosity achieved at VEPP-2M and then to increase the number of bunches (up to 120). To suppress multibunch instabilities electrons and positrons will circulate in two separate storage rings. They will collide at two interaction points. To reduce the minimum distance between the bunches they collide with a horizontal half angle of 10 mrad . To

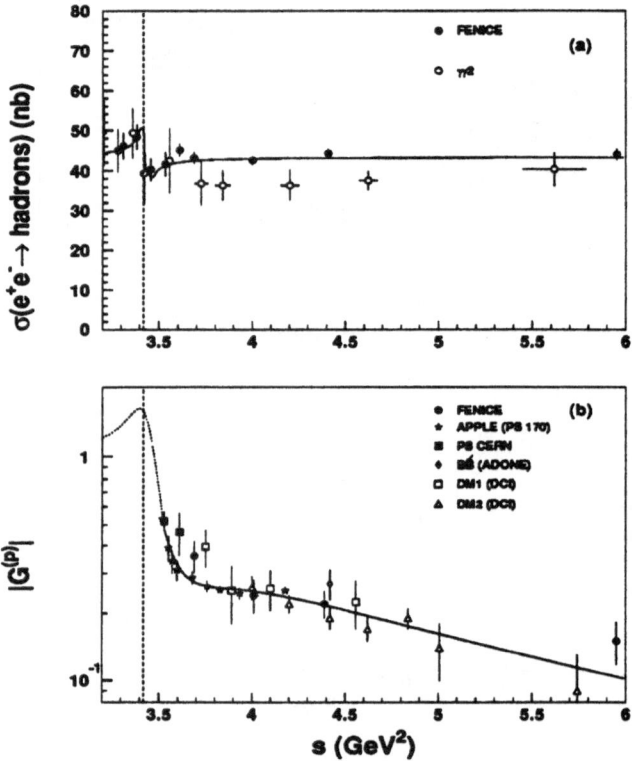

Figure 3: Total cross section $e^+e^- \to$ hadrons (a) and and Form Factor of the Proton (b) fitted supposing the interference of a narrow resonance.

avoid synchrobetatron oscillations the bunches will be very flat. In the interaction regions, equipped with two triplets of low-β quadrupoles, two experiments will be installed: KLOE, mainly for CP violation studies, and FINUDA, mainly for hypernuclei and low energy KN interactions studies.

With some loss in luminosity the maximum achievable total c.m. energy is 1.5 GeV. Some DAΦNE parameters at the Φ peak are reported in Table 2.

The KLOE [13] [14] detector, shown in Fig. 4 has been designed primarily with the goal of detecting CP violation in K^0 decays.

The main parts are, starting from outside, a superconductive solenoid that will provide a magnetic field of 6 KG, a lead-scintillator sampling electromagnetic calorimeter installed inside the coiland a very large drift chamber, filled with elium.

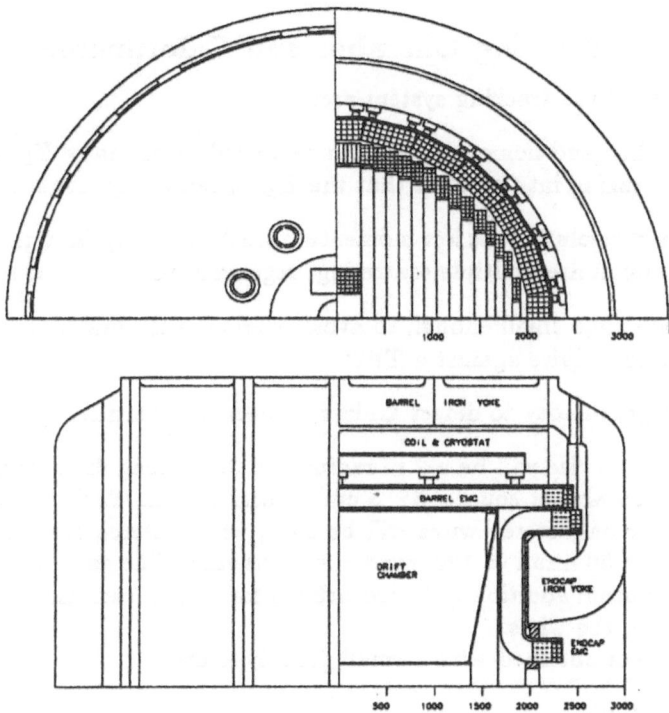

Figure 4: The KLOE detector.

Table 2: Main DAΦNE parameters.

Total CM energy (GeV)	1.0	Luminosity (10^{32} cm$^{-2}s^{-1}$)	~ 10
Bunches per ring per beam	120	Particles/bunch	10^{11}
Luminosity lifetime (hr's)	2	Single ring circumference (m)	98
Time between collisions (ns)	2.7	Crossing half-angle (mrad)	10
Hor. β-function at IP (m)	4.5	Ver. β-function at IP (cm)	4.5
Hor. r.m.s. size at IP (mm)	2.1	Ver. r.m.s. size at IP (μm)	21
Energy loss/turn (KeV)	9.3	Rel. nat. r.m.s. energy spread	4×10^{-4}
r.m.s. bunch length (cm)	3.0	Nat. emittance (mm $\times$ mrad)	1.0

DAΦNE and KLOE may be considered a Chiral Dynamics Factory for what concerns kaon decay, taking into account the large number of kaons produced by the Φ decay and the KLOE capability to fully reconstruct their various decay modes.

The order of magnitude of events per year ($\sim 10^7$ s) on tape are expected to be, taking also into account detection efficiencies, $\sim 2 \times 10^{10}$ $K^\pm$; $\sim 7 \times 10^9$ K_S; $\sim 2 \times 10^9$ K_L; $\sim 2 \times 10^8$ η; $\sim 5 \times 10^8$ $K_L \to \pi\mu\nu$; $\sim 5 \times 10^4$ $K_L \to \pi\pi e\nu$; $\sim 1 \times 10^5$ $\gamma\gamma \to \pi\pi$.

3.1 KLOE Tracking Chamber and Calorimeter

The main tasks of the tracking system are:

- Large radius and homogeneous volume to collect as many K_L decays as possible, taking into account that the K_L mean decay lenght is 3.5 m.

- Optimized resolution for low momentum tracks by using light gas and light walls to minimize multiple scattering, regeneration and γ's conversion.

- Collection time small enough to avoid overlap with Bhabha events. This requirement advise against a TPC.

- Enough granularity to detect kinks in a decaying particle.

The tracking device will be a 2 m radius and 3.5 m long drift chamber with cell configuration almost square, 3×3 cm^2 effective area. To fill uniformly the sensitive volume only stereo wires will be used, with a stereo angle that varies with radius from 50 mrad to 120 mrad going outward. The gas mixture will be elium with the usual addition of hydrocarbons for a total radiation lenght of $\sim$ 700 m, including the wires.

The performances obtained with a small prototype are:

- $\sigma_{r,\phi} < 200$ μm.

- Cell efficiency > 99 %.

- Momentum resolution $\sigma_{P_t}/P_t \sim 0.3\%$.

A large prototype has been built and will be tested next winter.

The main tasks of the calorimeter are:

- First of all full efficiency for γ's in the energy range 20–280 MeV, to discriminate between $K_L \to \pi^0\pi^0$ and $K_L \to \pi^0\pi^0\pi^0$. Of course hermeticity and good energy resolution are also demanded.

- The calorimeter should provide a fast trigger.

- Good time performance is also very useful to reconstruct the K_L neutral vertex with good resolution, taking into account that K_L are rather slow.

- Last but not least the calorimeter, which has to be very large in diameter, has to have a reasonable cost.

A lead-scintillator sampling calorimeter fulfils these requirements. It will consist of 0.5 mm lead layer, in which are embebbed 1 mm diameter scintillating fibers. The calorimeter will be composed of a barrel, with an inner radius of 2 m, a thickness of 23 cm (~15 radiation lengths), a length of 4.3 m and two end-caps that close hermetically the calorimeter. The total acceptance is 98 % of the full solid angle. The read-out is organized in five layers of twelve 4.5×4.5 cm cells coupled with two photomultipliers by light guides. The spatial resolution in reconstructing the shower apex is expected $\sigma_x \sim \sigma_y \sim 1.2$ cm and $\sigma_z \sim 2.5$ cm obtained by time difference.

A four meter long module has been built and tested with electrons, photons, muon and pion beams. The achieved performances are:

- $\sigma(E)/E \sim 4.4\%/\sqrt{E\ (GeV)}$.

- $\sigma(T) \sim 34$ ps$/\sqrt{E\ (GeV)}$.

- Very good linearity in the energy range of interest.

- No deterioration in the resolutions and in the response as a function of the incidence angle.

- Some capability for distinguishing between pions and muons.

4 $\gamma\gamma$ Physics at DAΦNE by $e^+e^- \to e^+e^-$ X

DAΦNE will offer an excellent possibility to study $\gamma\gamma$ interactions at low energy with high statistics [32]. Among the various topics the possibility of precise measurements of the $\gamma\gamma$ cross section and the study of e–π azimuthal correlations [33] will allow to test chiral perturbation theories and techniques.

The existing measurements on $\gamma\gamma$ interactions by means of high energy e^+e^- storage rings are affected by large statistical and systematic uncertainties at low $\gamma\gamma$ invariant masses. This is mainly due to the low available detection efficiencies and to the large background. Moreover the data from different experiments do not agree very well and the comparison with theoretical calculation is difficult. In Fig. 5 the experimental situation is reported for $\gamma\gamma \to \pi^+\pi^-$ and $\gamma\gamma \to \pi^0\pi^0$ together with theoretical predictions. For the $\pi^+\pi^-$ mode MARK II [16] data agrees quite well with a chiral prediction but for DM1 [17] and DM2 [18] the agreement is very poor. PLUTO [19] data are marginally consistent.

For the $\pi^0\pi^0$ mode only data from Crystal Ball [20] are available at present and in this case chiral calculations at tree level do not fit very well the data. Better agreement is found including two loop corrections [21] and in the case of Pennington's [22] calculation based on unitarity.

There are several advantages at DAΦNE due to:

- High $\gamma\gamma$ luminosity at low $\gamma\gamma$ invariant masses. It can be seen from Fig. 6 that already at the luminosity level DAΦNE will be competitive with LEP up to 600 MeV. Moreover, of course, detection efficiencies are negligible in the case of LEP experiments.

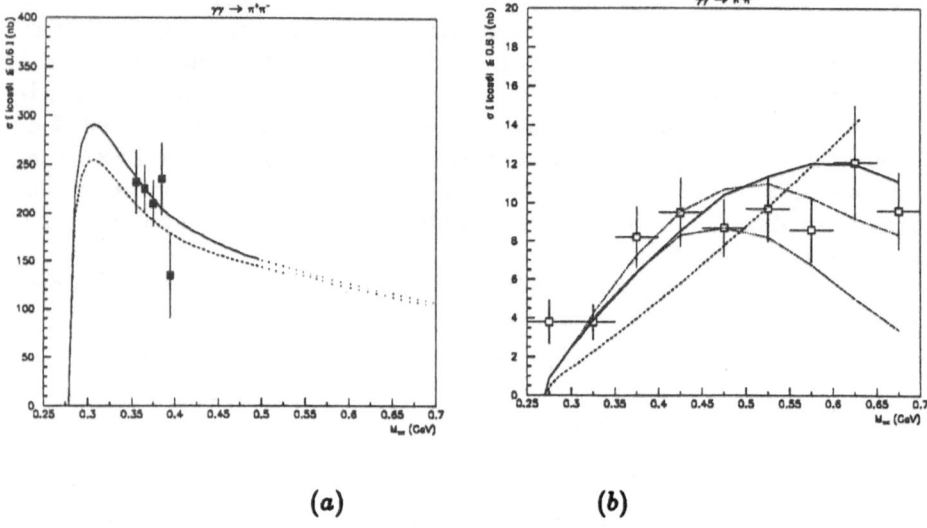

(a)   (b)

Figure 5: (a):$\gamma\gamma \to \pi^+\pi^-$ (solid line), (b):$\gamma\gamma \to \pi^0\pi^0$, χPT one loop (dashed l χPT two loop (solid line), unitarity predictions (dotted line)

- Good tagging efficiency due to the relatively high outgoing electron angle $\theta_e \sim m_e/E \sim 1$ mrad, to be compared with 10^{-2} mrad of PEP and PETRA. This will allow to work in an environment almost free of background, with a consequent reduction of systematic errors.

- To study azimuthal e–π correlations [33], that has not be performed at the existing colliders so far.

In conclusion the available mass range at DAΦNE for the study of these processes is from threshold up to $\sim$600 MeV . In one year running $\sim 10^4$ $\pi^0\pi^0$ and $\sim 10^6$ $\pi^+\pi^-$ will be produced for a luminosity L$\sim$5$\times$10^{32} cm^{-2}s^{-1}: a data sample two order of magnitude more than any present experiment. With DAΦNE running at the maximun available energy , 750 MeV per beam, the achievable mass range should be up to $\simeq 1$ GeV.

There is an important drawback in measuring $\gamma\gamma \to \pi\pi$ with DAΦNE sitting at the Φ. That is the presence of a large background from $\phi \to K_S K_L (K_S \to \pi\pi)$ when the K_L escapes the detector, neither decaying nor interacting with the electromagnetic calorimeter. The equivalent cross section for these events is $\sigma \sim$ 100 nb, to be compared with ~ 0.01 nb for $\gamma\gamma$ events.

So even if some coplanarity cut can help in reducing this background tagging of forward emitted electron (positron) is mandatory.

The proposed tagging system [15] consists in a small angle detector, SAT, installed downstream the Split Magnet, which is also a suitable magnetic analyzer for the outgoing electrons, and a wide angle detector, WAT, that is KLOE itself

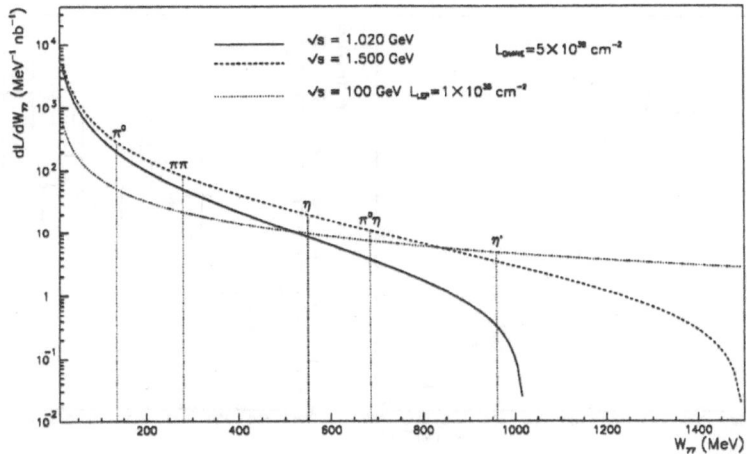

Figure 6: Photon-photon luminosity at DAΦNE as function of $\gamma\gamma$ invariant mass.

with eventually additional small detectors in front of the first low–β quadrupoles.

The angular acceptance is up to 10 mrad for the SAT and from 100 mrad on for the WAT.

Single and double tagging efficiencies for the SAT are $\sim$ 40 % and $\sim$ 15 % respectively. To avoid accidental coincidences in the SAT detector due to electrons from radiative Bhabha scattering an additional detector for forward emitted γ's detection is foreseen. However double tagging events are not affected by this problem.

Sensitivity has been evaluated performing an extensive Monte Carlo simulation, tracking the outgoing electrons from the interaction point to the various locations along the machine layout. The result [15] of a simulation is reported in Fig. 7(a) for events detected by the SAT in the single tag mode as a function of the $\gamma\gamma$ mass for one year running (10^7 s). Assuming the chiral two loop cross section and considering only statistical errors the significance of this measurement can be studied. In Fig. 7(b) the cross section is reported in a 25 MeV mass bin and compared with the Crystall Ball data. It is clear that the DAΦNE measurement will disentangle the two loop corrections near threshold. It has been pointed out several times that another way to get the pion polarizabilities is by means of the measurement of $\gamma\gamma \to \pi\pi$ at threshold [29][34]. Near threshold $\gamma\gamma$ interactions are supposed to provide mainly the $\pi\pi$ S wave, which is related to the difference, $\alpha_\pi - \beta_\pi$, between electric and magnetic polarizability.

In $\gamma\gamma$ interactions complications related to any nuclear target are avoided, but it has been demonstrated in the theoretical discussion that the extrapolation to the pion pole is much more difficult [36][21]. Therefore the conclusion

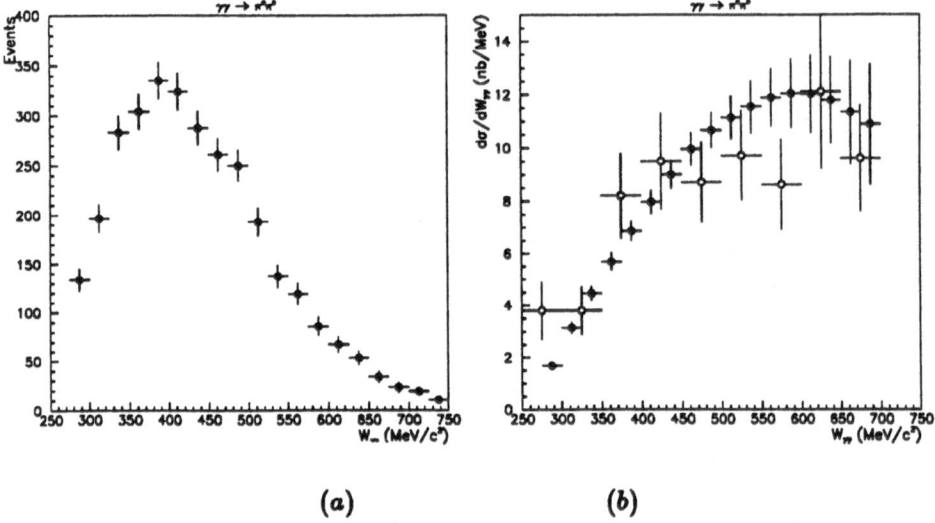

Figure 7: (a) Number of $\gamma\gamma \rightarrow \pi^0\pi^0$ events, (b) cross section (black dots) compared with Crystal Ball data (open dots)

has been achieved that $\gamma\gamma$ interactions are not the best way to get the pion polarizability. Nevertheless $\gamma\gamma$ interactions near threshold remain a very clean test of any theoretical description of strong interactions at low energies.

5 Other DAΦNE issues

The Φ decay into a pair of neutral kaons is a very suitable process to test Quantum Mechanics on a large scale. Many authors [23] have emphasized paradoxes related to this process, which are a good illustration of the celebrated Einstein, Podolsky, Rosen argument [24].

As an illustration two non intuitive Quantum Mechanics expectations are pointed out in the following and they may be exploited at a powerful Φ factory. Actually paradoxes arise because Φ and neutral kaons are both superposition of states.

Detection of CP violation provides a paradox in the time development of the two neutral kaon decays. Namely if the two final states are available to both K_S and K_L there is an interference term in their time evolution. Therefore the decay times are correlated, no matter how far one kaon is from the other. In particular Bose-Einstein statistics forbids symmetric decays in the same number of pions [23] [25].

Another striking paradox arises if there are thin regenerators upstream the decay points. To simplify the argument let ignore CP violation: therefore K_S

and K_L are identified by their decay products and only $K_S K_L$ events are produced in the Φ decay. However, if a thin regenerator is introduced on one side only, coherent regeneration will take place and a fraction of the events, decaying downstream the regenerator, will be either $K_S K_S$ or $K_L K_L$. Yet coherent regeneration cannot arise for a spherical regenerator [26] [27]: events where both kaons decay downstream the regenerator will be always $K_S K_L$. Quantum Mechanics is so that a neutral kaon, crossing a regenerator, knows what is occuring to the other kaon, even concerning a process at a microscopic level like regeneration. Again it happens no matter how far one kaon is from the other!

6   Conclusions

ADONE left a very wealthy legacy: the Neutron time-like form factor. Some heir should take it and fulfil the still unanswered questions on the Nucleon Form Factors. A new machine, DAΦNE , and new detectors, KLOE and FINUDA, will operate in the next future at Frascati. In addition to CP violation and Hypernuclei decay a lot of other interesting physics can be studied. In particular two-photon physics can be studied with high statistics and new tools.

All that will allow new, stringent, tests of Chiral Dynamics.

Acknowledgements

It is a pleasure to thank the organizers of this Workshop for a very stimulating working environment and for their patience.

References

[1] A. Lung et al., Phys. Rev. Lett. 70 (1993) 718.

[2] M. Castellano et a., Nuovo Cimento A14 (1973) 1;
G. Bassompierre et al., Phys. Lett. B64 (1976) 475;
B. Delcourt et al., Phys. Lett. B86 (1979) 395;
D. Bisello et al., Nucl. Phys. B224 (1983) 379;
M. Conversi et al., Nuovo Cimento 40 (1965) 690;
D. L. Hartill et al., Phys. Rev. 184 (1969) 1415;
C. Baglin et al., Phys. Lett. B163 (1985) 400;
G. Bardin et al., Phys. Lett. B255 (1991) 149; B257 (1991) 514.

[3] V. L. Chernyak and I. R. Zhitnitsky, Nucl. Phys. B246 (1984) 52;
T. Hyer, Phys. Rev. 47 (1993) 3875.

[4] N. Cabibbo and R. Gatto, Phys. Rev. 124 (1961) 1577;
V. Wataghin, Nucl. Phys. B10 (1969) 107;
J. G. Körner and M. Kuroda, Phys. Rev. D16 (1997) 2165;
P. Cesselli, M. Nigro and C. Voci, Proc. of Workshop on physics at LEAR, Erice (1982);
S. Dubnicka, Nuovo Cimento A104 (1991) 1075.

[5] M. E. Biagini, L. Cugusi, E. Pasqualucci, Zeit. Phys. C52 (1991) 631.

[6] J. Ellis, M. Karliner, H. Kowalski, Phys. Lett. B235 (1990) 341.

[7] A. Antonelli et al., Nucl. Instr. Meth. A337 (1993) 34.

[8] A. Antonelli et al., Phys. Lett B313 (1993) 283.

[9] E. Pasqualucci et al., presented at LEAP 94 (Bled, Slovenia), in printing.

[10] A. Antonelli et al., Phys. Lett. B334 (1994) 431.

[11] A. Antonelli et al., Phys. Lett. B301 (1993) 317.

[12] G. Vignola, Proc. of the XXVI Int. Conf. on High Energy Physics, ed. J. Sanford AIP (1992) 1941.

[13] The KLOE Collaboration, A general purpose detector for DAΦNE, LNF-92/019 (1992).

[14] The KLOE Collaboration, The KLOE detector, Technical Proposal, LNF-93/002 (1993).

[15] G.Alexander et al., The KLOE small tagging system at DAΦNE , KLOE-NOTE 1994-112 (1994).

[16] J.Boyer et al., Phys. Rev. D42 (1990) 1350.

[17] A.Courau et al., Phys. Lett. B96 (1980) 412.

[18] Z.Ajaltouni et al., Phys. Lett. B194 (1987) 573.

[19] Ch.Berger et ale., Z. Phys. C26 (1984) 199.

[20] H.Marsiske et al., Phys. Rev D41 (1990) 3324.

[21] S. Bellucci, J. Gasser and M.E. Sainio, Nucl. Phys. B423 (1994) 80.

[22] M.R. Pennington, THE DAΦNE PHYSICS HANDBOOK, L.Maiani, G.Pancheri, N.Paver, Eds. (1992) 379.

[23] H.J. Lipkin, Phis. Rev. 176 (1968) 1715.

[24] A. Einstein, B. Podolski, N. Rosen, Phis. Rev. 47 (1935) 777.

[25] I. Dunietz, J. Hanser, J. Rosner, Phis. Rev. D35 (1987) 2166.

[26] N.W. Tanner, R.H. Dalitz, Rev. Mod. Phys. 63 (1989) 1349.

[27] R. Baldini-Celio et al., LNF-90/007.

[28] A. Antonelli et al., to be published on Phis. Lett.

[29] B.R. Holstein, Comments Nucl. Part. Phys. 19 (1990) 221.

[30] M. E. Biagini, PANIC Proceedings, Perugia (1993) 763.

[31] FINUDA Collaboration Lab. Naz. di Frascati LNF-93/021 (1993).

[32] G. Alexander et al., Il Nuovo Cimento 107A (1994) 837.

[33] S. Ong, P. Kessler and A. Courau, Mod. Phys. Lett. A 4 (1989) 909.

[34] J.F. Donoghue, B.R. Holstein and Y.C. Lin, Phys. Rev. D3 (1988) 2423.

[35] The Mark II Collaboration (J. Boyer et al.) Phys. Rev. D42 (1990) 1350.

[36] D. Babusci, S. Bellucci, G. Giordano and G. Matone, Pys. Lett. B314 (1993) 112.

Experiments at the Electron Accelerator MAMI

Thomas Walcher

Institut für Kernphysik, University of Mainz, J.J. Becher-Weg 45, 55099 Mainz, Germany

1 Introduction

This workshop was about chiral dynamics and it became quite clear from the theoretical side that chiral perturbation theory is a fundamental approach to QCD. However, what is needed is the confrontation of this theory to experiment or in other words the finding of observables which represent a meaningful and valid test. Quite a few such observables using hadronic probes have been used for such tests, like e.g. the $\pi N - \Sigma$ term, $\pi\pi$ scattering, or $\kappa \to 2\pi$, 3π decays [MEI]. A less well exploited domain of such tests is the electroweak probe due to the lack of the appropriate experimental facilities. However, many calculations of such observables exist which are awaiting confrontation with experiments. Examples are: the electromagnetic and weak formfactors of the nucleon which can be investigated using parity violating and non–violating electron scattering [MEI], [BEI], electro– and photoproduction of pions at threshold [MEI], [BES], and polarizabilities of the nucleon and pion via real and virtual compton scattering [MEI], [DR3].

These examples show that the electroweak probe offers a rich field of experimental tests of chiral perturbation theory. One may, therefore, ask which are the right facilities to pursue these tests. The higher loop corrections limit the momentum transfer to $|q| \leq 0.2 MeV/c$ and invariant masses close to threshold, i.e. energies of about 1 GeV are required. Cross sections are small and make a high beam current $i \geq 50\mu A$ necessary. In order to make the high resulting luminosity usable, a duty factor of 1 is needed. Kinematics definition, background suppression etc. ask for a very good resolution ($\delta E/E \simeq 10^{-4}$) and clean beams with low emittances ($\Sigma \simeq \pi * n * m$). Polarized electrons and tagged polarized photons are a further request to the facilities. Luckily just three facilities AmPS–Amsterdam, Bates–MIT, and MAMI–Mainz are answering most of these requests and become just operational at a time when these exciting theoretical developments take place.

This contribution is about the Mainz Microtron MAMI. It presents some selected first results which show the performance of this facility and are related to overservables to test chiral dynamics.

The work at MAMI is organized in "collaborations" in the same sense as in high energy physics. These collaborations are oriented around one major facility and carry out a large number of different experiments in their respective framework. After a short review of the parameters of the MAMI accelerator, results of the different collaborations A1: "Coincidence Experiments with Electrons", A2: "Real Photons" and A3: "Electric Formfactor of the Neutron" are presented in turn. One more recent collaboration A4: "Parity Violating Electron Scattering" has not yet produced results and X1: "Coherent X–Ray Production" does not belong into the context of this workshop. A portrait of the MAMI laboratory appeared recently in ref. [VOL] .

2 The Mainz Microtron MAMI

The cw electron accelerator MAMI [HER] consists of three cascaded racetrack microtrons with a 3.5 MeV injector linac (see Fig. 1). The last stage delivers the beam from 180 to 855 MeV in 15 MeV steps. Normal conducting accelerating structures in connection with multiple beam recirculation represent an economical and reliable solution for stable acceleration of the intense electron beam. Owing to the extensive beam monitor and computer control system the accelerator complex is easy to operate, and several automatic optimizing routines serve for short beam setup and optimization times. The stability of the machine is excellent, the mean time between corrections by the operator normally being more than ten hours. Due to the highly stabilized rf amplitude in the acceleration sections, the beam intensity can easily be changed and even pulsed up to more than 110 μA with a rise time of 100 μs.

Recent measurements of the beam emittance along the 90 turns in the third stage have shown that the behaviour is very close to the theoretical calculations. In the horizontal plane, the beam emittance of $10^{-8}\pi * m * rad$ at maximum energy is mainly determined by stochastical emission of synchrotron radiation, whereas the vertical emittance is continuously decreasing to the low value of $0.7 * 10^{-9}\pi * m * rad$ due to pseudodamping. First measurements of the outer part of the beam profile behind the third microtron showed that there exists a faint halo, presumably generated by coulomb scattering off the residual gas atoms in the vacuum system. At maximum energy, the relative intensity outside a radius of 2.5 mm from the centre of a 0.5 mm beam spot (FWHM) is of the order of 10^{-6}.

The mean energy of the electrons in the third stage was measured with an absolute precision of ±160 keV by exact determination of the beam position on the accelerator axis and in several higher return paths. The long-term energy stability, limited by small thermal drifts of the bunch phase, is about 40 keV. Due to the stable non–isochronous longitudinal motion in microtrons, the residual rf–phase and amplitude fluctuations have only little influence on the beam energy, so that the spectrum at 855 MeV is mainly dominated by synchrotron radiation effects with a width of about 50 keV (FWHM).

A source of polarized electrons [HAR], [AUL] is attached to MAMI. The

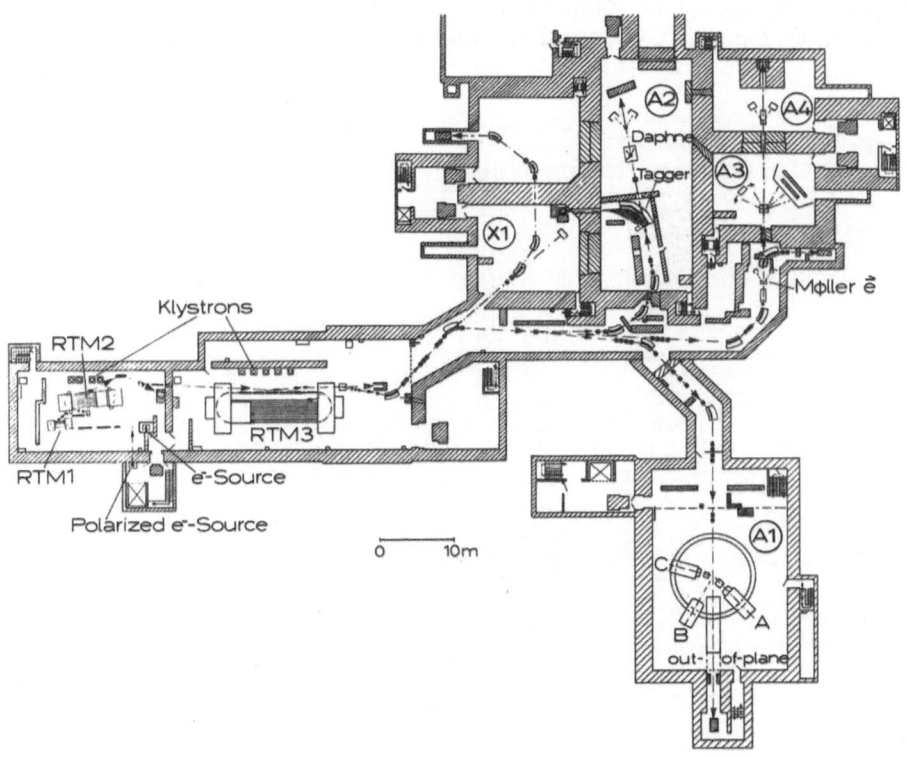

Fig. 1. Floor plan of the Mainz Microtron

spin direction can be oriented in any direction by means of a spin rotator [STF] in the 100 keV line behind the source. In January 1992, a polarized beam was accelerated to 855 MeV for the first time. The analysis by means of a Møller polarimeter showed that the initial degree of polarization of about 30%, given by the GaAsP photocathode, was preserved after more than 100 spin precessions during the acceleration.

3 Collaboration A1: Coincidence Experiments with Electrons

Electron scattering in coincidence with hadrons needs large solid angle, high resolution magnetic spectrometers. These tedious and costly instrumentations are mandatory to define the kinematics, resolve final states and suppress the high background typical for high luminosity electron experiments. The arrangement realized by the A1 collaboration is depicted in Fig. 2. The three spectrometers are rotatable around a common pivot and are installed in a new experimental hall with a size of 30m x 20m and a height of 17m from the floor to the crane hook.

The arrangements can be complemented with further magnetic or non–magnetic detectors, as, e.g., a BGO crystal ball and a neutron detector system.

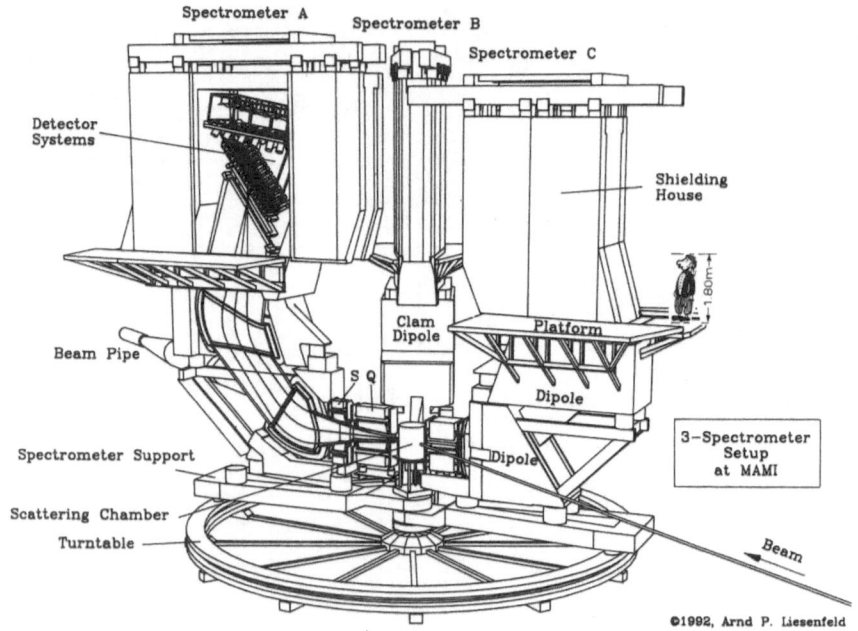

Fig. 2. 3-Spectrometer-setup at MAMI

The spectrometers A and C with point–to–point optics in the dispersive plane and parallel–to–point optics in the non–dispersive plane have large solid angles and large momentum acceptances, and each consist of a quadrupole, a sextupole and two dipole magnets (QSDD). The spectrometer B has a moderate solid angle and a reduced momentum acceptance, but it reaches a maximum momentum somewhat larger than that of the MAMI electron beam. It consists of a single clamshell dipole magnet with point–to–point optics in both planes, and due to its slim construction, it can be used at angles as small as 7°. The capability of reaching very small scattering angles allows a precise separation of longitudinal and transverse structure functions [DRE]. This separation needs as wide a variation of the transverse polarization ε at constant Q^2 as possible. The most essential parameters of the three spectrometers are compiled in Table 1.

A particularly important option in coincidence experiments with electrons is the capability for out–of–plane measurements in order to determine the longitudinal–transverse and transverse–transverse structure functions [DRE]. The interference–structure functions offer the possibility of amplifying small admixtures by the interference with a dominating contribution. A detailed study of the range of the out–of–plane angle demonstrated the particular need for small angles. For a range of 0° to 10°, Spectrometer B can be moved out–of–plane by means of a mechanical driving system.

Table 1. Main Parameters of the Three Spectrometers

Spectrometer configuration		A QSDD	B D*	C QSDD
Maximum momentum	$[MeV/c]$	735	870	551
Maximum induction	$[T]$	1.51	1.50	1.40
Momentum acceptance	$[\%]$	20	15	25
Solid angle	$[msr]$	28	5.6	28
Long-target acceptance	$[mm]$	50	50	50
Scattering angle range	$[°]$	18-160	7-62	18-160
Length of central trajectory	$[m]$	10.75	12.03	8.53
Dispersion (central trajectory)	$[cm/\%]$	5.77	8.22	4.52
Magnification (central trajectory)		0.53	0.85	0.51
Dispersion to magnification	$[cm/\%]$	10.83	9.64	8.81
Momentum resolution		$\leq 10^{-4}$	$\leq 10^{-4}$	$\leq 10^{-4}$
Angular resolution at target	$[mrad]$	≤ 3	≤ 3	≤ 3
Position resolution at target	$[mm]$	3–5	≤ 1	3–5

* clamshell dipole

Each spectrometer is equipped with a detector system consisting of four planes of vertical drift chambers, two planes of plastic scintillators and a threshold gas Cherenkov detector. Pions and protons are identified by comparing energy losses in the scintillators. This also discriminates against background of neutrons. The Cherenkov detector serves as a veto–detector for electrons. Particle momentum, emission angle and vertex are deduced by tracking trajectories back to the target using the four position measurements of the drift chambers.

In the summer of 1992, Spectrometer A was completed by mounting the shielding house and by installing the detector system. Detailed measurements with beam have proven that Spectrometer A exceeds the specifications.

In late fall 1992, a first coincidence experiment with Spectrometer A and a BGO ball detector [MOR] was performed. With this setup a rather complete measurement of multihadron final states in the quasielastic, dip and Δ resonance region was accomplished [A1R].

The second magnetic spectrometer, Spectrometer B, was ready for operation in June 1993. Again, detailed measurements with beam showed that Spectrometer B also exceeds the specifications, and that coincidence experiments with two magnetic spectrometers were now possible. Finally, the A1 collaboration could start a series of coincidence experiments. Measurements of the $(e, e'p)$ reaction on ^{12}C and ^{16}O were first performed at kinematics of earlier NIKHEF measurements [STE], subsequently kinematics, where up to now no experimental data exist, were investigated. Data with missing momenta $p_m = 700\ MeV/c$ and missing energies $E_m = 200\ MeV$ were taken. The data show an excellent ratio of true to accidental events (Fig. 3).

Some typical missing–energy–energy spectra are shown in Fig. 4. The missing energy resolution in this experiment was about 600 keV determined by a slight

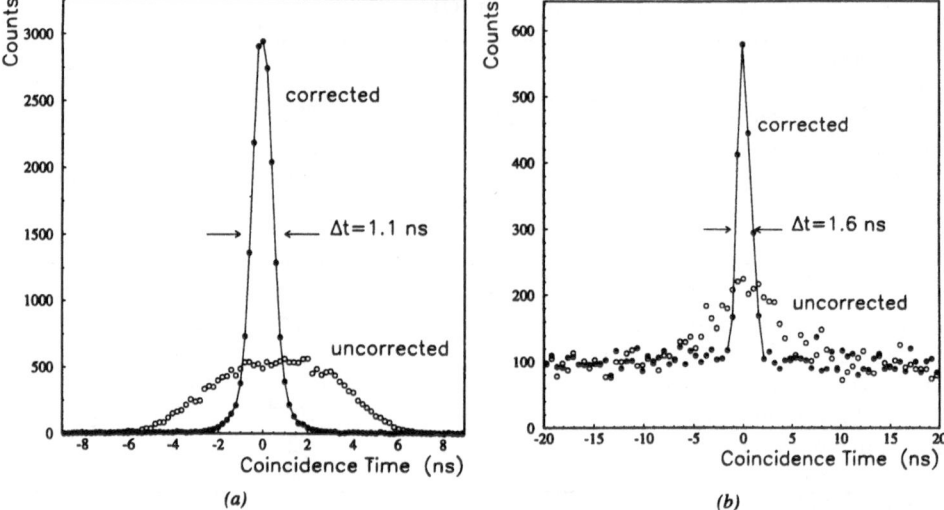

Fig. 3. Coincidence time spectrum for the $^{12}O(e, e'p)$ reaction at a missing momentum of 120 MeV/c (left) and 590 MeV (right), respectively. The measurements were performed with beam currents of 5 μA (left) and 24 μA (right), respectively, and with target thicknesses of 50 mg/cm^2 in both cases. The raw data are corrected for different path lengths of the electrons and protons relative to the central trajectory in each of the spectrometers

not corrected variation of the magnetic field in different runs. This resolution is, however, good enough to separate the $p_{1/2}$ and $p_{3/2}$ final states in ^{15}N and determine the reduced cross sections for these states as a function of missing momenta as depicted in Fig. 5. A detailed account of these results is about to appear [A1B].

A third spectrometer C (see Table 1) will be finished until the end of 1994 and allows the investigation of triple coincidences. This will open the study of short range correlations of nucleons through $A(e, e'pp)A'$ measurements or the study of the Δ resonance propagation in the nucleus serving as a spin–isospin filter.

Two experiments which have already produced results are the pion electroproduction measurements $p(e, e'\pi^+)n$ and $p(e, e'p)\pi°$. The theoretical framework is for instance developed in [DR2] which represents a standard reference.

The $p(e, e'\pi^+)n$ measurement has been performed at an invariant mass of W = 1125 MeV, i.e. 47 MeV above threshold. At a constant $q^2 = -0.117(GeV/c)^2$ a Rosenbluth separation could be done and the result is shown in Fig. 6. In order to extract some more direct physics quantities like the pion content of the nucleon from σ_L using the known pion formfactor or the axial formfactor of the nucleon from σ_T more q^2 values have to be measured. This work is in progress [LIE]. Another observable which could be for the first time extracted from the same data is the longitudinal-transverse interference cross-section σ_{TL}. Fig. 7 shows its dependence on the spherical angle $\Theta_{\pi CM}$ with respect to the momen-

NO radiative corrections applied

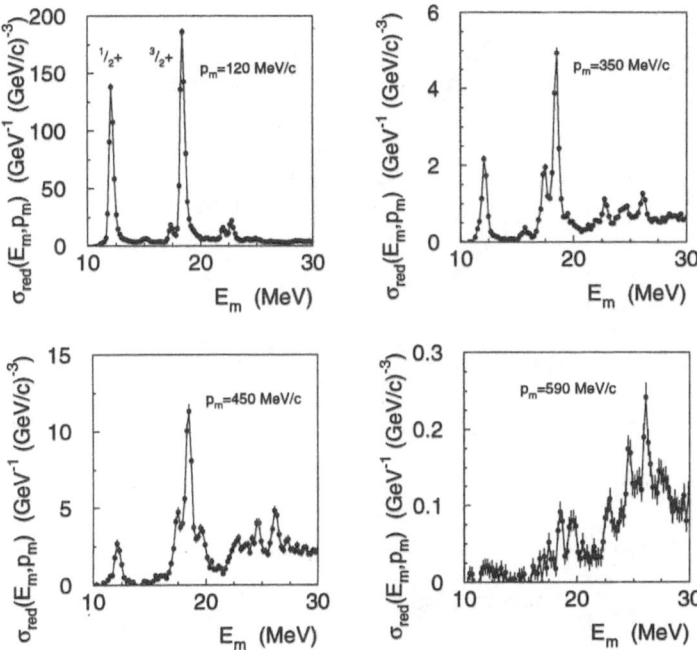

Fig. 4. Missing energy spectra in $^{16}O(e, e'p)$ reaction at 4 indicated, average missing momenta

tum transfer q. The measurement was done for an azimuthal angle between scattering and reaction plane of $\phi = 0°(\Theta_{\pi CM} > 0)$ and $\Phi = 180°(\Theta_{\pi CM} < 0)$.

A reaction which attracts particular interest is the threshold production of $\pi°$ via $p(e, e'p)\pi°$. At threshold chiral perturbation theory is making stringent predictions which deviate from the classical pseudovector coupling Born expansion of the production amplitude [BER]. In a first experiment taking up this new development an experiment was performed at NIKHEF [WEL]. At MAMI in a two days tune up experiment the $p(e, e'p)\pi°$ reaction was also investigated [A1D]. Due to the large solid angles of the spectrometers A and B, close to threshold ($\Delta W \leq 4~MeV$) the full CM–angular distributions can be measured in one setting. Fig. 8 shows a missing mass spectrum and demonstrates the cleanliness of the reaction identification due to the trajectory reconstruction in the spectrometers and the low rate of chance coincidences. The measurements had been taken at around $40\mu A$ with a 2cm LH_2 target so that a cw beam was mandatory. In Fig. 9 the measured CM–angular distribution is depicted together with the theoretical predictions [BER]. The exciting agreement with the chiral perturbation theory prediction is evident. In an intuitive picture one could interpret the one loop contributions which are added to the tree diagrams as the "pion cloud" of the proton. Such reactions close to threshold probe, therefore,

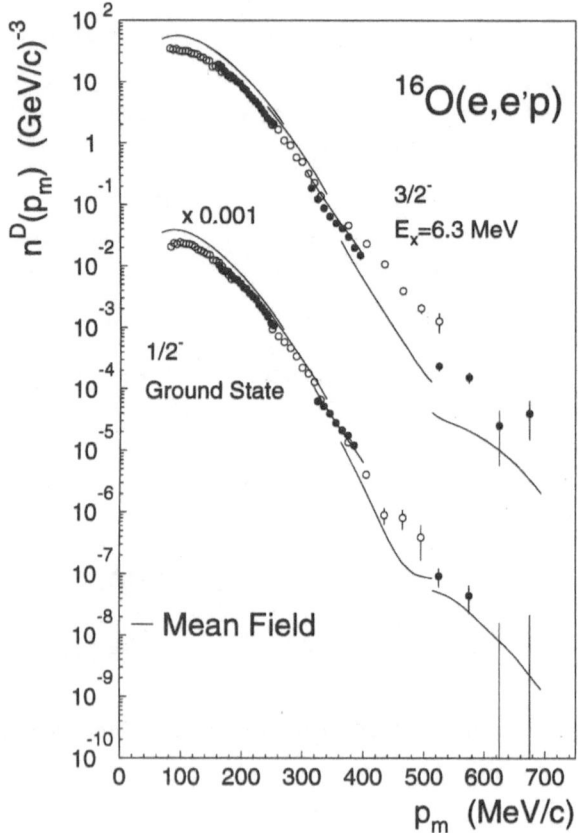

Fig. 5. Reduced cross section for the $1/2^-$ ground state and the $3/2^-$ excited final stale of ^{15}N as a function of the missing momentum p_m

the pion content of the nucleon as a natural effective degree of freedom. It is expected that the quantitative analysis in terms of partial waves as well as the determination of the interference structure functions σ_{TL}, σ_{TT} and $\sigma_{TL'}$ will pin down the pionic nature of the nucleon via chiral dynamics soon [1BW].

4 Collaboration A2: Experiments with Real Photons

This collaboration uses energy tagged Bremsstrahlungsphotons for a broad program devoted to experiments with real photons. The tagging is accomplished by means of a magnetic spectrometer with a large momentum bite (about 85%) which was designed and built by Glasgow University [ANT]. Using the 855 MeV beam of MAMI, tagged photons between 800 and 50 MeV with a resolution of 2 MeV are provided. If a diamond is used as the bremsstrahltarget linearly polarized photons are produced. In the future also circularly polarized photons using longitudinally polarized electrons will become possible. The total photon flux

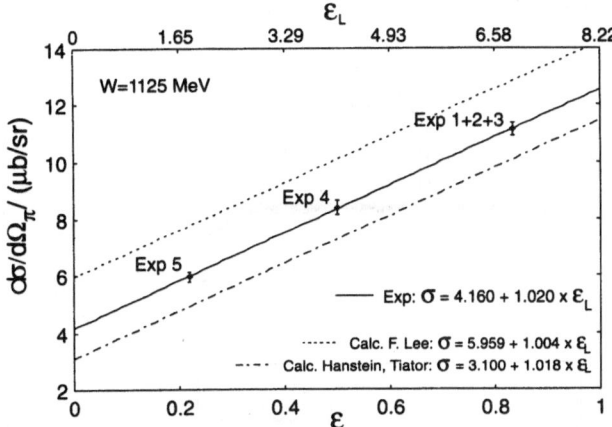

Fig. 6. Rosenbluth plot for π^+ production in the $p(e, e'\pi^+)n$ reaction. W = 1125 MeV, $q^2 = -0.117(GeV/c)^2$. Exp. 1 to 5 refers to different experimental runs. The full line is a fit of $\sigma = \sigma_T + \sigma_L \varepsilon_L$ to the data points. The dash pointed line shows the result of Hanstein and Tiator using a model based on Blomqvist–Laget operator [HAN]. The dashed line shows a calculation of Lee and Li [LEE]

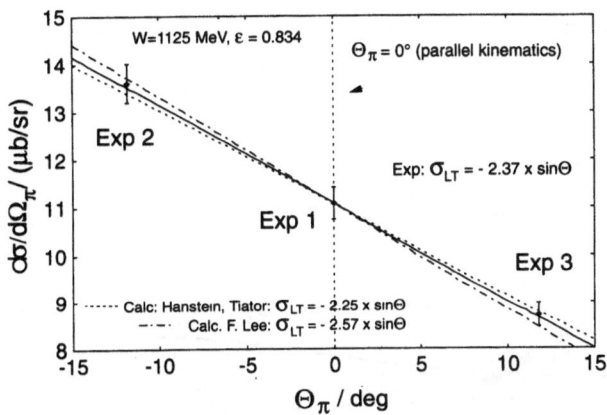

Fig. 7. The interference cross section σ_{TL} as a function of the polar angle $\Theta_{\pi CM}$ between the momentum transfer and the outgoing π^+

integrated over the momentum bite is about 10^8 photons/s. In order to use this moderate flux of energetically well resolved photons appropriately, a great many detectors, brought in by an international community, are used. Only a few can be mentioned here together with some selected results. Fig. 10 shows DAPHNE, a cylindrical detector consisting of a concentric stack of different detectors, which was built at Saclay [AUD]. This nonmagnetic detector was optimized to distinguish $p, n, \pi^\pm$ and π^0 and light fragments of nuclei after photoinduced reactions. Fig. 11 shows some typical results for the $p(\gamma, 2\pi)$ reactions. Some intriguing differences between the different channels are seen in this Δ resonance domi-

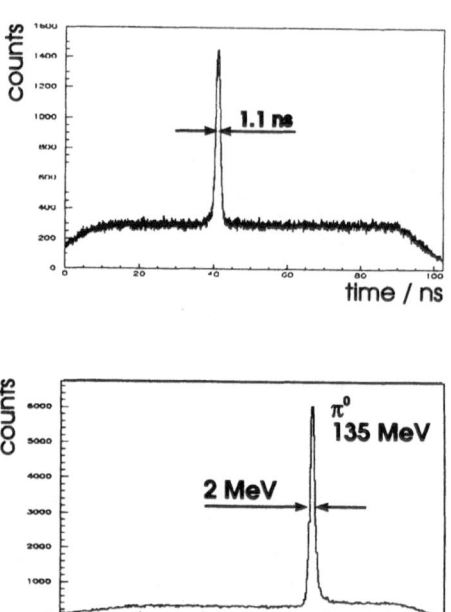

Fig. 8. Missing mass and coincidence time spectra as measured in the reaction $p(e, e'p)\pi^\circ$ [A1D] at a momentum transfer $q^2 = -0.1 (GeV/c)^2$

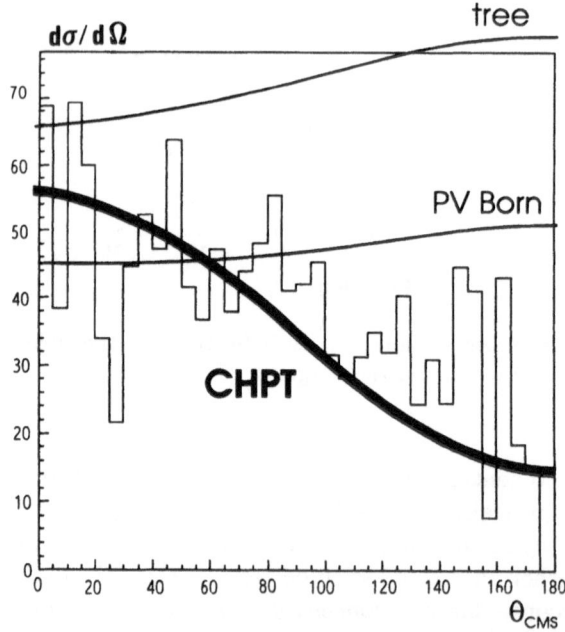

Fig. 9. Histogram of the measured CM–angular distribution [A1D] together with the PV–Born calculation (dashed), the tree diagram calculation (full) and the chiral perturbation theory prediction (broad band) [BER]

nated region which are awaiting theoretical explanations. This figure also shows a result of TAPS (the Three Arm Photon Spectrometer) which was particularly designed to look for photons in the $\pi^\circ \rightarrow 2\gamma$ and $\eta \rightarrow 2\gamma$ decays [NOV]. It delivered very precise data of the $p(\gamma, \eta)$ production at threshold which are being published [KRU]. It consists of 6 towers of 64 BaF-detectors which are grouped into different configurations depending on the reaction being investigated. Angular distributions for different photon energies are shown in Fig. 12.

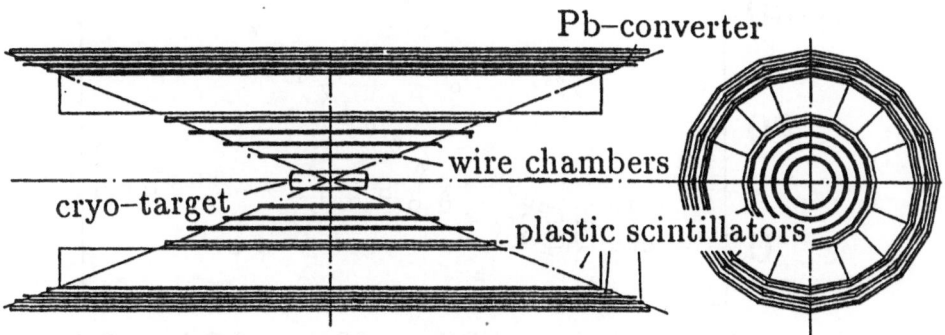

Fig. 10. The large–acceptance charged particle tracking detector DAPHNE [AUD]

A particular strength of TAPS is the possibility to go to energies close to the production threshold due to the measurement of the direct decay $\eta \rightarrow 2\gamma$. Here only a few partial waves contribute (see Fig. 12). This feature makes also a study of the $p(\gamma, \pi^\circ\pi^\circ)p$ process close to threshold accessible. Fig. 13 shows the result for a very preliminary measurement of 4 days [HAR]. It appears that the cross sections open directly at the threshold of $E_\gamma = 309\ MeV$ and than stays constant until the contribution from a resonance $N(1440)P_{11}(?)$ becomes dominant. This behaviour is at variance with chiral perturbation theory which predicts a very slow parabolic rise from threshold [BE2].

A final example of results in the framework of the A2 collaboration is the determination of the ratio of the E2/M1 transition amplitudes of the Δ resonance via the $p(\vec{\gamma}, \pi^\circ)p$ reaction [BEC]. The linearly polarized photons were produced by means of a diamond crystal as the bremsstrahltarget and for the $\pi^\circ \rightarrow 2\gamma$ decay measurement the ideally symmetric DAPHNE detector was used. Fig. 14 shows the differential cross section for a photon polarization parallel and perpendicular to the $p\pi^\circ$-decay plane. A systematic analysis of such data across the Δ resonance is in progress. A preliminary result indicates a result of around $E2/M1 = -2.5\%$.

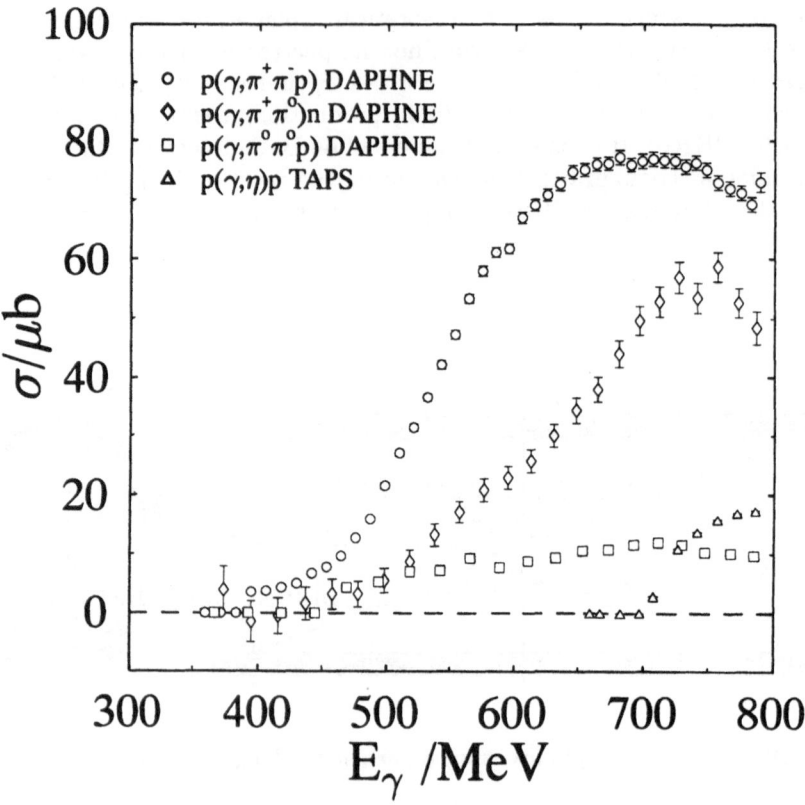

Fig. 11. Total cross section of the two π photoproduction in the Δ resonance region as measured with DAPHNE [DAP] and η photoproduction as measured with TAPS [KRU]

5 Collaboration A3: The Electric Form Factor of the Neutron

The A3 collaboration aims at a model–independent measurement of the electric form factor of the neutrons. This observable has attracted considerable interest because it provides a crucial test of quark models of the nucleon [EDE]. Due to the fact that the neutron as a target is bound, nuclear structure effects have always to be considered. Additionally, the electric form factor $G_{E,n}$ is much smaller than $G_{M,n}$ since the zero order term, the neutron charge vanishes. This means that the standard Rosenbluth separation of longitudinal ($G_{E,n}^2$) and transverse ($G_{M,n}^2$) is inaccurate because $G_{E,n}^2 \ll G_{M,n}^2$. However, elastic electron–deuteron scattering is sensitive to $G_{E,n}$ through interference with $G_{E,p}$ and the statistically most precise measurement stems from its measurement [PLA]. The model dependence due to different nucleon–nucleon amounts, however, to $\pm 100\%$. An alternative possibility which promises a model–independent measurement of $G_{E,n}$ is the use of polarization degrees of freedom [ARN]. Two

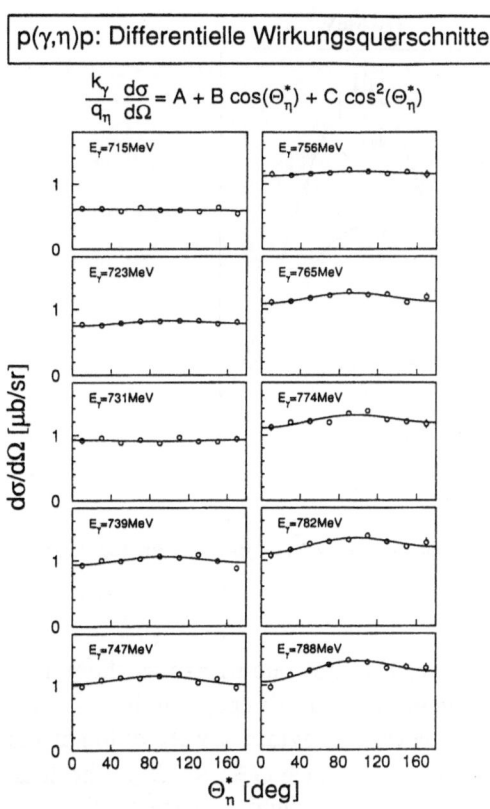

Fig. 12. Differential cross sections for the $p(\gamma, \eta)p$ reaction as measured with TAPS [KRU]

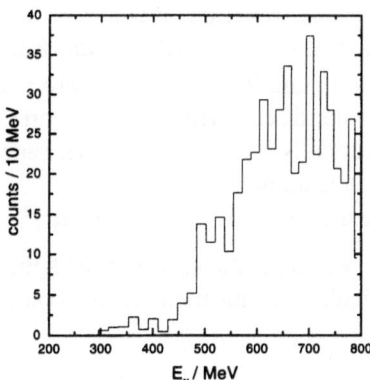

Fig. 13. Total cross section of the $p(\gamma, \pi^\circ\pi^\circ)p$ reaction close to threshold as measured with TAPS [HAR]

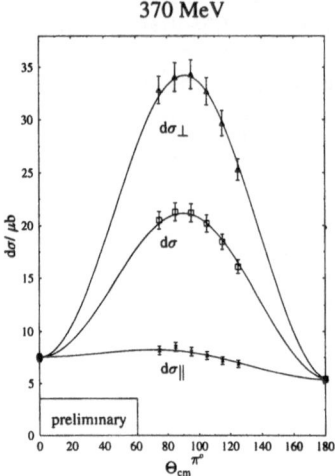

Fig. 14. Differential cross section of the $p(\vec{\gamma}, \pi^0)p$ reaction for $\perp$ and $\parallel$ polarized photons

equivalent possibilities exist. In the first one measures the polarization transfer P_x of longitudinally polarized electrons to the neutron in the $n(\vec{e}, e'\,\vec{n})$ reaction. In the second one measures the target asymmetry A_x of longitudinally polarized electrons in the $\vec{n}\ (\vec{e}, e'n)$ reaction. Since $G_{E,n} \ll G_{M,n}$ both asymmetries are proportional to $G_{E,n}/G_{M,n}$, i.e. they depend linearly on $G_{E,n}$.

The best approximation to a free neutron target is in the measurement of electron–neutron coincidences in quasifree kinematics. In a first measurement taking advantage of the 100% duty cycle and the polarized electron [HA2] of MAMI electron–neutron coincidences were observed in the $^3He\ (\vec{e}, e'n)2p$ reaction [MEY]. The particularity in this reaction is the need of a polarized ^3He target [ECK]. The second reaction used by the A3 collaboration is the polarization transfer in the $d(\vec{e}, e'\,\vec{n})p$ reaction. The difficulty here is the measurement of the neutron polarization by a second scattering in a neutron polarimeter. Since the quasifree kinematics are practically the same for both reactions, a common setup can be used. A scheme of it is shown in Fig. 15.

Using this apparatus the A3 collaboration determined at $q^2 = -0.31(GeV/c)^2$ the two asymmetries $A_\parallel$ and $A_\perp$, where in the first one the spin of $^3\vec{He}$ is parallel and in the second one perpendicular to the momentum transfer q. The result was

$$A_\parallel = (-7.40 \pm 0.73)\%, \qquad A_\perp = (0.89 \pm 0.30)\% \tag{1}$$

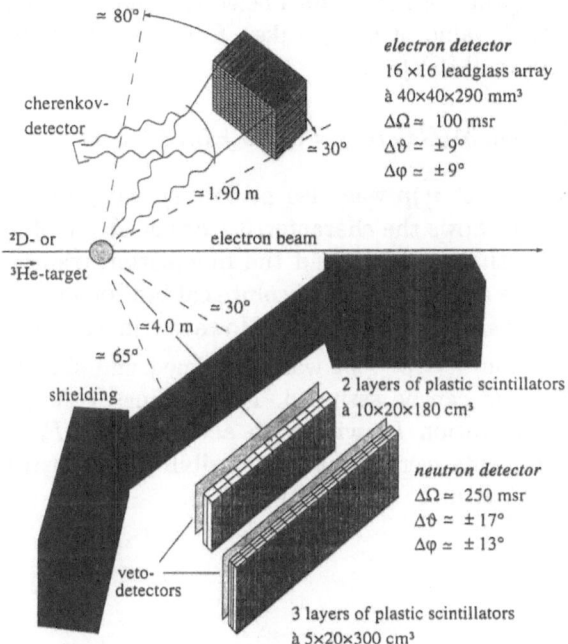

≈ 80°

cherenkov-
detector

electron detector
16 ×16 leadglass array
à 40×40×290 mm³
$\Delta\Omega \approx$ 100 msr
$\Delta\vartheta \approx \pm 9°$
$\Delta\varphi \approx \pm 9°$

≈ 30°

≈ 1.90 m

²D- or
³$\vec{\text{He}}$-target   electron beam

≈ 30°

≈ 4.0 m

≈ 65°

shielding

2 layers of plastic scintillators
à 10×20×180 cm³

neutron detector
$\Delta\Omega \approx$ 250 msr
$\Delta\vartheta \approx \pm 17°$
$\Delta\varphi \approx \pm 13°$

veto-
detectors

3 layers of plastic scintillators
à 5×20×300 cm³

Fig. 15. The setup for the $G_{E,n}$ measurements at MAMI

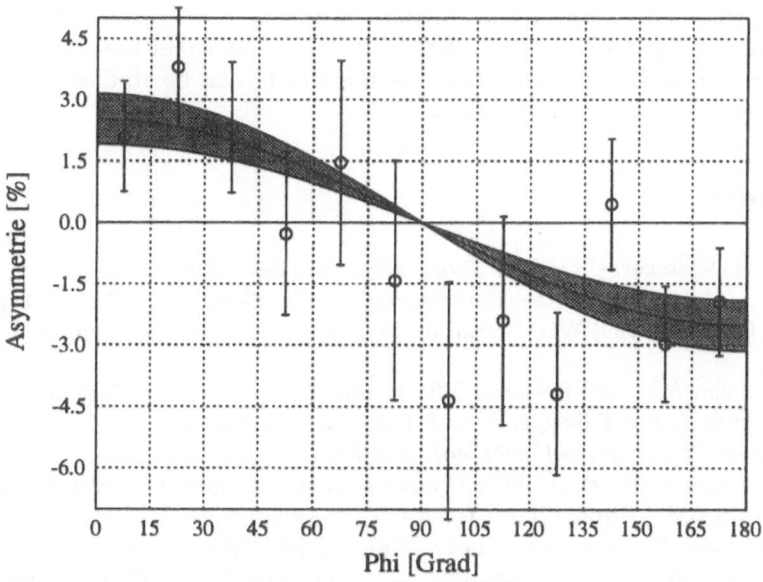

Fig. 16. Asymmetry of the recoil neutron in the $d(\vec{e}, e'\,\vec{n})p$ reaction as a function of the azimuthal angle phi [FRE]

Taking the ratio $A_\perp/A_\parallel$ the absolute value of electron and target polarization cancel and together with a value of $G_{M,n}$ taken from the dipole fit $G_{M,n} = \mu_n(1 + Q^2/0.71)^{-2}$ it follows [MEY]:

$$G_{E,n}(q^2 = -0.31(GeV/c^2) = 0.035 \pm 0.12 \pm 0.003 \tag{2}$$

The second reaction $d(\vec{e}, e'\ \vec{n})p$ was also performed with about the same statistical accuracy. Fig. 16 shows the characteristic cosine shape of asymmetry as a function of the azimuthal angle around the momentum transfer. For the extraction of $G_{E,n}$ from this asymmetry an absolute calibration of the analyzing power of the scintillators has still to be done. However, in another test measurement the mirror reaction $d(\vec{e}, e'\ \vec{p})n$ was performed and showed that the quasifree reaction mechanism can be assumed. The binding of the proton does not change the recoil polarization P_x within an accuracy of $\Delta P_x/P_x = 10\%$ [EYL]. All these measurements were meant as feasibility demonstrations and more precise measurements at more q^2 are in progress.

6 Conclusion

Though more work on the experiments presented has to be done, it became clear that the electroweak probe provides results which can be confronted to the calculations of chiral perturbation theory. This may be an incentive for theorists to proceed with the difficult calculations including two loop corrections, resonance effects etc. In this framework the guidance of theory is very much needed because no intuitive picture seems to exist. This may be inherent to QCD as a non linear field theory. On the other hand such a theory can be studied here with relatively modest experimental means.

References

[MEI] see e.g. Meissner, U.G.: Rep. Progr. Phys. **56** (1993) 903
[BEI] see eg., Proceedings of the Workshop on "Parity Violation in Electron Scattering", CalTech (1990) World Scientific, Singapore, Ed. Beise, E.J., and McKeown, R.D.
[BES] Bernstein, A.M. and Holstein, B.R.: Comm. Nucl. Part. Phys. **20** (1991) 197
[DR3] Drechsel, D. and Filkov, L.V.: Z. f. Phys. **A349** (1994) 177
[VOL] Nuclear Physics News (1994) **Vol. 4 No.2** 5-15
[HER] Herminghaus, H., et al.: First Operation of the 855 MeV CW Electron Accelerator "MAMI", Proc. LINAC Conf. (1990), Albuquerque, New Mexico
[HAR] Hartmann, W., et al.: Nucl. Instr. Meth. **A286** (1990) 1
[AUL] Aulenbacher, K., et al.: Operating Experience with the MAMI Polarized Electron Source, Contribution to the Workshop on Photocathodes for Polarized Electron Sources for Accelerators, SLAC-432 (1994)
[STF] Steffens, K.-H., et al.: Nucl. Instr. Meth. **A325** (1993) 378
[DRE] D. Drechsel and M.M. Giannin: Rep. Prog. Phys. **52** (1989) 1083

[MOR] Chris Morris, Los Alamos Meson Facility: contribution to this experiment (1994)

[A1R] A1 Collaboration; Blomqvist, I., et al.: (1994) to be published; Rosner, G.: contribution to this workshop

[STE] van der Steenhoven, G., et al.: Nucl. Phys. **A480** (1988) 547; Leuschner, M., et al.: Phys. Rev. **C49** (1994) 995

[A1B] A1 Collaboration; Blomqvist, I., et al.: accepted by Phys. Lett. B

[DR2] Drechsel, D. and Tiator, L.: J. Phys. G: Nucl. Part. Phys. **18** (1992) 449

[HAN] Hanstein, O. and Tiator, L.: private communication (1994) Hanstein, O.: Diploma Thesis, Mainz Univ. 1994; Blomqvist, I. and Laget, J.M.: Nucl. Phys. **A280** (1977) 405; Laget, J.M.: Nucl. Phys. **A481** (1988) 765

[LEE] Lee, F.: private communication; Li, Xiadong: Ph.D. Thesis, Chicago Univ. 1993

[LIE] Liesenfeld, A., Richter, A.R.: Doctorate Theses, Mainz Univ. 1994, and to be published

[BER] Bernard, V., Kaiser, N., Lee, T.-F.H. and Meissner, U.: Phys. Rep. **246** (1994) 315

[A1D] A1 Collaboration: Blomqvist, I., et al.: to be published; Distler, M.: Doctorate Thesis, Mainz Univ. 1995

[ABW] A1 Collaboration: Proposal A1/3-94; Bernstein, A. and Walcher, Th., Contactpersons

[WEL] Welch, T.P., et al.: Phys. Rev. Lett. **69** (1992) 2761

[ANT] Anthony, I., et al.: Nucl. Instr. Meth **A301** (1991) 230

[AUD] Audit, G., et al.: Nucl. Instr. Meth. **A301** (1991) 473

[NOV] Novotny, R.: IEEE, Transactions on Nucl. Sci. **38** (1991) 379

[KRU] Krusche, B., et al.: to be published in Z. f. Physik A

[DAP] Braghieri, A., et al.: submitted to Phys. Lett. B

[HAR] A2 Collaboration to be published; Härter, F.: Doctorate Thesis, Mainz Univ. 1995

[BE2] Bernard, V.: Contribution to this workshop

[BEC] A2 Collaboration; Beck, R. et al.: to be published

[EDE] see Eden, T. et al.: Phys. Rev. **C50**, 1749, and references therein (1994)

[PLA] Platchkov, S. et al.: Nucl. Phys. **A510** (1990) 740

[ARN] Arnold, R.G., Carlson, C.E. and Gross, F.: Phys. Rev. **C23** (1981) 363

[HA2] Hartmann, W. et al.: Nucl. Instr. Meth. **A286** (1990) 1; Aulenbacher, K. et al.: Contribution to the Workshop on Photocathodes of Polarized Sources for Accelerators, SLAC–432 (1994)

[MEY] Meyerhoff. M., Eyl, D., Frey, A. et al.: Phys. Lett. **B327** (1994) 201

[ECK] Eckert, G. et al.: Nucl Instr. Meth. **A320** (1992) 53

[FRE] Frey, A.: Doctorate Thesis, Mainz Univ. 1994

[EYL] Eyl, D., Frey, A. et al.: to be published in Z. f. Physik A; Eyl, D.: Doctorate Thesis, Mainz Univ. 1993

---

* supported by SFB 201, Deutsche Forschungsgemeinschaft

Conclusions

Aron M. Bernstein and Barry R. Holstein

Chiral dynamics [We79,Ga83] is fundamental because it represents a model-independent methodology for making QCD predictions at the confinement scale. It is important to confront these predictions with sensitive and accurate experiments, and this field has been growing rapidly due to increasing confidence in the theory and also as a result of a new generation of accelerators and detectors.

The foundations of chiral dynamics were reviewed by two of the creators of the field, Weinberg and Leutwyler. Chiral dynamics, or chiral perturbation theory, is an (low energy) effective field theory for QCD. At the confinement scale where the interaction between the quarks is very strong the relevant degrees of freedom are the baryons N, Σ, Δ, etc. and the Goldstone bosons π, η, K. A perturbative treatment is appropriate since the Goldstone interactions are relatively weak at low energies. The effective field theory obeys all of the underlying symmetries of the QCD Lagrangian. Of particular importance is chiral symmetry, which is broken both dynamically and also by the small but non-zero light quark masses (m_u, m_d, m_s).

Isospin is another important approximate symmetry which was discussed by both Weinberg and Leutwyler. At a superficial level one might expect large isospin violations due to the rather large ratio $r = (m_d - m_u)/(m_d + m_u) \sim 0.3$. However, the observed isospin violation is much smaller due to the fact that the up and down quark masses are small compared to the QCD scale and the relevant isospin breaking quantity is $(m_d - m_u)/\Lambda_{QCD} \sim 0.01$ [Leutwyler], which is of the same order as electromagnetic effects. Nevertheless there are some special cases which involve the interaction of π^0 mesons with nucleons for which this isospin violation can be larger [Weinberg], and this possibility, which is the focus of a new experimental initiative, is discussed below.

The price that must be paid for an effective, nonrenormalizable theory is the appearance of a number of priori unknown low energy constants, which must be obtained from experiment. It has been shown, however, that their magnitudes can be reliably estimated by saturating with the low lying resonances [Ecker] or very approximately determined by lattice gauge calculations [Negele]. Another approach, which appears promising, is based on a chiral version of the Schwinger-Dyson equation [Roberts].

The predictions of chiral perturbation theory are expressed as a power series in energy-momentum with a scale parameter $\sim 4\pi F_\pi \sim 1$ GeV. The convergence properties of the theory are determined by the structure of the amplitude under consideration and in a scattering problem, for example, will depend on the position of the lowest resonances and on the thresholds for particle production. The most accurately predicted quantities are the properties of the Goldstone bosons and their interactions. In this sector there is a unique connection between the number of loops and the corresponding power of the energy (momentum). For the case of $\gamma\gamma \to \pi\pi$ impressive calculations to order p^6 (two loops) were described by Gasser. In the relativistic treatment of the πN sector the simple relationship between the number of loops and the power of energy (momentum) is not valid due to the presence of the nucleon mass as an additional parameter with the dimensions of energy. However, in the heavy (non-relativistic) Fermion version of the theory the simple power counting is restored. There has been impressive progress in this arena in the past few years, as discussed by Meißner and Bernard.

From the previous discussion one can observe a "complementarity" about the experimental tests of chiral dynamics. On the one hand it is the Goldstone boson sector of the theory which is most amenable to reliable calculations. However, since one cannot make targets of these unstable particles more difficult, indirect methods, must be employed. The most mature example, summarized by Počanić, is the low energy $\pi - \pi$ interaction which was first predicted using current algebra and then refined with chiral perturbation theory. The original experimental technique used the nucleon as a source of virtual pions, with the $\pi - \pi$ interaction determined via extrapolation to the pole at $t = m_\pi^2$ (here t is the invariant four momentum transfer, which is negative in the physical region, hence the extrapolation). Accurate measurements of the final state $\pi - \pi$ interaction in the near-threshold $\pi N \rightarrow \pi\pi N$ reaction have more recently been performed. The largest error at the present time is due to the phenomenological method used by extract the $\pi - \pi$ scattering lengths from the data. The $\pi - \pi$ interaction has also been measured in the $K^+ \rightarrow \pi^+\pi^- e^+ \nu_e$ decay (K_{e4} decay, branching ratio $\sim 3.9 \times 10^{-5}$). At the present time there appears to exist a discrepancy between results obtained via different methods [Počanić]. However, due to possible systematic errors in the extraction of the scattering lengths from the $\pi N \rightarrow \pi\pi N$ reaction and the low statistics of the K_{e4} decay data, it is premature to draw definitive conclusions. At the present level of precision, it is clear that more needs to be done. We can look forward to improved theoretical analysis of the $\pi N \rightarrow \pi\pi N$ reaction and to vastly improved K_{e4} decay data from the Frascati Φ factory DAΦNE [Gasser and Sevior, Baldini and Pasqualucci]. In addition, there are plans to measure the $\pi - \pi$ scattering length by observation of pionium (bound $\pi^+\pi^-$ atoms) in experiments at CERN [Leutwyler] and Indiana [Vi92].

An additional area in which we can anticipate new experimental results is the study of the chiral anomaly. In this case parameter-free predictions can be made for threshold reactions. An anomaly is said to occur for a situation in which a symmetry of the classical Lagrangian is not obeyed when the theory is quantized. The most famous and successful example in particle physics is the prediction for the $\pi^0 \rightarrow \gamma\gamma$ decay rate. We can anticipate new results for the $\gamma\pi \rightarrow \pi\pi$ reaction. There is an approved experiment at CEBAF, which will utilize the virtual pions from a proton target, and another planned at Fermilab which will use a high energy pion beam and the virtual photons from a high Z nuclear target [Miskimen, Moinester].

Hadron polarizabilities $\bar{\alpha}$ and $\bar{\beta}$ (electric and magnetic) are important probes of internal structure. Again the "complementarity" between theory and experiment applies in that the predictions for the (hard to measure) pion are considerably more precise than for the nucleon. The nucleon situation was summarized by Nathan. For the proton, which has been studied for over 30 years, there has been considerable recent progress. At the present time the most precise results are obtained using the forward scattering (unitarity) sum rule constraint for $\bar{\alpha} + \bar{\beta}$ and fitting $\bar{\alpha} - \bar{\beta}$ from the Compton scattering measurements. For the neutron the only precise data has been obtained from elastic neutron scattering from Pb [Sc91]. The results are compared to theoretical predictions in Table 1. The theoretical results [Meißner] are shown with errors which take into account uncertainties associated with the relevant low energy constants. It can be seen that for $\bar{\beta}$ (the magnetic polarizability), where the contribution of the Δ is significant, the theoretical uncertainties are correspondingly large. Within errors, however, there is reasonable agreement between theory and experiment. Although this is encouraging, it is clear that more precise data, particularly on the neutron, is desirable. It is also very important to clarify the relationship between the formulations of baryon chiral perturbation theory with and without the Δ as an explicit participant, and to reduce the theoretical uncertainty [Butler and Nathan].

Table 1. Nucleon Polarizabilities (10^{-43} fm^3)

	Experimental[†]	Theory[*]
$\bar{\alpha}_p$	12.0 ± 0.9	10.5 ± 2.0
$\bar{\alpha}_n$	12.5 ± 2.50	13.4 ± 1.5
$\bar{\beta}_p$	2.2 ± 0.9	3.5 ± 3.6
$\bar{\beta}_n$	3.3 ± 2.7	7.8 ± 3.6
$\bar{\alpha}_p + \bar{\beta}_p$	14.2 ± 0.5	14.0 ± 2.2
$\bar{\alpha}_n + \bar{\beta}_n$	15.8 ± 1.0	21.1 ± 2.2

[*]CHPT [Meißner]   [†][Nathan]

For the case of the pion polarizabilities the theoretical calculations are now in excellent shape with a two loop result for the neutral pion already completed and a corresponding calculation for the charged pion in progress [Gasser]. On the other hand measurements of the pion polarizability are in their infancy. There have been preliminary experiments for which there exist highly divergent results. The working group on hadron polarizabilities [Baldini and Bellucci] discussed three active areas for future measurements. These are: 1) the $\pi^+ Z \rightarrow \gamma \pi^+ Z$ reaction from virtual photons in a high Z target (Primakoff effect) using high energy pions at Fermilab (K^+ and Σ^+ polarizabilities will be measured in a similar fashion) [Moinester]; 2) the $\gamma p \rightarrow \gamma \pi^+ n$ reaction (radiative photoproduction) extrapolated to the pion pole planned at Mainz; and 3) the $e^+ e^- \rightarrow e^+ e^- \pi^+ \pi^- (\pi^0 \pi^0)$ reaction planned at DAΦNE [Baldini and Pasqualucci]. It has been demonstrated that the latter reaction is an excellent probe of the dynamics but is unfortunately relatively insensitive to the pion polarizabilities [Gasser, Sec. 5.4, Baldini and Bellucci]. We can look forward to results from each of these experimental initiatives in the next few years.

The πN interaction at low energies is a fundamental testing ground for any theory of the strong interactions. Of particular interest for chiral dynamics is the sigma term which is predicted to be $\sigma = (35 \pm 5)$ MeV/(1 - y) where y is a measure of the strange quark content of the nucleon [Sainio]. There has been a long history of determining σ from the data which was reviewed by Sainio and Höhler. It is far too a complex a situation to summarize briefly. The complications include 1) an extrapolation must be made from the physical region to the Cheng-Dashen point by an analytic continuation of the empirically determined partial wave scattering amplitudes; 2) the relevant amplitude is isoscalar which is relatively small; 3) there exist systematic discrepancies in the data base; and 4) there is a rapidly varying t-dependence of the scalar form factor of the nucleon, which is determined by dispersion relations (lowest order chiral perturbation theory calculations are insufficient). To illustrate the problems associated with point 2 and 3 above, the sensitivity of the differential cross section to the sigma term for the elastic scattering of 30 MeV pions from the proton is shown in Fig. 1 [Sa94]. It can be seen that the variation of the cross section, for a range in the sigma term corresponding approximately to the present uncertainty in its determination, is about the same as the systemic error between the data sets. Despite all of these difficulties the value y = 0.2 ± 0.2, corresponding to a contribution of 130 MeV to the nucleon mass, has been determined [Sainio]. It is clearly desirable to reduce the 100% uncertainty in y. Of related interest is the suggestion by Leutwyler to measure the $\pi\pi$ sigma term from the observation of the $\pi - \pi$ scattering lengths in the pionium atom.

The interesting possibility of observing isospin violation due to the mass difference of the up and down quarks was discussed by Weinberg and Leutwyler. There was also a review of the present status of isospin violation in the Δ region [Höhler] as well as the presentation of several recent developments. In particular, new measurements of the π^-p and π^-p $\rightarrow \pi^0$n charge exchange scattering lengths have been performed at PSI by observing the transition energy and width of the 3p $\rightarrow$ 1s transition in pionic hydrogen [A. Badertscher et al. in the πN working group summary]. The values obtained show a possible violation of isospin symmetry [Achenauer et al. in the πN working group summary]. There is also a proposed new technique by which to measure π^+n $\rightarrow \pi^0$p and possibly π^0p scattering lengths in the γp $\rightarrow \pi^0$p reaction as a final state interaction effect [Be94]. All of the scattering lengths depend on the value of the πN coupling constant as was discussed by Pavan [πN working group summary] and Höhler.

Threshold electromagnetic pion production is an additional testing ground for chiral dynamics [Bernard, Meißner and Schoch]. The field dates back to current algebra derivations of "low energy theorems" for threshold pion photoproduction. However, it has only been in the past five years that accurate experimental data have been available. A flurry of activity was spurred by experiments at Saclay and Mainz on the threshold p(γ, π^0) reaction [Ma86,Be90].

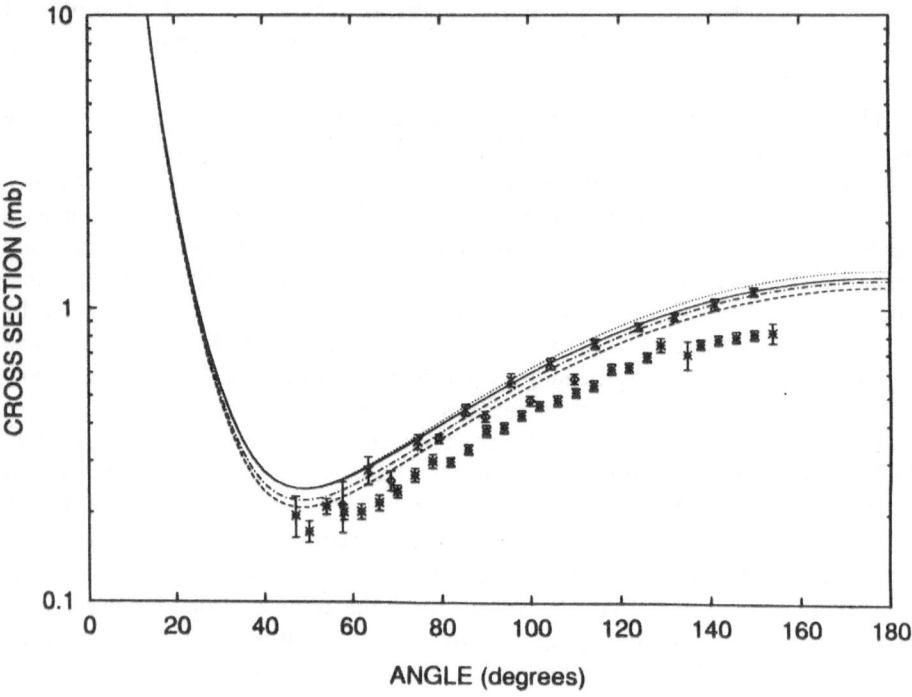

Fig. 1. Differential cross section for π^+p elastic scattering at $p_\pi = 97$ MeV/c (30 MeV pion kinetic energy). The curves show the variation of the predicted cross sections for a variation of Σ of 10 MeV which is the range of uncertainty in the πN sigma term. The points show the experimental data [Sa94].

When the dust settled down it was found that these experiments agreed with the soft pion predictions [Be91]. However, it was subsequently realized that the "low energy theorems" omitted important loop contributions and at the workshop Ecker demoted them to "low energy guesses." It now appears that the situation is more complicated with a substantial fraction of the prediction for the threshold electric dipole amplitude E_{0+} coming from an undetermined low energy constant which must be determined from experiment [Bernard]. Its value cannot be predicted with precision at the present time. Despite this situation, a *new* low energy theorem has been derived for two of the three p wave threshold multipoles [Bernard], and these predictions agree with the Mainz data. However, since an unpolarized cross section does not uniquely determine the multipoles this does not constitute a proof of these predictions are correct. In fact there is another empirical solution for the Mainz data which disagrees with the new p wave numbers [Be94]. A definitive determination of the multipoles awaits the results of more recent experiments for the (unpolarized) cross sections performed at Mainz [St90] and SAL [Bergstrom] as well as an experiment planned with linearly polarized photons at Mainz.

It was pointed out by Bernard that once the relevant low energy constants are determined by the data for the p(γ, π^0) reaction, there are no additional free parameters in the predictions for the corresponding p(e, e'π^0) process. It is therefore of increasing importance to make precision measurements of the threshold electroproduction reaction. Two preliminary results were reported at the workshop [Blok, Walcher] from NIKHEF and Mainz for $q^2 = -0.1$ GeV^2 and for CM energies W up to 15 MeV above threshold. The results are in qualitative agreement with the predictions of chiral perturbation theory [see Walcher, Fig. 9]. The forward peaking in the cross section indicates that the electric dipole amplitude $E_{0+}(q^2)$ has changed signs, as predicted, between the photon point, where it is negative, and $q^2 = -0.1$ GeV^2. As the magnitude of q^2 increases so does the relative contribution of the one loop contribution, reaching 50% of the total cross section by $q^2 = -0.1$ GeV^2. Therefore it is suggested that precision measurements of the p(e, e'π^0) reaction be carried out at a smaller magnitude of q^2 [Bernard]. An experiment at $q^2 = -.05$ GeV^2 is planned at Mainz during the next year.

The previous discussion has been focused on neutral pion production. For charged pions one expects the old low energy theorems to be more accurate since the leading (Kroll-Ruderman) term for the threshold electric dipole amplitude E_{0+} comes from a marriage of gauge and chiral invariance and is model-independent. Nevertheless there exist no accurate, modern experiments to test this prediction. There are plans to perform the p(γ, π^+)n reaction at Saskatoon [Bergstrom]. Also preliminary results for the p(π^-, γ)n reaction near threshold were presented by Kovash for a TRIUMF experiment. Finally we note the emergence of the ($\gamma, 2\pi$) threshold region as an interesting region of study [Bernard, Walcher].

The entire discussion of the πN sector is based on the physics of light (almost massless) up and down quarks. In the purely mesonic sector it is customary to treat the *three* light quarks (u, d, s). This SU(3) treatment has yet to be extended to the meson-baryon sector which would then include π, η, K as well as nucleons, hyperons (and Δ, Σ^* ?). Measurements in this extended sector are already in progress at Bates where the SAMPLE experiment is looking for the strange quark content of the nucleon [Kowalski]. Likewise experiments at Bonn and Mainz and planned experiments at CEBAF are investigating kaon and eta photo- and electro-production. In the mesonic sector we also look forward to exciting new data on the $\pi - \pi$ interaction from K$_{e4}$ decays and from the two photon production of pion pairs at DAΦNE [Baldini and Pasqualucci].

Although, in the interest of brevity, we have not discussed all of the new experimental and theoretical initiatives that are applicable to QCD studies at the confinement scale we hope that we have conveyed some of the sense of excitement and promise that was exhibited at the workshop. We anxiously look forward to future progress in this field.

References

[Be90] R. Beck *et al.*, Phys. Rev. Lett. **65**, 1841 (1990).

[Be94] A. M. Bernstein, to be published.

[Be91] A. M. Bernstein and B. R. Holstein, Comments Nucl. Part. Phys. **20**, 197 (1991), J. Bergstrom, Phys. Rev. **C44**, 1768 (1991), D. Dreschsel and L. Tiator, J. Phys. G: Nucl. Part. Phys. **18**, 449 (1992).

[Ga83] J. Gasser and H. Leutwyler, Phys. Lett. **125B**, 321 (1983); Ann. Phys. N.Y. **158**,142 (1984); Nucl. Phys. **250B**, 465, 517, 539 (1985).

[Ma86] E. Mazzucato *et al.*, Phys. Rev. Lett. **57**, 3144 (1986).

[Sa94] M. Sainio, private communication (1994).

[Sc91] J. Schmeidmeyer *et al.*, Phys. Rev. Lett. **66**, 1015 (1991).

[St90] Mainz experiment A2/7-90, TAPS collaboration, H. Ströher spokesman.

[Vi92] Proposal to the Indiana University Cyclotron Facility, S. E. Vigdor spokesman, April, 1992.

[We79] S. Weinberg, Physica **96A**, 327 (1979).

List of Participants

1. Jürgen Ahrens, Universität Mainz, Germany
2. Jose Enrique Amaro, MIT, Cambridge, MA USA
3. Hans Jürgen Arends, Universität Mainz, Germany
4. Andreas Badertscher, Institute for Particle Physics, Villigen PSI, Switzerland
5. Rinaldo Baldini, INFN, Frascati, Italy
6. Manoj K. Banerjee, University of Maryland, College Park, MD USA
7. Reinhard Beck, Universität Mainz, Germany
8. Stefano Bellucci, INFN, Frascati, Italy
9. Mohamed Benmerrouche, University of Saskatchewan, Saskatoon, SK Canada
10. John C. Bergstrom, University of Saskatchewan, Saskatoon, SK Canada
11. Veronique Bernard, CRN Strasbourg, France
12. Aron M. Bernstein, MIT, Cambridge, MA USA
13. William Bertozzi, MIT, Cambridge, MA USA
14. Henk Blok, Vrije Universiteit, The Netherlands
15. Anatoly Bolokhov, Sankt-Petersburg State University, Russia
16. Faustino Bonutti, INFN, Trieste, Italy
17. Edward C. Booth, Boston University, Boston, MA USA
18. Virginia R. Brown, LLNL, Livermore, CA USA
19. Urs Buergi, University of Berne, Switzerland
20. Paul Buettiker, University of Berne, Switzerland
21. Malcolm Butler, Saint Mary's University, Halifax, NS Canada
22. Juan Antonio Caballero, MIT, Cambridge, MA USA
23. Jian-ping Chen, MIT, Cambridge, MA USA
24. Csaba Csa'ki, MIT, Cambridge, MA USA
25. Giancarlo D'Ambrosio, INFN - Sezione di Napoli, Italy
26. Annalisa D'Angelo, INFN - Sezione Roma 2, Italy
27. Michael Otto Distler, Universität Mainz, Germany
28. T. William Donnelly, MIT, Cambridge, MA USA
29. Gerhard Ecker, University of Wein, Austria
30. Roudolph Eramzhyan, Institute for Nuclear Research, Russian Academy of Sciences, Russia
31. Harold W. Fearing, TRIUMF, Vancouver, BC Canada
32. Fabrizio Gabbiani, University of Massachusetts, Amherst, MA USA
33. Jürg Gasser, University of Bern, Switzerland
34. Howard Georgi, Harvard University, Cambridge, MA USA
35. Shalev Gilad, MIT, Cambridge, MA USA

36. José Luis Goity, CEBAF, Newport News, VA USA
37. Ralf Walter Gothe, Universität Bonn, Germany
38. Franz Gross, CEBAF, Newport News, VA USA
39. Thomas Hemmert, University of Massachusetts, Amherst, MA USA
40. Ross S. Hicks, University of Massachusetts, Amherst, MA USA
41. Gerhard Höhler, University of Karlsruhe, Germany
42. Barry R. Holstein, University of Massachusetts, Amherst, MA USA
43. Mahir S. Hussein, MIT, Cambridge, MA USA
44. Robert L. Jaffe, MIT, Cambridge, MA USA
45. Xiangdong Ji, MIT, Cambridge, MA USA
46. Joachim Kambor, University of Massachusetts, Amherst, MA USA
47. Sergey Kananov, Tel Aviv University, Israel
48. Melissa L. Kennedy, University of New Hampshire, Durham, NH USA
49. Mohammad A. Kermani, TRIUMF, Vancouver, BC Canada
50. Fritz Klein, Universität Mainz, Germany
51. Marc Knecht, Division de Physique Théorique – IPN, Faculté Des Sciences, France
52. Stefan Kopecky, Technischen Universität Wien, Austria
53. Michael A. Kovash, University of Kentucky, Lexington, KY USA
54. Stanley Kowalski, MIT, Cambridge, MA USA
55. Laird Kramer, MIT, Cambridge, MA USA
56. Hans-Jorg Leisi, Institute for Particle Physics, Villigen PSI, Switzerland
57. Heinrich Leutwyler, University of Berne, Switzerland
58. Wei Lin, MIT, Cambridge, MA USA
59. Aneesh Manohar, University of California at San Diego, La Jolla, CA USA
60. Victor Hugo Martinez, MIT, Cambridge, MA USA
61. Ulf-G. Meißner, CRN Strasbourg, France
62. Harald Merkel, MIT, Cambridge, MA USA
63. Richard G. Milner, MIT, Cambridge, MA USA
64. Rory A. Miskimen, University of Massachusetts, Amherst, MA USA
65. Nader Mobed, University of Regina, Regina, SK Canada
66. Murray A. Moinester, Tel Aviv University, Israel
67. Nimai C. Mukhopadhyay, Rensselaer Polytechnic Institute, Troy, NY USA
68. Euysoo Na, University of Massachusetts, Amherst, MA USA
69. Alan N. Nathan, University of Illinois at Urbana-Champaign, Urbana, IL USA
70. John Negele, MIT, Cambridge, MA USA
71. Reiner Neuhausen, Universität Mainz, Germany
72. Douglas Newton, MIT, Cambridge, MA USA
73. Martin Olsson, University of Wisconsin, Madison, WI USA
74. Stephen F. Pate, MIT, Cambridge, MA USA
75. Marcello Pavan, University of British Columbia/TRIUMF, Vancouver, BC Canada
76. Michael R. Pennington, University of Durham, United Kingdom
77. Antonio F. Perez, University of Massachusetts, Amherst, MA USA

78. Alexey A. Petrov, University of Massachusetts, Amherst, MA USA
79. Dinko Počanić, University of Virginia, Charlottesville, VA USA
80. Robert P. Redwine, MIT, Cambridge, MA USA
81. Craig Roberts, Argonne National Laboratory, Argonne, IL USA
82. Mikko Sainio, University of Helsinki, Finland
83. Adam J. Sarty, MIT, Cambridge, MA USA
84. Berthold Schoch, University of Bonn, Germany
85. Bernd Schreiber, MIT, Cambridge, MA USA
86. Christian Schütz, Forschungszentrum Jülich GmbH, Germany
87. Martin E. Sevior, University of Melbourne, Australia
88. Klara V. Shitikova, Moscow State University, Russia
89. Witold Skiba, MIT, Cambridge, MA USA
90. Dennis M. Skopik, University of Saskatchewan, Saskatoon, SK Canada
91. Victor Steiner, Tel Aviv University, Israel
92. Miloslav Svec, McGill University, Montreal, QB Canada
93. Eric Scott Swanson, MIT, Cambridge, MA USA
94. Antonios Tsapalis, MIT, Cambridge, MA USA
95. Christoph Tschalär, MIT, Cambridge, MA USA
96. Peter Unrau, MIT, Cambridge, MA USA
97. Eswara P. Venugopal, University of Massachusetts, Amherst, MA USA
98. Thomas Walcher, Universität Mainz, Germany
99. G. Wanders, Université de Lausanne, Switzerland
100. Steven Weinberg, University of Texas, Austin, TX USA
101. Xiaoming Xu, MIT, Cambridge, MA USA